T0390142

Accounting, Finance, Sustainability, Governance & Fraud: Theory and Application

Series Editor

Kıymet Tunca Çalıyurt, Iktisadi ve Idari Bilimler Fakultes, Trakya University Balkan Yerleskesi, Edirne, Turkey

This Scopus indexed series acts as a forum for book publications on current research arising from debates about key topics that have emerged from global economic crises during the past several years. The importance of governance and the will to deal with corruption, fraud, and bad practice, are themes featured in volumes published in the series. These topics are not only of concern to businesses and their investors, but also to governments and supranational organizations, such as the United Nations and the European Union. Accounting, Finance, Sustainability, Governance & Fraud: Theory and Application takes on a distinctive perspective to explore crucial issues that currently have little or no coverage. Thus the series integrates both theoretical developments and practical experiences to feature themes that are topical, or are deemed to become topical within a short time. The series welcomes interdisciplinary research covering the topics of accounting, auditing, governance, and fraud.

Kasım Kiracı · Kıymet Tunca Çalıyurt
Editors

Corporate Governance, Sustainability, and Information Systems in the Aviation Sector, Volume I

 Springer

Editors
Kasım Kiracı
Department of Aviation Management
İskenderun Technical University
İskenderun, Turkey

Kıymet Tunca Çalıyurt
Trakya University
Edirne, Turkey

ISSN 2509-7873   ISSN 2509-7881   (electronic)
Accounting, Finance, Sustainability, Governance & Fraud: Theory and Application
ISBN 978-981-16-9275-8   ISBN 978-981-16-9276-5   (eBook)
https://doi.org/10.1007/978-981-16-9276-5

This Springer imprint is published by the registered company Springer Nature Singapore Pte Ltd.
The registered company address is: 152 Beach Road, #21-01/04 Gateway East, Singapore 189721,
Singapore

Foreword

I am delighted to write this foreword, not only because Kıymet Tunca Çalıyurt and Kasım Kiracı have been good friends and colleagues for many years but also because I strongly believe sustainability improves the quality of our lives, protects our ecosystem, and preserves natural resources for the future generations. Kıymet and Kasım are particularly well suited for tackling this challenging problem by producing such an inspiring and informative book. Their knowledge and experience are self-evident in this book.

For more than two decades, the concept of sustainability has maintained a high position in public policy and international consciousness. Sustainability studies whether something will be able to continue as it currently exists into the future without significant adverse impacts. Hence, ensuring sustainability and long-term growth is an urgent problem for the aviation industry, which is an integral part of the global business. There are three main categories of sustainability: environmental, financial, and social.

Modern use of the term sustainability commenced with the United Nations Commission on Environment and Development, also known as the Brundtland Commission, in 1983. As far back as 2009, the airline industry pledged to cut emissions from flying to half of 2005 levels by 2050. Transitioning to net zero is increasingly a top priority for all industries across the globe. Many industries are now looking at how they can reformulate their operations for a sustainable future. Sustainable Aviation Fuel (SAF) is a clean substitute for fossil jet fuels. Using SAF results in a reduction in carbon emissions compared to the traditional jet fuel over the lifecycle of the aircraft. SAF is produced from sustainable resources such as waste oils from a biological origin and agricultural residues. SAF, made from renewable biomass and waste resources, has the potential to deliver petroleum-based jet fuel with a fraction of its carbon footprint. The International Air Transport Association (IATA) considers SAF to be one of the critical elements to reduce the carbon footprint within the environmental impact of aviation.

Kasım and Kıymet have written an exciting book about Corporate Governance, Sustainability, and Information systems in the Aviation Sector. They offer a unique perspective, combining the expertise of trained academicians with practical knowledge of the practitioners. In this book, operational resilience through disruption management in airlines was examined in the context of sustainability. Different critical factors in the ecosystem were evaluated in tandem to provide a clearer understanding of the relationship between sustainability and value co-creation related to the sector. Furthermore, they have identified different variables to measure the sustainability of the financial performance of airlines. Utilizing Data Envelopment Analysis (DEA), the financial performance of 20 airlines in the context of sustainability was addressed.

Readers will indeed find the materials in this book to be worthwhile and interesting, and I hope that the response to this book be such that Kıymet and Kasım will continue to write more books in this area.

Dr. Bijan Vasigh
President and Founder
Aviation Consulting Group, LLC
Professor, Embry-Riddle Aeronautical
University
Daytona Beach, USA

Contents

Editors and Contributors

About the Editors

Dr. Kasım Kiracı received his first B.Sc. degree in Aviation Management in 2010 from Kocaeli University and second B.Sc. degree in Economics in 2015 from Anadolu University. He obtained his first M.Sc. degree in Economics from Gebze Technical University and second MSc degree in Aviation Management from Anadolu University. He completed his Ph.D. in Aviation Management with the dissertation titled "Determinants of The Capital Structure in Different Business Models: A Panel Data Analysis On Low Cost and Traditional Airlines" at the Anadolu University, Eskişehir, in 2017. He joined Faculty of Aeronautics and Astronautics, Iskenderun Technical University in 2018. He has been working as a faculty member at the Department of Aviation Management at İskenderun Technical University since 2018. He received his associate professorship in finance in 2021. He has many articles published in web of science journals on aviation economy and airline financing. In addition, there are many book chapters published on different topics of aviation.

Prof. Dr. Kıymet Tunca Çalıyurt graduated from the Faculty of Business Administration at Marmara University, Istanbul, Turkey. Her Masters and Ph.D. degrees are in Accounting and Finance Programme from the Social Graduate School, Marmara University. She has worked as auditor in Horwath Auditing Company, manager in McDonalds and finance staff in Singapore Airlines before positioning in academia. After vast experience in private sector, she has started to work in Trakya University as lecturer in 1999. She has been as visiting researcher in Concordia University, Canada (2001), Amherst Business School, Massachussetts University, USA (2014), and UNWE, Sofia, in 2019.

In 2009, she has founded of the International Group on Governance, Fraud, Ethics and Social Responsibility (IGonGFE&SR) and International Women and Business Group (IWBG), which organizes a global, annual conferences. She published papers, book chapters and books both nationally and internationally on fraud, social responsibility, ethics in accounting/finance/aviation disciplines in Springer and Routledge. She is a book series editor with following titles in Springer and Routledge.

- Accounting, Finance, Sustainability, Governance & Fraud: Theory and Application (Springer Nature indexed by Scopus)
- Women and Sustainable Business (Routledge)

She is acting as associate or editorial board member in following titles:

- Journal of Financial Crime (ESCI)
- International Journal on Law and Management
- Journal of Money Laundering Control
- International Journal of Climate Change Strategies and Management (SSCI)

She is a regular speaker at International Economic Crime Symposium in Jesus College, Cambridge University and partner of Herme Consulting in Trakya University Technopark.

Contributors

Leyla Adiloğlu-Yalçınkaya Faculty of Aviation and Aeronautical Sciences, Özyeğin University, Istanbul, Turkey

Şahap Akan Department of Civil Aviation Management, Anadolu University, Eskişehir, Turkey

Muhammet Mikdat Akbaba Turkish Airlines Technic, Bursa, Turkey

Seda Arslan Faculty of Aeronautics and Astronautics, Iskenderun Technical University, Hatay, Turkey

Veysi Asker Department of Aviation Management, Civil Aviation High School, Dicle University, Diyarbakır, Turkey

Mutlu Yuksel Avcilar Faculty of Economics and Administrative Sciences, Osmaniye Korkut Ata University, Osmaniye, Turkey

Mahmut Bakır Department of Aviation Management, Samsun University, Samsun, Turkey

K. Gülnaz Bülbül Department of Flight Training, Faculty of Aeronautics and Astronautics, Eskisehir Technical University, Eskişehir, Turkey

Kıymet Tunca Çalıyurt CFE, CPA, Accounting—Finance Division, Department of Business Administration, Faculty of Business Sciences and Economics, Trakya University, Edirne, Turkey

Onur Çetin Faculty of Economics and Administrative Sciences, Department of Business Administration, Trakya University, Edirne, Turkcy

Ümit Doğan Department of Civil Aviation Management, Anadolu University, Eskişehir, Turkey

Dilek Erdoğan Department of Aviation Management, Tarsus University, Mersin, Turkey

Nuriye Günebakan Faculty of Aeronautics and Astronautics, Iskenderun Technical University, Hatay, Turkey

Hilal Inan Faculty of Economics and Administrative Sciences, Cukurova University, Adana, Turkey

Corina Joseph UiTM Cawangan Sarawak, Samarahan, Sarawak, Malaysia

Kasım Kiracı Department of Aviation Management, İskenderun Technical University, İskenderun, Turkey

Yeşim Kurt Lüleburgaz Faculty of Aeronautics and Astronautics, Department of Civil Aviation Management, Kırklareli University, Kırklareli, Turkey

Emircan Özdemir Department of Aviation Management, Eskişehir Technical University, Eskişehir, Turkey

Inci Polat Süleyman Demirel University, Isparta, Turkey

Metin Reyhanoğlu Faculty of Economics and Administrative Sciences, Hatay Mustafa Kemal University, Hatay, Turkey

Nisa Seçilmiş Faculty of Aeronautics and Aerospace, Aviation Management Department, Gaziantep University, Gaziantep, Turkey

Gökhan Tanrıverdi Department of Aviation Management, Ali Cavit Çelebioğlu School of Civil Aviation, Erzincan Binali Yıldırım University, Erzincan, Turkey

Akansel Yalçınkaya Faculty of Tourism, Istanbul Medeniyet University, Istanbul, Turkey

Harun Yılmaz Faculty of Aeronautics and Astronautics, Iskenderun Technical University, Hatay, Turkey

List of Figures

List of Tables

Part I
Introduction

Chapter 1
Introduction: Towards a New Management Approach in the Aviation Industry After the COVID-19 Pandemic

Kasım Kiracı and Kıymet Tunca Çalıyurt

Abstract We are announcing a new sub-series in our book series titled "Corporate Governance, Sustainability and Information System in the Aviation Sector". This series, as editors, we will discuss sustainable management issues in aviation sector. This series which we have been planning to open for a long time, has become more meaningful with the COVID-19 pandemic, which has hit the aviation industry since March 2019.

Keywords Airlines · Aviation · Management · COVID-19 · Pandemic

1.1 Introduction

The COVID-19 pandemic has significantly affected the airline industry. In order to reduce the spread of coronavirus, many countries have issued travel restrictions, especially air travel, to limit the travel volume. Many organizations have moved to working from home and using Zoom meetings to avoid any interruption. People chose to travel by car whenever possible. (Truong 2021) (Fig. 1.1).

Hanson et al. stated that the impact of COVID-19 on air travel could last longer than its impact on economic growth (Hanson et al. 2021).

The air transportation industry benefits society in a variety of ways, both economically and socially. It allows societies to interact culturally with one another. It makes it easier for countries and regions to develop trade relations. In addition to all of these benefits, the airline transportation industry interacts with a variety of organizations. The general public, trade unions, non-governmental organizations, consumers, and environmental organizations all play a role in air transportation regulation.

K. Kiracı
Department of Aviation Management, İskenderun Technical University, İskenderun, Turkey

K. T. Çalıyurt (✉)
CFE, CPA, Accounting—Finance Division, Department of Business Administration, Faculty of Business Sciences and Economics, Trakya University, Edirne, Turkey
e-mail: kiymetcaliyurt@trakya.edu.tr

K. Kiracı and K. T. Çalıyurt (eds.), *Corporate Governance, Sustainability, and Information Systems in the Aviation Sector, Volume I*, Accounting, Finance, Sustainability, Governance & Fraud: Theory and Application, https://doi.org/10.1007/978-981-16-9276-5_1

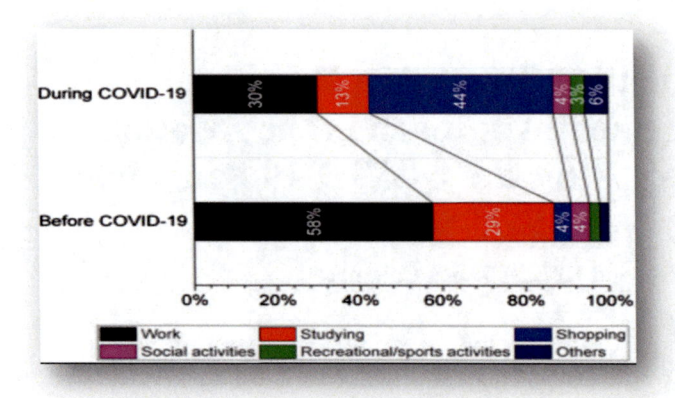

Fig. 1.1 Primary purpose of traveling before and during COVID-19 pandemic (Abdullah et al. 2020)

Climate change and global warming have recently become more important than ever. The global warming problem, which is caused by rising greenhouse gas emissions, as well as the environmental disasters that result from it, have a negative impact on all countries around the world. In this context, the airline transportation industry requires a structure that is both sustainable and environmentally friendly. In order for air transportation to become sustainable, all stakeholders, particularly aircraft manufacturers and airlines, must work together. Issues related to sustainability that are academically relevant to air transport industry stakeholders are discussed in "Corporate Governance, Sustainability, and Information Systems in the Aviation Sector–Volume 1". As a result, the subject of sustainability in the airline transportation industry has been examined in the book in the context of various disciplines and topics.

In the *second* chapter, operational resilience through disruption management in airlines, was examined in the context of sustainability. In this context, chapter briefly presents the concept of resilience and sustainability and introduces the airline operations and scheduling process. Afterwards, it provides detailed information on airline disruption management along with studies in the existing literature and methods used during the process.

In the *third* chapter, the sustainability process based on value co-creation was evaluated within the framework of the DART model. To provide a clearer understanding of the relationship between sustainability and value co-creation related to the sector, the actors in the ecosystem were evaluated separately.

The subject of chapter *fourth* is the corporate social responsibilities of aviation organizations. The chapter aims to reveal the relationship between aviation organizations and CSR initiatives and the effects of the COVID-19 pandemic on this relationship. The research area has been created with a global approach, from airport and airline organizations in different parts of the world. The research utilizes the literature review and secondary sources obtained from the websites of the relevant aviation organizations. According to the results of the chapter, airport organizations mostly carry out ethical and legal responsibilities related to health, hygiene, and safety measures against the pandemic. In addition to the aforementioned measures, airline organizations carry out voluntary corporate social responsibilities such as donating and transporting medical protective equipment, carrying the COVID-19 vaccine, free transport of pandemic groups, especially healthcare workers, or gifting flight miles to these groups.

In the *fifth* chapter, the relationship between sustainability and financial performance is examined for the airline industry. In the first stage of the chapter, authors have reached the last 10 years of financial data of top 20 airlines operating globally. In the second stage, authors determined variables to measure the sustainable financial performance of airlines. Authors used the studies in the literature for the determination of the variables. Within the scope of the study, we created four different models in order to reveal the financial performance of airlines.

The *sixth* chapter focused on sustainable aviation based on innovation. The aim of chapter is to reveal the importance of the technological steps taken in the fight against climate change within the framework of sustainable aviation. Patent data was used as a technology development indicator to measure the success of political action. The findings of the chapter confirm that there is a consensus among all industry stakeholders that the most effective policy tool in this struggle is sustainable innovation. In this context, research and development on aircraft and engine technology and alternative jet fuels has been considered a priority for sustainable aviation in order to reduce aviation-related GHG emissions.

In the *seventh* chapter, intention to buy and willingness to pay more for green airlines are analyzed in the context of Theory of Planned Behavior (TPB). The chapter is one of the first attempts in the airline industry to associate the purchase intention for green airlines with TPB. The outputs of the TBP model reveal the green purchase intention of the airline passengers and their willingness to pay more for green airlines.

In the *eight* chapter, factors affecting passengers' avoidance of using airline mobile applications were examined. results of the chapter; perceived irritation is another factor affecting passengers' avoidance of using mobile apps of airlines. From this point of view, if individuals perceive an irritation towards the mobile apps themselves or their content, they may avoid to use the app with the motivation to protect themselves. Therefore, airline companies should ensure that their mobile apps are attractive and not distracting passengers from using their mobile apps.

In the *ninth* chapter, supplier performance was evaluated by using cluster analysis and artificial neural networks in maintenance and repair overhaul (MRO) in aviation industry. The chapter is important in that the supplier evaluation studies related to the MRO sector are generally limited in the literature and very limited in the domestic literature. The results of the chapter indicate that from a managerial point of view, classifying suppliers with CA instead of average score may be an alternative to the company's existing supplier evaluation system.

In the *tenth* chapter, the issue of strategic outsourcing in the airline transport industry was examined. Outsourcing has become a vital source of competitive advantage because it provides several benefits for organizations, such as reducing the cost of ownership of products/services, resolving technical problems without increasing the number of staff, and enabling the company to focus on its core business. Airline companies outsource in many different functions such as ticket sales and distribution, aircraft leasing, maintenance training, information system and technology, pilot training, advertising, engine overhaul, ground handling operation. The results of the chapter indicate that outsourcing the reservation system, which affects the main business of the airline, is a strategic decision. Because the reservation system is customized according to the company, that is, it is a process with high asset specificity, which increases the dependency of the buyer on the supplier.

In the *eleventh* chapter, the importance of organizational behavior model applications in air transportation industry was examined. The aim of the chapter is to emphasize the importance of applying organizational behavior models in the air transportation industry. Thus, it will be possible to identify the problems experienced by individuals working in the aviation industry and to produce solutions. The results of the chapter indicate that companies applying organizational behavior models will be one step ahead in achieving sustainable growth besides gaining advantage in competition.

In the *twelfth* chapter, role of strategic alliances in achieving sustainable competitive advantage in the airline industry is examined from the perspective of Resource Dependency Theory (RDT). According to Resource Dependency Theory (RDT), strategic alliances are one of the political actions and behaviors that organizations operating in intense competitive environments resort to gain access to necessary resources and avoid resource uncertainty. In the chapter, authors argue how strategic alliances provide sustainable competitive advantage to airlines from resource dependence theory perspective in this systematic and bibliometric review. The findings of the chapter confirm that strategic alliances are seen as a network and that airlines achieve a sustainable competitive advantage through access to network resources.

The aim of chapter *thirteenth* is to understand and investigate the historical change of the board composition of Turkish Airlines from 1956 to 2020. For this purpose, the members of the board have been analyzed from the political economy perspective and resource dependence theory and the data obtained from the various secondary data. Research findings of the chapter showed that there are three periods which can be named as Militarization period (1956–1982), Demilitarization period (1983–2002), and Politicization period (2003–2020).

Fourteenth chapter aims to investigate the extent of responsiveness plans and actions undertaken by three Malaysian aviation companies in battling the COVID-19 pandemic. The analysis includes the examination of online news which was captured from a Google search from a period 18 March to early October 2020. There were 82 online news captured involving three aviation companies, i.e., AirAsia, Malaysian Airline System and Malindo. The results of the chapter indicate that the online news is not used extensively to communicate about the strategic responsiveness plan in mitigating the COVID-19. Perhaps both airlines are legitimizing their strategic responsiveness plan via other communication platforms such as websites or internal bulletins.

In this section, the effects of the COVID-19 pandemic on the aviation industry were discussed. Then, the fourteen chapters in the book were summarized. We expect book chapters and book studies from all academicians and practitioners working in the aviation industry to our series. We wish you pleasant reading.

References

Abdullah M, Dias C, Muley D, Shanin M (2020) Exploring the impacts of COVID-19 on travel behavior and mode preferences. Trans Res Interdiscip Perspect. https://doi.org/10.1016/j.trip.2020.100255

Hanson D, Delibasi TT, Gatti M, Cohen S (2021) How do changes in economic activity affect air passenger traffic? The use of state-dependent income elasticities to improve aviation forecasts. J Air Trans Manag 98:102147. https://doi.org/10.1016/j.jairtraman.2021.102147

Truong D (2021) Estimating the impact of COVID-19 on air travel in the medium and long term using neural network and Monte Carlo simulation. J Air Trans Manag 96:102126. https://doi.org/10.1016/j.jairtraman.2021.102126

Kasım Kiracı received his first BSc degree in Aviation Management in 2010 from Kocaeli University and second BSc degree in Economics in 2015 from Anadolu University. He obtained his first MSc degree in Economics from Gebze Technical University and second MSc degree in Aviation Management from Anadolu University. He completed his PhD in Aviation Management with the dissertation titled "Determinants of The Capital Structure in Different Business Models: A Panel Data Analysis On Low Cost and Traditional Airlines" at the Anadolu University, Eskişehir in 2017. Dr. Kiracı joined Faculty of Aeronautics and Astronautics, Iskenderun Technical University in 2018. He has been working as a faculty member at the Department of Aviation Management at İskenderun Technical University since 2018. He received his associate professorship in finance in 2021. Kasım Kiracı has many articles published in web of science journals on aviation economy and airline financing. In addition, there are many book chapters published on different topics of aviation.

Part II
CSR and Sustainability in Airlines

Chapter 2
A Necessity for Sustainability: Operational Resilience Through Disruption Management in Airlines

K. Gülnaz Bülbül

Abstract In today's turbulent, complex, ever-growing, global system organizations face various challenges such as terrorist attacks, pandemic diseases, natural disasters, economic recession, human error, equipment failure, besides minor disruptions. These challenges pose severe threat to the operational continuity of organizations. This vulnerability brings resilience forward as an important issue since it is a necessary precondition for sustainability. Air transportation, which is one of the crucial components of this global system, has also widened its network while gaining more and more importance every day. Due to its scale and complexity, it has also become more vulnerable to disruptions caused by internal and external factors such as weather conditions, strikes, political reasons, aircraft mechanical problems, sickness of crew. These factors can break operational continuity of airlines—one of the main components of air transportation—by causing flight disruptions. A single flight disruption may trigger a snowball effect causing delays or even cancellations on several other flights legs. As today, most of the large airlines serve on a tight hub-and-spoke network, a single glitch in a hub can swiftly affect the whole schedule of an airline. Considering that the notable growth in air transportation has resulted in congested airports and airspace, serving on a large network with limited resources can easily contribute to the propagation of each disruption, costing airlines billions of dollars each year. Thus, effective disruption management is crucial for airlines, as it aims to predict the occurrence of disruptions and to find feasible plans, considering the costs, that allow the airline to recover from these disruptions and their associated delays. In the scope of disruption management, robust and dynamic operational research methodologies have gained popularity. However, due to high system complexity, disruption management is usually covered in different problems such as robust aircraft scheduling, robust crew scheduling, aircraft recovery, crew recovery, and passenger recovery problems. In this context, this chapter, first, briefly presents the concept of resilience and sustainability and introduces the airline operations and

K. G. Bülbül (✉)
Department of Flight Training, Faculty of Aeronautics and Astronautics, Eskisehir Technical University, Eskişehir, Turkey
e-mail: kgbulbul@eskisehir.edu.tr

scheduling process. Afterwards, it provides detailed information on airline disruption management along with studies in the existing literature and methods used during the process.

Keywords Sustainability · Resilience · Airline planning and scheduling · Airline disruption management · Airline recovery problem · Robust airline scheduling

2.1 Introduction

Due to globalization, today, economies, populations and cultures are now linked in a large, complex and ever-changing network. In this volatile global system, organizations can face a wide variety of challenges. Besides minor uncertainties or disturbances, terrorist attacks, pandemic diseases, natural disasters, economic recession, human error, equipment failure are some of the main, unpredictable factors that can pose significant threat to the operational continuity of an organization (Bhamra et al. 2011). This vulnerability brings resilience forward as an important issue. Although the term resilience is used in various different fields, in each context the concept is closely related to the capacity of an element to return to a stable state subsequent to a disruption and continue to function (Anderies et al. 2013; Holling 1973; Derissen et al. 2011; Fiksel 2006). In business context, it is defined as the capacity for an enterprise to survive, adapt, and grow in the face of turbulent change (Fiksel 2006).

Resilience is identified as a necessary precondition for sustainability such that, without it a system can only possess insubstantial sustainability (Derissen et al. 2011; Marchese et al. 2018). For a sustainable development, it is essential to develop adaptive strategies to cope with abrupt challenges, instead of holding on to"steady-state" models (Fiksel 2006).

Air transportation is one of the crucial components of today's world. With globalization, air transportation system has also widened its network while gaining more and more importance every day. However, it has also become more vulnerable to disruptions caused by internal and external factors. Adverse weather conditions, strikes, political reasons, aircraft mechanical problems, sickness of crew, and other different causes can impact the long-term success of an airline by causing flight disruptions and thus breaking operational continuity (Jimenez Serrano and Kazda 2017; Ball et al. 2007).

Flight disruption occurs when a scheduled flight leg is canceled, or delayed for 2 h or more, within 48 h of the original scheduled departure time (ACI 2013). Since a flight leg is a component of various schedules, a disturbance in timing of a single leg may trigger a snowball effect causing delays or even cancellations on several other legs (Jimenez Serrano and Kazda 2017; Ball et al. 2007). Considering most of the large airlines serve on a tight interconnected hub-and-spoke network, a single glitch in a hub can swiftly affect the whole schedule of a carrier. Moreover, resource shortages can worsen this impact. The notable growth in air transportation has resulted in congested airports and airspace (Guimarans et al. 2018; Jimenez Serrano and Kazda 2017). Serving at maximum capacity increases operational disruption risk since limited resources can easily contribute to propagation of one single disruption. Disruptions cost airlines and their costumers up to $60 billion yearly, which is approximately 8% of worldwide airline revenue (Gershkoff 2016). Hence, disruption management, which aims to predict the occurrence of disruptions and to find feasible, cost-minimizing plans that allow the airline to recover from these disruptions and their associated delays, is crucial for airlines (Ball et al. 2007; Guimarans et al. 2018). This raised attention to airline disruption management both in the literature and airline industry, where robust and dynamic operational research methodologies have gained popularity. However, due to high system complexity disruption-management is usually covered in different problems such as robust aircraft/crew scheduling, aircraft/crew/passenger recovery problems (Guimarans et al. 2018; Ball et al. 2007).

In the scope of this study, first a brief definition of resilience and its connection with sustainability is presented. Then airline operations and scheduling process is introduced. Finally, detailed information on airline disruption management is provided along with studies in the existing literature and methods used during the process.

2.2 A Brief Description of Sustainability

For almost over four decades, importance and thus awareness of sustainability have increased prominently among both government and industry. Sustainable development has become one of the main strategic agendas for companies. This led an increasing amount of research on concept of sustainability in various perspectives. Although the definition can alter according to the context, the main focus of sustainability is enhancing the quality of an entity, considering environmental, economic and social aspects. With the inclusion of these three pillars, sustainability is a compound concept. The most common and cited definition of this concept belongs to United Nations World Commission on Environment and Development (UCLA Sustainability Committee, n.d.). The term is defined as "development that meets the needs of the present without compromising the ability of future generations to meet their own needs" in 1987 (Emas 2015).

Today's global system consists of highly complex, dynamic, nonlinear, and interdependent sub-systems. This presents a need for a comprehensive system approach for achieving sustainability, though collaborative decision making processes. It is crucial to develop adaptive and dynamic models, instead of holding on"steady-state" ones of static optimization, in order to attain sustainability. This is only possible by an improved insight into complex systems, their dynamic, adaptive behavior and resilience against disruption (Fiksel 2006; Anderies et al. 2013). Correspondingly, UCLA Sustainability Committee defines sustainability as "the integration of environmental health, social equity and economic vitality in order to create thriving, healthy, diverse and resilient communities for this generation and generations to come. The practice of sustainability recognizes how these issues are interconnected and requires a systems approach and an acknowledgement of complexity."(UCLA Sustainability Committee, 2021).

Fiksel (2006), mentions that according to the U.S. Environmental Protection Agency's Office of Research and Development accomplishing sustainable systems involves a set of essential challenges (Fiksel 2006):

- Addressing multiple scales over time and space.
- Capturing system dynamics and points of leverage or control.
- Representing an appropriate level of complexity.
- Managing variability and uncertainty.
- Capturing stakeholder perspectives in various domains.
- Understanding system resilience relative to foreseen and unforeseen stressors.

Without a doubt, sustainability is becoming more and more crucial and challenging as systems get even more complex and interconnected. These challenges can only be overcome by addressing sustainability as a systems problem that demands collaborative solutions. This necessity, of the decision-making for sustainability, leads to the need of a set of sustainability science tools. As one of these tools resilience will be discussed in Sect. 2.3.

2.3   The Concept of Resilience

The concept of resilience was first introduced by Holling (1973). Holling (1973) defined resilience as the size of the stability domain around stable time-invariant equilibria (point attractors) or stable oscillations (periodic attractor), in the context of ecological systems (Holling 1973; Lele 1998). Ever since it was first described, the term has been employed by a wide variety of fields such as engineering, psychology, economics, strategic management, supply chain management, and transportation (Bhamra et al. 2011; Wan et al. 2018). In each of these various contexts, the main focus of resilience is the ability to return to a stable state after a disruption and continue to function (Holling 1973; Fiksel 2006; Derissen et al. 2011; Anderies et al. 2013). It encompasses not only individual but also organizational responses to unexpected disturbance and disruptions (Bhamra et al. 2011).

Carpenter et al. (2001) define three main properties of resilience (Carpenter et al. 2001):

- The amount of change that a system can undergo while retaining the same controls on structure and function.
- The degree to which the system is capable of organizing itself without disorganization or force from external factors.
- The degree to which a system develops the capacity to learn and adapt in response to disturbances.

Considering these main properties, resilience can be expressed as a function of the vulnerability of a system and its adaptive capacity. Vulnerability is commonly defined as being subject to a range of disruptions where adaptive capacity indicates the learning aspect of system behavior regarding a disruption (Carpenter et al. 2001; Bhamra et al. 2011). Adaptive cycle theory states that dynamic systems journey through four stages; rapid growth and exploitation, conservation, creative destruction, and renewal or reorganization, rather than being stable (Ponomarov and Holcomb 2009).

Both communities and organizations are dynamic and highly complex systems that are in the form of networks composed of interconnected and interdependent entities. This type of interaction is more prone to disruptions caused by the uncertain environment. This is mainly because of it is challenging to have control over complex systems and facilitate constructive communication, between elements, in times of disruptions. Thus, as a system becomes more complex, resilience gains more importance and becomes even a harder goal to achieve.

Ponomarov and Holcomb (2009) describe three phases of resilience:

- readiness and preparedness,
- response and adaption, and
- recovery or adjustment.

It is clear that developing and implementing procedures in order to enhance resilience is a must, since systems are drifting to increased complexity with globalization. Each phase of resilience should be addressed punctiliously while considering the mutual relationship between each phase.

2.4 Relationship Between Sustainability and Resilience

The concepts of sustainability and resilience are presented in previous sections. From the definitions given in these sections, it can be seen that the concepts resemble one another, where a clear distinction is fairly difficult. They are both used to describe a system, which can be almost any one (Carpenter et al. 2001). Both resilience and sustainability address the state of a system focusing on the durability of that system under either ordinary or abrupt conditions. Because of the aspect of"continuity" that they both contain, in some studies they are used interchangeably. However, in spite of the similarities, resilience and sustainability are two distinct concepts, which are frequently misused.

To begin with, sustainability focuses on longer time scales, while the need of resilience belongs to a more specific point in time (Derissen et al. 2011; Marchese et al. 2018). Furthermore, they dissociate in the way that they are used in decision making. To a greater extent, sustainability is often a concern of institutional level decision-making, where goals range from a single process enhancement to constructing community well-being through development.

Resilience, on the other hand, generally is a weapon used to defuse interruptions (Marchese et al. 2018). Marchese et al. (2018) present a literature review in their study where they focused on the conflict about the relationship between resilience and sustainability. They concluded that there are three management frameworks for regarding the relation between sustainability and resilience (Marchese et al. 2018):

1. resilience as a component of sustainability,
2. sustainability as a component of resilience, and
3. resilience and sustainability as separate objectives.

In Framework 1, resilience harmonizes with formerly established institutional boundaries of sustainability. Framework 2 suggests that being economically, socially, and ecologically sustainable makes a system more resilient. According to Framework 3 resilience and sustainability are two separate concepts without a hierarchical relationship. Considering these different points of view, the similarities and differences between sustainability and resilience is framework-dependent (Derissen et al. 2011).

In the scope of this chapter, we define resilience as a necessity for sustainability. Fiksel (2006) argues that providing sustainability in the face of ever-changing global environment depends upon resilience. From this perspective, development of modeling and decision-making tools that encourage adaptive and dynamic management, instead of static optimization, should be prior concern of sustainability studies. In order to achieve that, alternatives and their relative competence in improving system resilience should be analyzed thoroughly (Fiksel 2006).

Anderies et al. (2013) suggest that resilience provides a rich set of ideas that can contribute to sustainable management in the face of turbulent environment. They define sustainability as an indicator of system performance, a goal, while resilience is a substantial concept in the success of the processes that leads the way to this goal (Anderies et al. 2013). Thus, it is essential to comprehend the skill set for achieving resilience, since it affects the sustainability of a system (Bhamra et al. 2011). In other words, sustainability is a fundamental concept for systems to maintain, where resilience is a necessary precondition for sustainability (Derissen et al. 2011).

2.5 Operational Resilience

Commonly organizations are, too, complex systems that consists of interconnected elements that need to be adaptive in order to survive. Today all kinds of organizations in each industry, government, and academia are operating in an uncertain environment that is subject to sudden unexpected events (Caralli et al. 2010). Natural disasters, economic crises, pandemic diseases, terrorist attacks, political crises present a great source of risk to the operational continuity of organizations (Bhamra et al. 2011; Essuman et al. 2020). Accordingly, in light of the threat that these disruptive events present, resilience has become more of an issue among organizations. This is where the concept of operational resilience is introduced.

Operational resilience focuses on the ability of an organization to cope with unwanted events that affect its core operational capacities (Stolker et al. 2008; Caralli et al. 2010). These events can have a high or low impact and a high or low probability of occurrence.

Stolker et al. (2008) define the capabilities of an organization related to operational resilience as (Stolker et al. 2008):

- The ability of an organization to prevent disruptions from occurring.
- The ability to quickly respond to and recover from a disruption when is occurs.

Operational resilience provides the ability of operations to achieve their mission. An operationally resilient organization can attain its mission under disruptive conditions. On the other hand, an organization is not operationally resilient if it is not capable of returning to normal after being disrupted (Allen and Davis 2010). As the definition suggests, operational resilience is a multidimensional concept that encompasses two specific components; disruption absorption, and recoverability. Each component has a particular positive effect on efficiency under different disruptive conditions (Essuman et al. 2020). Effective disruption management plays a key role in operational efficiency. Organizations that have high disruption absorption capacity are more successful in preserving their structure and functioning of operations. This allows such organizations to sustain their normal performance levels, eliminating the risk of inefficiency caused by disruptions. Besides, organizations with high recoverability are more competent at restoring operations after disruptive events. This contributes to reducing the risk of propagated inefficiency (Essuman et al. 2020).

It is suggested that recoverability is mostly associated with flexibility, while disruption absorption is related with investment in redundancies and buffer capacities. Thus, resilience comes with a cost (Essuman et al. 2020). The trade-off between resilience building and the investment needed to succeed that should be considered meticulously. In situations that present a low operational disruption risk, it can be counterproductive and thus inefficient to increase operational resilience.

It is evident that understanding the concept of operational resilience along with its behavior is vital in the efficient execution of operations. Therefore, organizations should define their operational processes in detail and determine the need of resilience carefully in order to achieve competitive advantage.

2.6 Airline Operations and Planning

There are two main components of air transportation service; airports and airline companies. Composed of a gigantic network of dynamic processes, air transportation requires detailed planning at every level. In order for all processes to operate efficiently, detailed planning and scheduling is crucial for both airports and airline companies. In airlines, scheduling is a complex process that starts a year ahead of the operation (i.e., flight), and continues until the moment of flight. An airline schedule is a tool that represents the basic plan of the company, matching available resources with demand (Grosche 2009). An airline schedule considers the operational, strategic, and marketing objectives of the company, in a way that maximizes the profits. There are three main components of an airline schedule; flight, aircraft, and crew schedules (Barnhart et al. 2003; Belobaba 2009). Due to its complexity, it is highly challenging to model airline scheduling problem en-masse. Thus, the common approach is to divide it into sub-problems and solve each sub-problem separately. The main sub-problems are flight scheduling, aircraft scheduling and crew scheduling.

Flight scheduling is a long-term planning activity. After a flight schedule is constructed, aircraft scheduling is made based on this schedule. Aircraft scheduling is composed of two main sub-problems as fleet assignment and aircraft routing. Following the aircraft scheduling, crew schedules are determined according to the aircraft flight assignments. Crew scheduling process also contains two main sub-problems; crew pairing and crew assignment (Fig. 2.1).

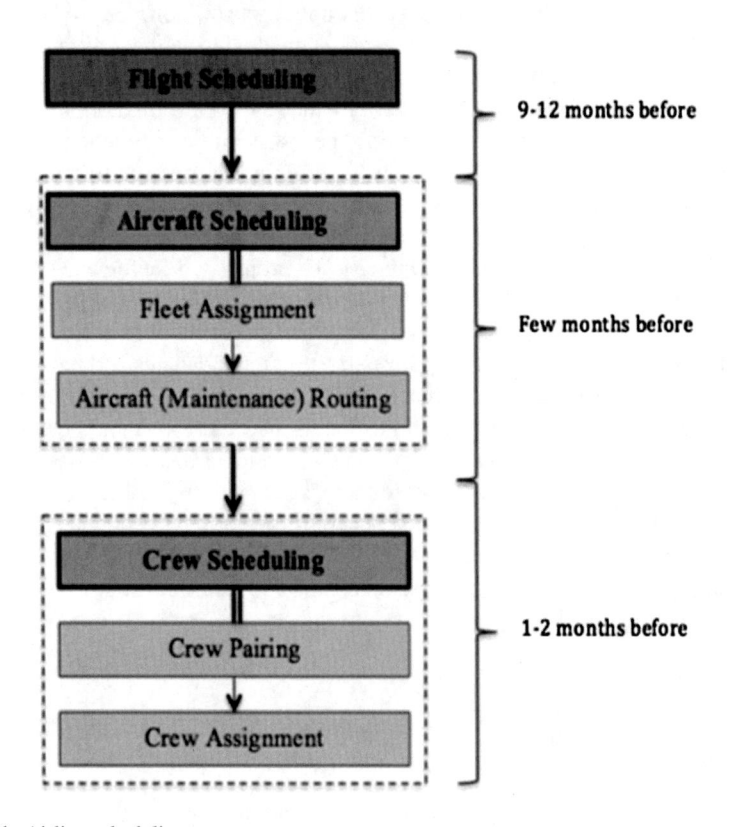

Fig. 2.1 Airline scheduling process

2.6.1 Flight Scheduling

A flight schedule is the output of the process in which the city pairs (i.e., flight legs) to be flown and the flight times are determined for a certain time interval, considering the passenger demands. It is the first phase of the airline scheduling process. Flight scheduling plays a key role on managing and creating passenger demand. It is also a crucial step, as it leads and has a great effect on the other planning steps (Patty and Diarnond 1998). The purpose of this step is to create a timetable that contains information on arrival–departure airports and times, flight frequency, and days. The main limitations of flight scheduling process are size and structure of fleet and other resources, regulations and restrictions, and traffic rights.

Three sub-problems can be defined within the scope of flight schedule process (Grosche 2009):

- Network design.
- Frequency assignment.
- Schedule generation.

Network design focuses on identifying origin–destination city pairs that the airline intends to serve and the network structure. Two common network structures are point-to-point and hub and spoke network structures. In point-to-point networks, there are direct flights between all cities (i.e., point) pairs. Hub-and-spoke networks, on the other hand, are composed of collection and distribution airports. There are only direct flights between larger cities (hubs), while smaller cities (spokes) are connected via hub airports. It is possible to serve a wider market with hub and spoke network (Patty and Diarnond 1998). Point-to-point network structure is more appropriate for markets where the distances between cities are very short (long), demand is high (low), and the number of cities are low (high) (Lederer and Nambimadom 1998).

Frequency assignment is the phase to determine flight frequencies for each city pairs. Frequency together with the ticket prices has a major effect on a passenger's preference of a flight or airline. Thus, it plays a key role in creating a competitive advantage (Grosche 2009).

In schedule generation phase, exact departure times of each flight leg is determined. Departure times mainly depend on the distribution of demand and connection possibilities. Times that the demand is relatively higher (peak times), are identified and considered during the schedule generation process, while fleet, crew, airport constraints, and operational costs are also taken into account.

2.6.2 Aircraft Scheduling

The aim of aircraft scheduling is to assign a tail number to each flight leg in the schedule generated in flight scheduling process. As mentioned earlier, aircraft scheduling process is usually considered as two sub-problems.

First fleet assignment problem is solved. The fleet assignment problem aims to assign an aircraft type to each flight leg with the objective of maximizing the profit. In the most general definition, it is the effort to match the demand (passengers) with the resource (number of seats every flight legs) (Bazargan 2010). The main factors to be considered in the fleet assignment process are cruise speed, fuel consumption, capacity of the aircraft, noise restrictions, requirements related to flight crew, ground equipment and maintenance, and gate restrictions (Grosche 2009; Bazargan 2010). In the fleet assignment problem, there are two types of costs that must be minimized; operation cost and the spill cost (Barnhart et al. 2003; Grosche 2009). The main constraints that need to be considered in the scope of fleet assignment are (Barnhart et al. 2003; Grosche 2009; Bazargan 2010):

- Fleet size: The total number of aircraft assigned of a type should not be greater than the fleet size.
- Flight coverage: Only an aircraft type should be assigned to each flight.
- Flow balance: The number of arrivals and departures for each aircraft type should be equal for each airport.
- Regulatory restrictions: Aerodrome's operational limitations for each aircraft type should be considered (noise limitation, gate sizes, length of runway, etc.)

Following the fleet assignment, in aircraft routing phase, a physical aircraft (tail number) is assigned to flight legs in a fleet schedule. The aircraft routing problem is generally solved for each fleet separately. At this stage; feasible sequences of flight legs, where the arrival point for one flight leg is the departure point of the next one, are generated and assigned to an individual aircraft. There are maintenance requirements for each type of aircraft, specified by regulations. These maintenance requirements are also considered in aircraft routing. Hence, this sub-problem is also commonly defined as aircraft maintenance routing. Minimizing number of aircraft used, maximizing through value, and maximizing robustness are some main objectives that are considered in the scope of aircraft maintenance routing problem.

2.6.3 Crew Scheduling

The purpose of the crew scheduling problem is to assign crew members (cockpit and cabin) to flight legs. Crew scheduling is done after aircraft scheduling mainly because flight crew are licensed to serve in different aircraft types. Crew costs is the second highest cost for airlines, following the fuel costs (Barnhart et al. 2003; Grosche 2009). In this regard, an optimal crew schedule has a crucial role in reducing costs. Crew scheduling problem is highly complex as it considers regulations, union agreements, and operating policies and thus divided into two sub-problems as crew pairing and crew assignment.

In crew pairing stage, multi-day work schedules with the lowest cost and highest team utilization rate, called pairing, are created. The limitations to be addressed within the scope of the crew pairing problem are as follows (Barnhart et al. 2003; Grosche 2009);

- Maximum permitted working hours per day;
- Maximum flight hours allowed;
- Maximum allowed working hours on consecutive days;
- Restrictions for the shortest and longest rest time between duties;
- Maximum number of flight duties in a pairing;
- Restrictions for the shortest and longest rest time between flight legs;
- Maximum number of days the team can stay away from their home base.

Main cost items to be considered in crew scheduling can be listed as follows (Grosche 2009):

- Cost per flight hour.
- Accommodation expenses.
- Total elapsed work time.
- On-duty time costs.
- Deadhead costs.
- Minimum guarantee costs.

After crew pairings are generated, individual crew members are assigned to these pairings on a monthly basis. The objective of this step is to make efficient and fair assignments. There are two main approaches used in crew assignment; bidline generation and rostering. In bidline generation, crew members choose the pairings they want to be assigned. Seniority is considered during this process. The senior the crew member, the higher the bidding priority. Rostering, on the other hand, is a relatively complex method. Crew members are assigned directly to flight legs, without any bid. However, while doing this, the following restrictions on crew members are taken into account (Grosche 2009):

- Annual leave.
- Sick leave.
- Training or observer flights.
- Visa regulations.
- Medical appointments.
- Flight and rest periods determined by regulations.

2.7 Operational Resilience in Airlines

Air transportation is an indispensable component of global business and tourism as it enables worldwide rapid transport. Along with the globalization, in the last century, both the capacity and the network of air transportation system has expanded enormously. In this period of time, air transportation network evolved to a highly connected network that is capable of serving over. 4.3 billion passengers in 2018 (ICAO 2018). In 2019, there was an increase of 4.2% in demand (revenue passenger-km(RPK)) compared to the previous year. Morever, the capacity (available seat-km (ASK)) increased by 3.4%, while the load factor increased by 0.7 points, reaching a record high of 82.6% (IATA 2020c). The International Air Transport Association (IATA) stated that airline passenger traffic, which grew by 7.3% worldwide in 2018, grew by 4.2 % in 2019, despite difficulties such as economic recession, weak global trade, political and geopolitical tensions, Boeing 737 MAX crisis, strikes, Brexit uncertainty (IATA 2020c).

As air transportation system is probably the most intricate person-made system beyond being possibly the most complex transportation system, it has been a focus of interest among researchers (Ball et al. 2007; Cardillo et al. 2013). There are numerous studies on the properties of airline transportation systems by means of network theory which aims to reveal its structural characteristics and behavior. Air transportation system is also a multi-layered system whose nodes commonly belong to different layers at the same time, just like other real-world systems (Cardillo et al. 2013). Airports are the fixed nodes of the system where aircraft are the components that provide the transportation service in terms of moving an entity from a point to another. There are passengers who demand to travel between various origins and destinations at a specific time. There are also flight crew members who carry out the flight service. These are the most obvious components of air transportation system which has numerous other sub-elements. All elements, in different layers, have different kinds of interactions, coordinated by flight schedules, in this complex network (Ball et al. 2007; Janić 2015).

As a result of the augmenting complexity and uncertainty in worldwide systems, air transportation, maybe more than other modes of transportation, is vulnerable to a wide range of disturbances. Besides minor disruptions; terrorist attacks, pandemic diseases, natural disasters, economic recession, strikes, human error, equipment failure, sickness of crew and other different causes pose significant threat to the operational continuity of airlines (Janić 2015; Jimenez Serrano and Kazda 2017).

The disruptions caused by the Icelandic eruption of Eyjafjallajökull volcano, in 2010, have led airlines to lose about $ 1.7 billion of revenue globally. It also estimated that 1.2 million passengers a day were affected (BBC 2010). In 31 December 2015, due to the expectation of strong southwester and rain in Marmara, Turkish Airlines, has canceled a total of 100 domestic and international flights from and to its hub, Atatürk International Airport (Onedio 2015). Brussels Airport shut down, and suspended all operations due to the terrorist attacks, 2016 for 4 days (Vora 2016). Cabin crew of Lufthansa Airlines went on a strike in 2019, and a 48-h strike resulted 1,300 Lufthansa flights to be cancelled (DW 2019). In 7 February 2020, an aircraft of Pegasus Airlines skiddedoff the runway of Sabiha Gökçen International Airport in forcing the runway to be closed for almost a day (Charpetreau 2020). The most recent and the most outspread disruptive event is COVID-19 pandemic. Due to the pandemic Global RPKs decreased by 94.1%, while ASK decreased by 87%, and saw the bottom, in April 2020, when it is compared to the year before (IATA 2020a). It is still recovering but with cautious baby steps. In IATA's report, that represents the August 2020 numbers, RPKs fell by 75.3% and ASKs fell by 63.8% year-on-year (IATA 2020b).

As the examples unveil, air transportation operations only rarely develop according to the plan. There is almost no day without a disruptive event in air transportation, since a flight leg is a constituent of various schedules (i.e., flight and crew schedules and passenger itineraries). A tiny disturbance in a single leg can lead to a snowball effect, resulting a wider system disturbance and it is a known fact that flight disruptions are very common (Jimenez Serrano and Kazda 2017; Ball et al. 2007).

Flight disruption occurs when a scheduled flight leg is cancelled, or delayed for two hours or more, within 48 h of the original scheduled departure time (ACI 2013). Serving at maximum capacity, in congested airports and airspace, increases operational disruption risk since limited resources can easily contribute to propagation of one single disruption. Disruption cost airlines and their costumers up to $60 billion per year, or about 8% of worldwide airline revenue (Gershkoff 2016). Moreover, failure of air transport system brings a socio-economic burden.

This brings the operational resilience forward in the agenda of airlines, as an important issue. It is mentioned that there are four main characteristics of a resilient transportation system; redundancy, robustness, and recoverability. These characteristics are claimed to determine the overall performance of the system on how long it is able to maintain operations without collapsing, what processes it will follow during a disruption and how long will it take to go back to normal after being disrupted (Wan et al. 2018). In the light of these, airline operational resilience is the ability of an airline to maintain its planned function during disruptions and its agility in recovering up to the specified state after facing one. Hence, disruption management, which aims to predict the occurrence of disruptions and to find feasible plans to airline to recover from these disruptions and their associated delays, is crucial for airlines (Ball et al. 2007; Guimarans et al. 2018). In the following sections description of airline disruption management is presented along with studies in the existing literature and methods used during the process.

2.8 Airline Disruption Management

A variety of events were mentioned, in the previous section, which can affect the operational continuity of airlines and cause disruptions by pointing out that disruption management is crucial for operational resilience of airlines. As described in Sect. 2.5 operational resilience is composed of two merits; preventing/absorbing disruptions and respond to and recover from them expeditiously. Hence, disruption management in airlines concerns with minimizing disruptions and their impacts as well as responding to and recovering from them. From this point of view, airline disruption management encompasses two approaches; robust planning as a proactive approach and airline recovery as a reactive approach (Barnhart 2009; Lee 2020). Robust planning concerns with preventing disruptions and enhancing the ability of airline to absorb them by reducing airlines vulnerability to future disruptions. Airline recovery, on the other hand, focuses on re-planning and recovering of airline resources with the objective of minimizing the costs related to disruption. In this section an overview of robust planning and airline recovery problem is presented along with successful applications.

2.8.1 Airline Recovery

Main concern of an airline during irregular operations is bringing operations back to plan. In order to achieve that, airline operation controller center (AOCC) re-allocate the resources of an airline with the goal of finding a minimum cost, feasible plan that enables the airline to recover from the disruption. When disruption occurs, AOCC alters scheduled operations by the actions listed below (Rapajic 2016; Arıkan et al. 2017; Marla et al. 2017; Marla et al. 2020).

- Delaying flights
- Cancelling flights
- Diverting flights
- Swapping aircraft
- Introducing additional flights
- Calling up reserve crew
- Deadheading crew
- Ferrying aircraft
- Reassigning crew
- Re-accommodating passengers

Alterations, that arise from irregular operations, should also satisfy all constraints included in airline scheduling process (See Sect. 2.6). Airline recovery planning may have different objectives, which some of them sometimes considered together. Minimizing reserve crew cost, passenger recovery cost, loss of passengers' goodwill and total time spend on bringing operations back to plan are some possible objectives of schedule recovery planning (Barnhart 2009). The major difference between airline scheduling process and recovery process is the time span. Regardless of the objective, most of the times, recovery problems must be solved within a short time period or the solution may become obsolete. Considering time limit along with the scale of the airline scheduling problems, as one would expect, it is even more difficult to solve a recovery problem as a whole. Thus, most airline recovery processes are sequential, consisting of three steps that address recovery of single resource; aircraft, crew, or passengers. The first step is aircraft recovery which encompasses decisions on flight leg delays, aircraft swaps and flight leg cancellations. The second step is crew recovery. Crew recovery focuses on assigning crews to uncovered flight legs. Cew recovery; comprises reassigning of crews and assigning of reserve crews. The last step is passenger recovery where disrupted passengers, whose planned itinerary is broken and impossible to execute during irregular operations, are re-accommodated (Ball et al. 2007).

2.8.1.1 Aircraft Recovery

One of the critical resource in airline operations is aircraft. Thus, aircraft recovery is crucial, during schedule disruptions, for an airline's operational continuity. Aircraft recovery problem focuses on determining altered departure times and cancellation of flights, and re-routings for affected aircraft while addressing a specific objective (e.g., minimizing cost, minimizing delay). Aircraft re-routing encompasses several options (Ball et al. 2007);

- diverting-landing to an alternate airport
- swapping-re-assignment of flights among different aircraft
- ferrying- relocating a vacant aircraft to another airport, where it can be utilized.
- over-flying-flying to another scheduled destination

The actions to be taken for aircraft recovery must comply required rules such as maintenance and aircraft balance requirements and station departure curfew restrictions. Aircraft should be positioned accordingly in order to resume operations as planned at the end of the irregular period. It is usually not possible to only cancel a single flight leg since it requires reserve aircraft of the same type assigned to the destination of the cancelled leg. Delaying departures, making cancellations, swapping aircraft are decisions that have a snowball effect. They usually require significant changes to crew schedules, passenger itineraries and maintenance plan (Barnhart 2009). All these requirements and consequences of aircraft recovery decision add up to problem complexity. Consequently, it has been a focus of interest among researchers, who presents a wide variety of problem and model descriptions and solution approaches.

Researches on aircraft recovery problem dates back to 80's. In 1984, Tedorovic and Guberinic published a pioneer paper on aircraft recovery problem, where the objective is to minimize total passenger delay while determining routes for aircraft. This study considers only single fleet without considering maintenance requirements. They use branch and bound method to solve the model they proposed (Tedorovic and Stojković 1984). This study is than extended by study of Tedorovic and Stojković (1990), which adds minimizing total number of cancelled flights to the objective (Tedorovic and Stojković 1990). Network flow models are widely used in researches considering aircraft recovery problem (Jarrah et al. 1993; Rakshit et al. 1996; Mathaisel 1996; Cao and Kanafani 1997c, b, a). These studies, except (Tedorovic and Stojković 1990) consider both cancellation and re-timing. While some of them take into account only one aircraft type, a majority consider different fleets. Some of the studies define the aircraft recovery problem as a time-line network (Yan and Yang 1996; Thengvall et al. 2000; Thengvall et al. 2001; Løve et al. 2001; Thengvall et al. 2003, Vos et al. 2015). There are also studies based on time-band network, which was first introduced by Argüello et al. (1997). Both studies of Argüello (1998) and Bard et al. (2001) propose a model based on time-band network, to regenerate aircraft routings subsequent to a delay. Eggenberg et al. (2007), also model aircraft recovery problem on time-band network, considering multi-fleet structure with reserve aircraft and maintenance requirements (Eggenberg et al. 2007). Connection network is also widely used in the scope of the studies on aircraft recovery problem (Tedorovic and Stojković 1984; Rosenberger et al. 2003; Andersson and Värbrand 2004; Liu et al. 2008; Hu et al. 2017; Liang et al. 2018). Most of the studies in the literature consider aircraft swap, delay and cancellation as recovery options. Swapping planned maintenance of aircraft, if maintenance requirement is met, is an other option that can be considered during aircraft recovery. However, studies that consider maintenance swap is limited in the literature (Eggenberger et al. 2010; Liang et al. 2018). Considerable amount of studies use set-partitioning formulation in modelling aircraft recovery problem and use column generation as solution method (Rosenberger et al. 2003; Andersson and Värbrand 2004; Eggenberger et al. 2010; Liang et al. 2018). Metaheuristics are also popular in aircraft recovery literature, since they present advantage in solution time of complex problems. Argüello et al. (1997) introduce a greedy algorithm for aircraft recovery problem (Argüello et al. 1997). Methods based on local search metaheuristic are presented in Løve et al. (2001), Løve et al. (2001). Liu et al. (2006) and Liu et al. (2008) propose multi-objective Genetic Algorithm to construct new aircraft routings (Liu et al. 2006; Liu et al. 2008).

2.8.1.2 Crew Recovery

Reparations done on broken aircraft schedules, during aircraft recovery process, mostly leads disruptions in crew schedules. Cancelling, delaying or diverting a flight or swapping aircraft results in the absence of flight crew at the station needed. Thus, aircraft recovery follows with crew recovery problem where several options for handling crew disruption are available. These options include (Ball et al. 2007);

- deadheading-flying crews in passenger seats from the point of disruption to the location of the later flight leg that they are assigned to.
- reassigning crew-reassign a crew from its original schedule to an alternative schedule
- calling up reserve crew-assigning reserve crew to the flight legs that are left crewless due to the disruption

All of these actions must satisfy the requirements defined in crew scheduling problem (see. Sect. 2.6.3). With the options mentioned below and restrictions described, crew recovery problem is to create new schedules, for both the disrupted and reserved crews, that covers all flights at minimum cost (Barnhart 2009). Crew cost is the second highest cost in airline operating costs, following fuel cost. However crew recovery has received much less attention since it is significantly harder. Besides the need to solve the crew recovery problem in minutes, the objective function of the problem is compound. Studies often consider both minimizing the incremental crew costs and the number of crew schedule changes made together (Ball et al. 2007). In addition, information regarding the current location and flight history of every crew member should always be known so that the assignments can be made according to regulations. Crew recovery literature is relatively limited because of the challenges mentioned below. Even though, crew recovery refers to both cabin and cockpit crew most of the studies in this area focus on cockpit crew, as they are more costly and restrictive compared to the cabin crew.

Wei et al. (1997) represent the crew recovery problem on a connection network and model it as an integer multi-commodity network flow problem. The objective is to return to the original schedule as soon as possible while minimizing the operational cost (Wei et al. 1997). Stojković et al. (1998) also use multicommodity network flow formulation and present an integer nonlinear model. They decompose the model where the master problem is a set partitioning problem and the sub-problem is a resource contrained shortest path problem (Stojković et al. 1998). Guo (2004) model the crew recovery problem as a set partitioning problem, aiming to minimize the deviations from the planned schedule (Guo 2004). A different modeling approach is proposed by Nissen and Haase (2006). They construct a duty-based formulation, which is specifically well suited for crew recovery in European airlines, which commonly employ fixed monthly crew rates (Nissen and Haase 2006; Castro et al. 2014). Medard and Shawney (2007), propose a model that is a flight-based equivalent of the original pairing-based rostering model. In the model re-assigned flights corresponds the flight pairings (Medard and Sawhney 2007). All the aforementioned studies use fixed flight schedules as input. There are also studies that consider flight delays as well as cancellations during the crew recovery process. There are a number of the studies consider cancelattion as a recovery option (Lettovský et al. 2000; Yu et al. 2003; Chang 2012; Liu et al. 2013). Lettovský et al., (2000) extended the classical formulation of the problem by introducing a set of decision variables that enables considering flight cancellations. Yu et al. (2003), implement a crew recovery decision support system in Continental Airlines, where flight cancellations are also considered (Yu et al. 2003). Stojković and Soumis (2001) extend the study of Stojković et al. (1998), by explicitly considering to delay scheduled flights in their formulation (Stojković and Soumis 2001). Abdelghany et al. (2004), present a decision support tool that proactively recovers crew problems ahead of time before their occurrence (Abdelghany et al. 2004). Zhao et al. (2007) transform the formulation proposed by Abdelghany et al. (2004) into a gray programming model (Zhao et al. 2007).

2.8.1.3 Passenger Recovery

Disrupted passenger is a passenger who must be re-accommodated on a different itinerary than planned due to missed connections caused by flight delays and flight cancellations (Barnhart 2009). Thus, both the aircraft and the crew decision can lead passenger disruptions. Accordingly the next phase in airline recovery process is passenger recovery. In this phase disrupted passengers are assigned to alternative itineraries that start at the location they disrupted and end at their destination or a nearby location. Passenger recovery problem also presents some challenges. To begin with, it is challenging to estimate the cost associated with passenger recovery because of its complex structure. The main reason of this complexity is that the number of seats available at the time of recovery and the passenger demand for these seats are dependent on the previous disruptions and recovery actions taken (Barnhart 2009).

Bratu and Barnhart (2006) present two models, considering passenger recovery problem. The models allow flight cancellations and delay decisions as well as decisions on assigning reserve crew and aircraft to the flight legs (Bratu and Barnhart 2006). Zhang and Hansen (2008) propose an intermodal solution to airline passenger recovery problem, where disrupted passengers are allowed to use ground transportation modes (Zhang and Hansen 2008). Jafari and Zegordi (2011), consider the problem of recovering disrupted passengers after aircraft recovery decisions (Jafari and Zegordi 2011). Sinclair et al. (2016) formulate the integrated aircraft and passenger recovery problem as a mixed-integer programming problem (Sinclair et al. 2016). Zhang et al. (2016), propose a heuristic approach that solve schedule recovery, aircraft recovery and passenger recovery sequentially (Zhang et al. 2016). Marla et al. (2017) also consider, aircraft and passenger recovery problems together. They propose an approach where speed changes during flight is an option. Their formulation is based on (Bratu and Barnhart 2006; Marla et al. 2017).

In most of the studies that consider passenger recovery, at least one other phase (i.e., aircraft recovery, crew recovery) is also considered. There are also studies that consider different combination of recovery problems in an integrated manner. In the scope of this section a brief review of airline recovery problems is presented. For more review a detailed reader can refer to (Barnhart 2009; Clausen et al. 2010; Castro et al. 2014; Hassan et al. 2021).

2.8.2 Robust Airline Scheduling

Robust optimization is an approach that addresses uncertainty, by modelling future uncertainties. The main idea behind robust planning is to determine operational plans that are "good" for most scenarios and acceptable for the worst case scenario. The main goal is to keep the original plan intact in an uncertain operating environment (Yu and Qi 2004). Yu and Qi (2004) describe a robust planning process with the following steps;

- Identifying the potential disruptive scenarios
- Determining a robustness criterion appropriate for the decision maker
- Incorporate the determined criterion in generating a robust plan
- Carrying out the plan

As mentioned above, a robust plan can handle worst case scenarios, but that virtue may come with costs; average performance and sunk cost. When the probability of disruptive event is low, actions taken in order to be robust may cause settling in a lower performance plan. Moreover investment in robustness may generate sunk-costs, which are redundant. However, robust planning is an important approach in disruption management and thus, an important tool that contributes to resilience.

As a proactive approach in airline disruption management, robust scheduling aims to create flight and crew schedules and aircraft rotations that are less vulnerable to disruptions by creating flexible plans. The main purpose of robust scheduling is to provide a large range of recovery options for aircraft, crew and passengers or generate plans that are able to absorb disruptions and remain feasible (Ball et al. 2007; Clausen et al. 2010).

Robust airline scheduling is also consists of sub-problems, as robust fleet assignment, robust aircraft routing, robust crew scheduling. The studies on robust airline scheduling can be categorized under two approaches; recovery robustness and absorption robustness (Clausen et al. 2010).

Research on recovery robustness focuses on minimizing the impact of disruptions by generating schedules that are able to respond effectively to disruptions. Recovery robustness addresses schedules that align well to the current recovery strategies in case of a disruptions. A schedule that is recovery robust, is costly and may have opportunities for aircraft and crew swapping (Clausen et al. 2010). Rosenberger et al. (2003) use short cycle strategy in fleet assignment in order to minimizing the propagation of cancellations (Rosenberger et al. 2003). A study that considers swapping opportunities is done by Smith and Johnson (2006). They limit the number of aircraft types at each airport in order create swapping possibilities (Smith and Johnson 2006). Arıkan et al. (2013) present robust airline scheduling strategies, under delay propagation uncertainty (Arıkan et al. 2013). Froyland et al. (2014) focuses on aircraft routing considering the uncertainty of future disruptions and resulting recovery (Froyland et al. 2014). Schaefer et al. (2005) focus on heuristic solution methods, that are tractable, for crew scheduling problems with uncertainty (Schaefer et al. 2005). Schaefer and Nemhauser (2006) present a framework for perturbing scheduled departure and arrival times, after a crew schedule has been found, while keeping the schedule legal without increasing the planned cost of the crew schedule (Schaefer and Nemhauser 2006). Shebalov and Klabjan (2006), propose a model for bi-objective robust crew scheduling problem where the objectives are minimizing the crew cost and maximizing the number of move-up crews (Shebalov and Klabjan 2006). Yen and Birge (2006) propose a stochastic integer programming model for the airline crew scheduling problem (Yen and Birge 2006).

The main aim of absorption robustness it to maintain the feasibility of plans when smaller disruptive events occur and prevent the ripple effects. Thus, studies on absorption robustness focus on minimizing the occurrence of propagated delays. For providing robustness, airlines may use buffer times when constructing schedules. However there is a trade-off between buffers and resource efficiency. The questions of "how much buffer time should be used" and "where the buffer times be placed" should be addressed meticulously (Clausen et al. 2010). Lan et al. (2006) propose an approach to reduce delay propagation through using buffers in aircraft routing and schedule re-timings. They formulate the problem as a mixed- integer programming problem with stochastic data (Lan et al. 2006). Ahmadbeygi et al. (2010), present an approach to reduce delay propagation by redistributing existing slack in the planning process by rearranging existing slacks in the schedule (AhmadBeygi et al. 2010). Borndörfer et al. (2010), propose a column generation method for the robust aircraft rotation problem that aims to minimize the probability of delay propagation (Borndörfer et al. 2010). Dunbar et al. (2012) consider aircraft routing and crew pairing problems in an integrated manner in order to minimize propagated delay (Dunbar et al. 2012). Yan and Kung (2018) propose a robust optimization approach that addresses the uncertainty in primary delays and incorporates it with aircraft routing. Marla et al. (2018), focus on a data-driven approach and compare the performances of robust optimization, stochastic programming and chance programming based approaches in robust aircraft routing problem (Marla et al. 2018).

2.9 Summary

The modern global framework has led to an interconnected world. In such a complex and linked structure organizations face even more complex challenges. Terrorist attacks, pandemic diseases, natural disasters, economic recession, human error, equipment failure are some of the main, unpredictable factors that can pose significant threat to the operational continuity of an organization. Besides, it is not only the major disasters but also some minor uncertainties or disturbances can also cause a great deal of challenge to organizations. The ever-changing operation environment and the fragility it induces, highlights resilience as a critical issue.

Resilience is identified as a component of sustainability such that, a system that is not resilient can only posse tenuous sustainability. Thus, it is a must to have adaptive capabilities in today's interconnected, complex, volatile world in order to cope with challenges.

The importance of resilience holds also for air transportation industry, which is a crucial component of today's world. There are a wide variety of factors that can cause flight disruptions and thus breaking operational continuity, that in return can impact the long-term success of an airline. A disruption in a single flight leg can have a ripple effect, leading delays or even cancellations and other disruptions in airline operations. This is why, airline disruption management has been a focus of interest both in the research and airline industry. Disruption management in airlines concerns with minimizing disruptions and their impacts as well as responding to and recovering from them.

Airline disruption management comprises two methods; robust planning and airline recovery. Robust planning is a proactive method which aims to prevent disruptions and enhance the absorption capability of an airline. Differently, airline recoveryaddresses re-planning and recovering of airline resources and minimizing the costs related to disruption, as a reactive method.

There exists a broad range of research on airline disruption management and the number of studies is only increasing. Research on robust planning is rather new compared to airline recovery. Nonetheless, it has already received a significant attention, which is promising for airline disruption management.

References

Abdelghany A, Ekollu G, Narasimhan R, Abdelghany K (2004) A proactive crew recovery decision support tool for commercial airlines during irregular operations. Ann Oper Res 127(1–4):309–331. https://doi.org/10.1023/B:ANOR.0000019094.19940.41

ACI (2013) Passenger protection under cases of flight disruption. 4–6. Retrieved from www.icao.int/meetings/atconf6

AhmadBeygi S, Cohn A, Lapp M (2010) Decreasing airline delay propagation by re-allocating scheduled slack. IIE Trans 42(7):478–489. https://doi.org/10.1080/07408170903468605

Allen JH, Davis N (2010) Measuring operational resilience using the CERT resilience management model. Tech Note 84

Anderies JM, Folke C, Walker B, Ostrom E (2013) Aligning key concepts for global change policy: Robustness, resilience, and sustainability. Ecol Soc 18 (2). https://doi.org/10.5751/ES-05178-180208

Andersson* T, Värbrand P (2004) The flight perturbation problem. Transp Plan Technol 27(2):91–117

Argüello MF, Bard JF, Yu G (1997) A GRASP for aircraft routing in response to groundings and delays. J Comb Optim 1(3):211–228. https://doi.org/10.1023/A:1009772208981

Argüello MF, Bard JF, Yu G (1998) Models and methods for managing airline irregular operations. In: Operations research in the airline industry (pp. 1-45). Springer, Boston, MA

Argüello MF (1998) Framework for exact solutions and heuristics for approximate solutions to airlines irregular operations control aircraft routing problem (Unpublished doctoral dissertation). USA

Arıkan M, Deshpande V, Sohoni M (2013) Building reliable air-travel infrastructure using empirical data and stochastic models of airline networks. Oper Res 61(1):45–64. https://doi.org/10.1287/opre.1120.1146

Arıkan U, Gürel S, Aktürk MS (2017) Flight network-based approach for integrated airline recovery with cruise speed control. Transp Sci 51(4):1259–1287. https://doi.org/10.1287/trsc.2016.0716

Ball M, Barnhart C, Nemhauser G, Odoni A (2007) Chapter 1 air transportation: irregular operations and control. Handbooks Operat Res Manag Sci 14(C):1–67. https://doi.org/10.1016/S0927-0507(06)14001-3

Bard JF, Yu G, Argüello M (2001) Optimizing aircraft routings in response to groundings and delays. IIE Trans 33:931–947

Barnhart C (2009) Irregular operations: schedule recovery and robustness. In: Belobaba P, Odoni A, Barnhart C (eds) The global airline industry. John Wiley and Sons Ltd., Publication, Wiltshire, pp 253–274

Barnhart C, Belobaba P, Odoni A (2003) Applications of operations research in the air transport industry. Transp Sci 37(4):368–391

Bazargan M (2010) Airline operations and scheduling. Ashgate, Farnham

BBC (2010) BBC website Flight disruptions cost airlines $ 1.7bn, says IATA. Retrieved October 17, 2020, from http://news.bbc.co.uk/2/hi/business/8634147.stm

Belobaba P (2009) Airline schedule optimization. In: Belobaba P, Odoni A, Barnhart C (eds) The global airline industry. John Wiley and Sons Ltd., Publication, Wiltshire, pp 183–211

Bhamra R, Dani S, Burnard K (2011) Resilience: the concept, a literature review and future directions. Int J Prod Res 49(18):5375–5393. https://doi.org/10.1080/00207543.2011.563826

Borndörfer R, Dovica I, Nowak I, Schickinger T (2010) Robust tail assignment (Tech. Rep. No. May)

Bratu S, Barnhart C (2006) Flight operations recovery: new approaches considering passenger recovery. J Sched 9(3):279–298. https://doi.org/10.1007/s10951-006-6781-0

Cao JM, Kanafani A (1997a) Multifleet routing and multistop flight scheduling for schedule perturbation. Eur J Operat Res (103):155–169

Cao JM, Kanafani A (1997b) Real-time decision support for integration of airline flight cancellations and delays part II: Algorithm and computational experiments. Transp Plan Technol 20(3):201–217. https://doi.org/10.1080/03081069708717589

Cao JM, Kanafani A (1997c) Real-time decision support for integration of airline flight cancellations and delays part I: mathematical formulation. Transp Plan Technol 20(3):183–199. https://doi.org/10.1080/03081069708717588

Caralli RA, Curtis PD, Allen JH, White DW, Young R (2010) Improving operational resilience processes: the CERT R resilience management model. In: Proceedings–SocialCom 2010: 2nd IEEE international conference on social computing. pp 1165–1170. https://doi.org/10.1109/SocialCom.2010.173

Cardillo A, Zanin M, Gòmez-Gardeñes J, Romance M, Garcíadel Amo AJ, Boccaletti S (2013) Modeling the multi-layer nature of the European air transport network: resilience and passengers re-scheduling under random failures. Eur Phys J Spec Top 215(1):23–33. https://doi.org/10.1140/epjst/e2013-01712-8

Carpenter S, Walker B, Anderies JM, Abel N (2001) From metaphor to measurement: resilience of what to what? Ecosystems 4(8):765–781. https://doi.org/10.1007/s10021-001-0045-9

Castro AJ, Rocha AP, Oliveira E (2014) A new approach for disruption management in airline operations control, vol 562. Springer. https://doi.org/10.1007/978-3-662-43373-7

Chang SC (2012) A duty based approach in solving the aircrew recovery problem. J Air Transp Manag 19:6–20

Charpetreau C (2020) Page 1 of 16. Retrieved October 17, 2020, from https://www.aerotime.aero/clement.charpentreau/24420-istanbul-airport-closed-after-pegasus-flight-skids-off-runway

Clausen J, Larsen A, Larsen J, Rezanova NJ (2010) Disruption management in the airline industry-concepts, models and methods. Comput Oper Res 37(5):809–821. https://doi.org/10.1016/j.cor.2009.03.027

Derissen S, Quaas MF, Baumgärtner S (2011) The relationship between resilience and sustainability of ecological-economic systems. Ecol Econ 70(6):1121–1128. https://doi.org/10.1016/j.ecolecon.2011.01.003

Dunbar M, Froyland G, Wu CL (2012) Robust airline schedule planning: minimizing propagated delay in an integrated routing and crewing framework. Transp Sci 46(2):204–216. https://doi.org/10.1287/trsc.1110.0395

DW (2009) 1,300 Lufthansa flights cancelled as courts approve strike. Retrieved October 17, 2020, from https://www.dw.com/en/1300-lufthansa-flights-cancelled-as-courts-approve-strike/a-51131875

Eggenberg N, Bierlaire M, Salani M (2007) A column generation algorithm for disrupted airline schedules

Eggenberg N, Salani M, Bierlaire M (2010) Constraint-specific recovery network for solving airline recovery problems. Comput Oper Res 37(6):1014–1026

Emas R (2015) The concept of sustainable development: definition and defining principles. In: Brief for GSDR. Florida International University, pp 1–3

Essuman D, Boso N, Annan J (2020) Operational resilience, disruption, and efficiency: Conceptual and empirical analyses. Int J Prod Econ 229

Fiksel J (2006) Sustainability and resilience: toward a systems approach. Sustain Sci Pract Policy 2 (2):14–21. https://doi.org/10.1080/15487733.2006.11907980

Froyland G, Maher SJ, Wu C-L (2014) The recoverable robust tail assignment problem. Transp Sci 48(3):351–372. https://doi.org/10.1287/trsc.2013.0463

Gershkoff I (2016) Airline disruption management. Amadeus Whitepaper. https://doi.org/10.1016/j.eneco.2008.08.004

Grosche T (2009) Airline scheduling process. In: Grosche T (ed) Computational intelligence in integrated airline scheduling. John Wiley and Sons Ltd., Publication, Berlin, pp 7–46

Guimarans D, Arias P, Tomasella M, Wu CL (2018) A review of sustainability in aviation: a multidimensional perspective. Sustain Trans Smart Logist Decis-Mak Models Solut 91–121. https://doi.org/10.1016/B978-0-12-814242-4.00004-1

Guo Y (2004) A decision support framework for the airline crew schedule disruption management with strategy mapping. In: Fleuren H, Hertog D, Kort P (eds.), Operations research proceedings 2004. operations research proceedings, vol 2004. Springer

Hassan LK, Santos BF, Vink J (2021) Airline disruption management: A literature review and practical challenges. Comput Oper Res 127:105137.

Holling CS (1973) Resilience and stability of ecological systems. Annu Rev Ecol Syst 4:1–23

Hu Y, Liao H, Zhang, S, Song Y (2017) Multiple objective solution approaches for aircraft rerouting under the disruption of multi-aircraft. Expert Syst Appl 83:283–299

IATA (2020a) IATA april air passenger market analysis (Tech. Rep. No. April)

IATA (2020b) IATA august air passenger market analysis (Tech. Rep. No. August)

IATA.(2020c) IATA–slower but steady growth in 2019 (No. February). Retrieved October 17, 2020c, from https://www.iata.org/en/pressroom/pr/2020c-02-06-01/

ICAO (2018) The world of air transport in 2018. Retrieved October 17, 2020, from https://www.icao.int/annual-report-2018/Pages/the-world-of-air-transport-in-2018.aspx

Jafari N, Hessameddin Zegordi S (2011) Simultaneous recovery model for aircraft and passengers. J Franklin Inst 348(7):1638–1655. https://doi.org/10.1016/j.jfranklin.2010.03.012

Janić M (2015) Modelling the resilience, friability and costs of an air transport network affected by a large-scale disruptive event. Trans Res Part A Policy Pract 71:1–16. https://doi.org/10.1016/j.tra.2014.10.023

Jarrah AI, Yu G, Krishnamurthy N, Rakshit A (1993) Decision support framework for airline flight cancellations and delays. Transp Sci 27(3):266–280. https://doi.org/10.1287/trsc.27.3.266

Jimenez Serrano FJ, Kazda A (2017) Airline disruption management: yesterday, today and tomorrow. Trans Res Proc 28:3–10. https://doi.org/10.1016/j.trpro.2017.12.162

Lan S, Clarke JP, Barnhart C (2006) Planning for robust airline operations: optimizing aircraft routings and flight departure times to minimize passenger disruptions. Transp Sci 40(1):15–28. https://doi.org/10.1287/trsc.1050.0134

Lederer P, Nambimadom R (1998) Airline network design. Oper Res 46(6):785–804

Lee J (2020) New approaches to airline recovery problems (Unpublished doctoral dissertation). Graduate College of the University of Illinois at Urbana-Champaign

Lele S (1998) Resilience, sustainability, and environmentalism. Environ Dev Econ 3(2):249–254

Lettovský L, Johnson EL, Nemhauser GL (2000) Airline crew recovery. Transp Sci 34(4):337–348

Liang Z, Xiao F, Qian X, Zhou L, Jin X, Lu X, Karichery, S (2018) A column generation-based heuristic for aircraft recovery problem with airport capacity constraints and maintenance flexibility. Transp Res B: Methodol 113:70–90

Liu TK, Jeng CR, Chang YH (2008) Disruption management of an inequality-based multi-fleet airline schedule by a multi-objective genetic algorithm. Transp Plan Technol 31(6):613–639. https://doi.org/10.1080/03081060802492652

Liu TK, Jeng CR, Liu Y.-Y, Tzeng J.-y (2006) Applications of multi-objective evolutionary algorithm to airline disruption management. In: 2006 ieee international conference on systems, man and cybernetics. p 4274546

Liu Q, Zhang X, Chen, X, Chen X (2013) Interfleet and intrafleet models for crew recovery problems. TTransp Res Rec 2336(1):75–82

Løve M, Sørensen KR, Larsen J, Clausen J (2001) Using heuristic to solve the dedicated aircraft recovey problem (Tech. Rep.). In: Informatics and Mathematical Modelling. Technical University of Denmark

Marchese D, Reynolds E, Bates ME, Morgan H, Clark SS, Linkov I (2018) Resilience and sustainability: similarities and differences in environmental management applications. Sci Total Environ 613–614:1275–1283. https://doi.org/10.1016/j.scitotenv.2017.09.086

Marla L, Jacquillat A, Lee J (2020) Dynamic disruption management in airline networks under airport operating uncertainty. In: Lee J, Marla L, Jacquillat A (eds) Dynamic airline disruption management under airport operating uncertainty, Transportation science, Forthcoming. https://doi.org/10.2139/ssrn.3082518

Marla L, Vaaben B, Barnhart C (2017) Integrated disruption management and flight planning to trade off delays and fuel burn. Transp Sci 51(1):88–111. https://doi.org/10.1287/trsc.2015.0609

Marla L, Vaze V, Barnhart C (2018) Robust optimization: Lessons learned from aircraft routing. Comput Oper Res 98:165–184. https://doi.org/10.1016/j.cor.2018.04.011

Mathaisel DF (1996) Decision support for airline system operations control and irregular operations. Comput Oper Res 23(11):1083–1098. https://doi.org/10.1016/0305-0548(96)00007-X

Medard CP, Sawhney N (2007) Airline crew scheduling from planning to operations. Eur J Oper Res 183(3):1013–1027. https://doi.org/10.1016/j.ejor.2005.12.046

Nissen R, Haase K (2006) Duty-period-based network model for crew rescheduling in European airlines. J Sched 9(3):255–278. https://doi.org/10.1007/s10951-006-6780-1

Onedio (2015) Beklenen Lodos THY'nin 103 Seferini İptal Ettirdi. Retrieved October 17, 2020, from https://onedio.com/haber/beklenen-lodos-thy-nin-103-seferini-iptal-ettirdi-444938

Patty B, Diarnond J (1998) The complex configuration model. In: Yu G (ed) Operations research in the airline industry. Springer, Boston, pp 370–403

Ponomarov SY, Holcomb MC (2009) Understanding the concept of supply chain resilience. Int J Logist Manag 20(1):124–143. https://doi.org/10.1108/09574090910954873

Rakshit A, Krishnamurthy N, Yu G (1996) System operations advisor: a real-time decision support system for managing airline operations at United Airlines. Interfaces 26(2):50–58. https://doi.org/10.1287/inte.26.2.50

Rapajic J (2016) Beyond airline disruptions, 2nd edn. Routledge, New York

Rosenberger J, Johnson E, Nemhauser G (2003) Rerouting aircraft for airline recovery. Trans Sci (37):408–421

Schaefer AJ, Nemhauser GL (2006) Improving airline operational performance through schedule perturbation. Ann Oper Res 144(1):3–16. https://doi.org/10.1007/s10479-006-0003-1

Schaefer AJ, Johnson EL, Kleywegt AJ, Nemhauser GL (2005) Airline crew scheduling under uncertainty. Transp Sci 39(3):340–348

Shebalov S, Klabjan D (2006) Robust airline crew pairing: move-up crews. Transp Sci 40(3):300–312. https://doi.org/10.1287/trsc.1050.0131

Sinclair K, Cordeau J-F, Laporte G (2016) A column generation post-optimization heuristic for the integrated aircraft and passenger recovery problem. Comput Oper Res 65:42–52. https://doi.org/10.1016/j.cor.2015.06.014

Smith BC, Johnson EL (2006) Robust airline fleet assignment: imposing station purity using station decomposition. Transp Sci 40(4):497–516. https://doi.org/10.1287/trsc.1060.0153

Stojković M, Soumis F (2001) An optimization model for the simultaneous operational flight and pilot scheduling problem. Manage Sci 47(9):1290–1305

Stojković et al., 1998 Stojković M, Soumis F, Desrosiers J (1998) The operational airline crew scheduling problem. Transp Sci 32(3):232–245. https://doi.org/10.1287/trsc.32.3.232

Stolker R, Karydas D, Rouvroye J (2008) A comprehensive approach to assess operational resilience. In: Proceedings of the third resilience engineering symposium(2005). pp 247–253

Tedorovic D, Stojković G (1984) Optimal dispatching strategy on an airline network after a schedule perturbation. Eur J Operat Res (15):178–182

Tedorovic D, Stojković G (1990) Model for operational daily airline scheduling. Transp Plan Technol (14):273–285

Thengvall BG, Bard JF, Yu G (2000) Balancing userpreferences for aircraft schedule recovery during irregular operations. IIE Trans 32(3):181–193

Thengvall BG, Yu G, Bard JF (2001) Multiple fleet aircraft schedule recovery following hub closures. Transp Res Part A: Policy Pract 35(4):289–308

UCLA Sustainability Committee (2021) What is sustainability? Retrieved October 10, 2020, from https://www.sustain.ucla.edu/what-is-sustainability/

Vora S (2016) With Brussels airport shut down, what travelers need to know. Retrieved from https://www.nytimes.com/2016/03/23/travel/with-brussels-airport-shut-down-what-passengers-need-to-know.html

Vos HWM, Santos BF, ve Omondi T (2015) Aircraft schedule recovery problem–a dynamic modeling framework for daily operations. Transp Res Procedia 10:931–940

Wan C, Yang Z, Zhang D, Yan X, Fan S (2018) Resilience in transportation systems: a systematic review and future directions. Transp Rev 38(4):479–498. https://doi.org/10.1080/01441647.2017.1383532

Wei G, Yu G, Song M (1997) Optimization model and algorithm for crew management during airline irregular operations. J Comb Optim 1(3):305–321. https://doi.org/10.1023/A:1009780410798

Yan S, Yang DH. (1996) A decision support framework for handling schedule perturbation. Transp Res B: Methodol 30(6):405–419

Yen JW, Birge JR (2006) A stochastic programming approach to the airline crew scheduling problem. Transp Sci 40(1):3–14

Yu G, Argüello M, Song G, McCowan SM, White A (2003) A new era for crew recovery at Continental Airlines. Interfaces 33(1):5–18. https://doi.org/10.1287/inte.33.1.5.12720

Yu G, Qi X (2004) Introduction. Disruption management: feamework, models and applications. World Scientific Publishing Co., Pte. Ltd., Singapore, pp 1–30

Zhang Y, Hansen M (2008) Real-time intermodal substitution: Strategy for airline recovery from schedule perturbation and for mitigation of airport congestion. Transp Res Rec 2052:90–99. https://doi.org/10.3141/2052-11

Zhang D, Yu C, Desai J, Lau HH (2016) A math-heuristic algorithm for the integrated air service recovery. Transp Res Part B Methodol 84:211–236. https://doi.org/10.1016/j.trb.2015.11.016

Zhao XL, Zhu JF, Guo M (2007) Application of grey programming in irregular flights scheduling. In: IEEM 2007: 2007 IEEE International Conference on Industrial Engineering and Engineering Management. pp 164–168. https://doi.org/10.1109/IEEM.2007.4419172

K. Gülnaz Bülbül was graduated from the Department of Industrial Engineering, Faculty of Engineering, Başkent University. She received her master's and PhD degrees in Civil Aviation Management at the Anadolu University. She is currently a faculty member of Faculty of Aeronautics and Astronautics of Eskisehir Technical University. Her researches are mainly on Operations Research in Air Transportation. Her current studies focus on modelling and optimization of airline operations.

Chapter 3
Value Co-Creation in Airline Ecosystem: Framework Integrating Sustainability and Dart Model

Inci Polat

Abstract The value concept which has been defined as produced by the firm and consumed by the customer in the traditional Goods-Dominant logic perspective is discussed in the marketing literature based on the Service-Dominant logic for two decades. Service-Dominant logic led to radical changes in the basic arguments of value by conceptualizing the value created in the interaction between actors. It is important to eliminate or reduce obstacles such as boundaries and asymmetry in the exchange of information between the firm and the customer in the process of value co-creation. The DART model consists of dialogue, access, risk, and transparency components, which are important in making the flow between the actors faster and more efficient. The model creates interdependent conditions for creating an organizational climate conducive to co-creating value. The DART approach, which provides the basic conditions for value co-creation among the actors, also plays an active role in ensuring long-term and sustainable firm–customer relations. Efforts to integrate actors into value co-creation processes also lead to formulating innovation strategies of firms. The existence of innovative approaches in the service concepts perceived by the customers enables the firm to be perceived as creative and progressive. Value co-creation also leads to the emergence of a firm perception that differs from its competitors by customers. The active participation of actors in the creation processes of value causes customer loyalty and therefore to plays an active role in the formation of sustainable relationships by providing interaction. The airline firms play an active role (with increasing their bilateral interactions and information sharing) in the process of value co-creation. The fact that the information about benefits and costs in the relationship between actors (firms and passengers) is more clear and understandable, will increase the passengers' participation in the value co-creation process for sustainability.

Keywords Value co-creation · DART model · Sustainability · Airline ecosystem · S-D Logic

İ. Polat (✉)
Süleyman Demirel University, Isparta, Turkey
e-mail: incisesliokuyucu@sdu.edu.tr

© The Author(s), under exclusive license to Springer Nature Singapore Pte Ltd. 2022
K. Kiracı and K. T. Çalıyurt (eds.), *Corporate Governance, Sustainability, and Information Systems in the Aviation Sector, Volume I*, Accounting, Finance, Sustainability, Governance & Fraud: Theory and Application, https://doi.org/10.1007/978-981-16-9276-5_3

3.1 Introduction

The value concept which has been defined as produced by the firm and consumed by the customer in the traditional goods-dominant (G-D) logic perspective has been discussed in the marketing literature based on the service-dominant (S-D) logic for two decades. S-D logic led to radical changes in the basic arguments of value by conceptualizing the value created in the interaction between actors (Vargo and Lusch 2004). As such, the creation of shared value is a value co-creation process that includes not only companies themselves but also all actors in their ecosystem. Providing unique value to customers has become a must for businesses, especially in the high-tech era (Fernando and Chukai 2018). Although value is principally expressed in monetary terms (i.e., cost/sacrifice-type value), research shows other forms of value that appear to be relevant to the concept of value chain integrity. Namely, relational synergy occurs when a long-term relationship between two organizations provides more collective value than independently acting organizations can provide (Borys and Jemison 1989; Thorelli 1986). The approach to the co-creation process of value states that leveraging the stakeholder capabilities of social, commercial, and civil communities within the business network will lead to better governance, infrastructure, development, and sustainability (Aquilani et al. 2016), and joint value creation can create a competitive advantage for firms (Grandy and Levit 2015). Any positive difference that a firm can maintain its superiority will enable the firm to perform better than its competitors (Porter and Van der Linde 1995).

It is important that institutional actions must create value for society by consulting stakeholders (Clarkson 1995), and that value is created interactively with each actor rather than embedded in objects (Ramaswamy 2008; Vargo 2008). Therefore, businesses' actions may not support sustainability unless beneficiaries are included in the early stages of social responsibility programs (Biggemann et al. 2014). Although there are many definitions of sustainability, the basic principle in sustainability is to shape human systems by saving, producing, and keeping the capabilities of all ecosystems alive. Sustainability deals with the activities of companies to implement sustainable and social-ecological requirements throughout the entire value chain (Arnold 2017). Sustainability strategies can be developed by all actors that contribute to value in the service ecosystem. The business in the service ecosystem can plan its goals and operate accordingly, but these activities may depend on the actions and reactions of other actors with which it interacts. Thinking about sustainability recognizes the importance of defining and understanding the needs and interests of all stakeholders and the value created in interacting with them. However, value assessment is inherently difficult because of the importance attached to the economic value that most research focuses on (Salzmann et al. 2005) because economic value only takes place in the long run. However, identifying those who benefit from value is also a problem. Thus, the problem is not only about how much value is created but also how it will be distributed (Biggemann et al. 2014). Increasing global sustainability pressures have increased the importance of sustainable business models in collaboration between firms and other key stakeholders. In common value creation processes, how companies manage their organizational processes and which stages will be transferred to which departments is not an issue that is heavily discussed. Due to the increasing number of actors in co-creation processes, coordination in the production of long-term products and/or services requires better coordination (Arnold 2017).

While there are many scientific studies conducted by theorists and practitioners on the creation of shared value (Vargo et al. 2008; Payne et al. 2008; Grönroos 2011; Edvardsson et al. 2011; Saarijärvi 2012; Grönroos and Voima 2013; Yi and Gong 2013; Gummesson et al. 2014; Ranjan and Read 2016; Cossío-Silva et al. 2016; Assiouras et al. 2019) examining the relationship between shared value creation and sustainability studies (Biggemann et al. 2014; Lacoste 2016; Ma et al. 2019) have not been discussed much in the academic literature. In this context, the study aims to examine the sustainability process in the airline ecosystem based on value co-creation. The framework of the study was established through the literature review on the airline ecosystem with a sustainability-focused value co-creation perspective. After evaluating the value co-creation process, Dialogue–Access–Risk Assessment–Transparency variables are handled in terms of sustainability within the scope of the DART model, and in line with these findings, the sustainability-oriented value co-creation perspective is evaluated specifically for the airline ecosystem.

3.2 Value Co-Creation

In the industrial value approach, value is treated as the equivalent of a price, and customers are expressed as the destructor of the value created by producers. This approach reflects the traditional perspective in which suppliers or producers are seen as the sole producers of value, and consumers as only consuming units (Payne et al. 2008). S-D logic changes contemporary marketing thinking, where marketing is seen as the facilitator of voluntary exchange processes that continue through collaborative, value-creating relationships between actors. S-D logic sets out eleven basic propositions aiming to create a framework for a service-oriented perspective (Vargo and Lusch 2016).

If S-D logic is explained based on its basic premises (Vargo and Lusch 2008; Williams and Aitken 2011), it presents a marketing perspective created by operant sources such as knowledge and abilities and realized through value creation processes, based on service, the main cause of exchange. The operant capabilities that facilitate and develop value co-creation processes with the S-D logic perspective are strategic resources that assume a central role for an organization to gain competitive advantage (Karpen et al. 2012, 2015). In this context, in today's world, the differentiation in the communication of consumers with suppliers is defined more broadly. The propositions handled by S-D logic are not a set of rules, but rather propositions that represent an evolving and collaborative effort to create a better marketing-based understanding of value and change. "*Value is co-created by multiple actors, always including the beneficiary.*" and "*Value is always uniquely and phenomenologically determined by the beneficiary.*" The co-creation of value approach addressed based on these propositions states that customers are involved in the co-creation and re-creation process of value with organizations in the new paradigm (Ramirez 1999).

While sustainability-focused value creation strategies use vertical supplier relationships, suppliers provide sustainable value with their customer networks (direct customers and customers of their customers) to increase the benefits of themselves and their customers by re-providing sustainability advantage (Lacoste 2016). Value re-creation is a way of sharing, combining, and innovating each other's resources and capabilities between firms and active users to create value through new forms of interaction, service, and learning mechanisms (von Hippel 2005; Zwass 2010; Arnold 2017). Nidumolu et al. (2009) present it as a five-step process that goes beyond the product lifecycle to move towards sustainable superiority:

Seeing compliance with environmental norms as an opportunity: It is to ensure that compliance with norms becomes an innovation opportunity. Competencies are required to anticipate and shape regulations and to work with competitors and other firms to implement creative solutions. As a result of this situation, the company and its partners in its ecosystem are encouraged to experiment with sustainable processes and/or technologies.

Making value chains sustainable: Productivity is increased along the value chain. Technical specialization competencies are required in evaluating outputs (carbon management, life cycle, etc.). By designing operations to produce less harmful output (such as less energy, water use, emission generation, and waste release), partners are directed to environmentally friendly production.

Designing sustainable products and services: It is the process of developing sustainable offerings or redesigning existing ones to make them environmentally friendly. However, this process reveals competency requirements regarding which products and services cause the least damage to the environment, the support of the actors in the ecosystem, and the sustainable material supply and production. Compact and sustainable products and/or services are developed as a result of the application of new techniques.

Developing new business models: It is the process of introducing new processes of value creation and acquisition that will fundamentally change competition. It requires the development of talents to increase the interaction between consumers and to increase their shared value. Accordingly, it is ensured that value chain relationships are changed and new business models are developed by combining digital and physical infrastructures.

Creating new application platforms: It is the process of questioning the dominant logic underlying today's business world from a sustainability perspective. In this process, competencies are required to reveal how renewable/non-renewable resources affect business processes and ecosystems, and to synthesize applications, technologies, and regulations in different sectors. In line with this, partners in the ecosystem are enabled to create sustainable innovations on the platforms.

The value created in the interaction between actors in the service ecosystem may arise from the sharing of resources, knowledge, and technology, and may include sustainability within members of the chain on the road to final consumption. An alternative concept is that value is not fully included in a product or service, but rather in use (Biggemann et al. 2014). Prahalad and Ramaswamy (2004a, b) propose that unique value can be co-created by involving the end-user in the value creation process, then the value is in the experience of co-creation rather than merely in physical or service delivery. The interaction between the firm and the customer, which has become the focus of value creation, places the relationship between the two actors at the center of the application (Prahalad and Ramaswamy 2004a, b). Provision of services is offered as part of an eco-design strategy where environmental impact can be minimized by providing appropriate service activities throughout the product life cycle or by performing the desired function themselves, rather than providing tangible products (Geum and Park 2011; Laperche and Picard 2013; Rantala et al. 2018). Furrer (2010) states that improving services means reducing cost, risk, and uncertainty, saving time, increasing knowledge, and increasing the social status and prestige of the image of the firm. The emergence of new business models explains how companies create and own value (Björkdahl and Holmén 2013). While business model innovation means the implementation of a new model for the firm (Björkdahl and Holmén 2013), it is also called an innovation application consisting of new combinations of existing factors (McKelvey and Holmén 2006). Value co-creation processes refer to the active participation of the actors in the ecosystem in the development of new services/goods or the improvement of existing ones (service/good). In this context, the DART model, which focuses on both contributing to the co-creation process and interacting with stakeholders to gain more benefits within these processes, addresses the creation of common value in terms of Dialogue, Access, Risk Assessment, and Transparency.

The key point in value co-creation for actors through common forums or forms of interaction is the central dialogue, understanding the benefits of access, transparency, and risk.

3.2.1 Dialogue

Dialogue, which is one of the most important building blocks in value creation processes, expresses "interaction, deep participation, and the ability and willingness to act on both sides". For the dialogue to take place between the parties, that is, to develop an active dialogue and a common solution process, the actors should be made equal and common problem solvers. This gathering should be developed by focusing on issues that concern all actors and within clearly defined rules (Prahalad and Ramaswamy 2004a, b). If the subjects and rules are clearly defined, it will ensure that problems arising from dialogue in the value co-creation process are prevented from the beginning. The rules in the process should be clear and understandable, as the dialogue includes the possibility of disclosure of the ideas that will remain confidential for the actors (especially the companies) as a result of the transfer to other actors. If confidentiality principles are not followed in new product/service development processes, it can be stated that companies that are sensitive to secrets will not lean towards active and intensive participation in value co-creation processes (Prahalad and Ramaswamy 2004a, b; Hoyer et al. 2010).

Creating symmetrical relationships between actors in a successful value co-creation process will lead to the dynamic participation of the actors involved in the cooperation (Lindström and Polsa 2016; Sesliokuyucu and Polat 2020). In particular, it is vital for symmetrical communication to anticipate the differing expectations of the actors involved in the sustainability programs of companies and/or respond quickly to these expectations. Scandelius and Cohen (2016) revealed in their study findings that the actors (employees, suppliers, and colleagues in the sector) aim at symmetrical communication within the ecosystem and that these groups facilitate the company's sustainability programs. Communicating sustainability programs to the actors through forums by creating simplification, flexibility, and emotional commitment will ensure that communication is placed in a consistent framework (Scandelius and Cohen 2016).

3.2.2 Access

Access refers to facilitating the creation of shared value between actors by using the right communication tools. It also covers how the interaction strengthens the client's access to information, tools, and experience, thus enabling value creation to occur (Russo-Spena and Mele 2012). In the traditional sense, firms have operated by using the information asymmetry between them and the actors in their ecosystems. However, in today's world, actors can access mutual information more effectively due to the existence of many connection points. The existence of effective access offers increased freedom of choice for customers and brings marketing solutions with it (Prahalad and Ramaswamy 2004a, b). The opportunities offered within the scope of access will lead actors to find innovative solutions by creating an interactive space that can tell stories and create shared experiences (Russo Spena and Mele 2012).

Access, as the second component of the DART model, facilitates dialogue and enables companies to optimize when, where, and how to provide the opportunity to create value with actors. Instead of seeing actors outside the value creation process, firms should try to facilitate customers' access to the processes and resources used to create and deliver their product or service offerings, taking into account the law and competitors. This thus facilitates the meaningful participation of actors in the value co-creation process (Albinsson et al. 2016). The desire to deliver superior value involves the participation of a large number of parties (Anderson et al. 2007) in the shared value creation process, which increases the difficulties of coordinating interaction between actors. The development of information technologies brings with it the participation of more actors in these processes and an increase in their access to information. This situation causes some problems to arise. Increased access to information leads to the emergence of new consumer groups who proactively seek information and research actor behaviors in the ecosystem, and to examine the behavior of other actors (Biggemann et al. 2014).

3.2.3 Risk Assessment

Stating that access and dialogue also hold actors partially responsible for the results of the shared value creation process, the risk assessment provides actors with information on their contributions. Efficient risk assessment facilitates conscious decisions about the risks of co-creation of value (Albinsson et al. 2016). The basic idea is that actors make conscious choices and have the right to be aware of the risks as well as the benefits they create. And actors can make an in-depth assessment (Mazur and Zaborek 2014). Co-creation processes bring along problems that may arise in intellectual property rights. The willingness of the actors in the ecosystem to receive responses for their skills and efforts (as well as voluntarily transferring them to other actors without any approval) is one of the most important risks (Hoyer et al. 2010). For this reason, determining the rules at the beginning of value co-creation processes should be the basic approach in risk assessment. For traditional goods-oriented companies, detailed and understandable rules have strategic importance in healthy customer relationships. In this context, companies should treat actors as honorable and reliable partners working towards a common value optimization goal (Albinsson et al. 2016).

3.2.4 Transparency

Transparency, which emphasizes the elimination of information asymmetry and openness of information between actors, facilitates collaborative dialogue with those actors. Providing continuity with access and risk assessment can reveal new business models and functions for the parties (Prahalad and Ramaswamy 2004a, b). Firms should update business-related information on their side, such as disclosing pricing-related information for transparency. Transparency of company information in the value co-creation process will increase the willingness of actors to accept the good/service quality (Prahalad and Ramaswamy 2001a, b; Taghizadeh et al. 2016). The effects of integrity on achieving more profitable business results are such that companies find more motivation to demonstrate and maintain integrity as well as the whole value chain. Even a mining company should be proud of its stakeholders not only because of the product they market but also because of the beneficial relationships that can be established, the value created by this activity, and therefore the benefits obtained with others (Biggemann et al. 2014).

Firms that are successful in creating shared value share information that is previously registered and/or can be considered strategically dangerous to reveal outside the company's internal environment. The information shared can be diverse (e.g., transaction fees, security transactions, profit margins, product development details), and at first sight can be counterproductive in gaining customer loyalty and market advantage. However, this transparency reveals the integrity of the firm and its commitment to openness (Albinsson et al. 2016). Despite its risks, transparency paves the way for a better perception of needs (through dialogue and access) and active participation of different actors in the processes through the participation of actors (Hatch and Schultz 2010).

3.3 Sustainability and Value Co-Creation in Airline Ecosystem

In service ecosystems, whose focal point is a chain of activities based on a symbiotic relationship, value is created through the application of competencies, unlike the sequential activities of goods-oriented ecosystems (Lusch and Nambisan 2015; Vargo and Lusch 2004, 2008; Agarwal et al. 2016). While value definitions in G-D approaches are based on the interaction between subject and object. In the S-D approach, it is used to explain the value that arises when a service is used (Biggemann et al. 2014). S-D logic is concerned with its role in marketing, helping customers create valuable experiences at all stages of the "consumption" process, including selection, purchasing, consumption, and recycling. Service providers are in this dynamic nature of customers by continuing to innovate in ways that make it easier to create value with customers (Flint 2006). Service innovation provided through this shared value creation plays a key role in companies whose sustainable competitive advantages depend on the development and implementation of new services (Gebauer et al. 2011).

The value created jointly within a mutually beneficial relationship network based on S-D logic (Vargo et al. 2008) shows that the aircraft offered to the passenger is a tool used in the process of providing seat service when considered from the perspective of the airline business and its customers. The airline operator should include the passenger in this process by focusing on the passenger's service procurement process, not the passenger on the plane or seat. In the airline ecosystem, the effectiveness of the airline can be evaluated in the best way by including the customer in the service production process as well as the quality of service it provides to the customer. Every customer contributes to the service process as the co-creator of value by being included in the production process of the service they need or expect. The intangible service offered to the airline by each actor in an airline ecosystem is provided by humans. Therefore, proper management of human resources is important for every actor in the ecosystem. Taneja (2016) states that airlines have a critical success factor linked to organizational structure and culture. Taneja (2016) states that cooperation with actors in the airline ecosystem and greater use of data to differentiate value propositions to provide a total travel experience.

In sustainability processes, which have various dimensions such as economic, ecological, and social, businesses interact with many actors as they are an important part of expanded networks within their ecosystem. Considering that most of the actors in the ecosystem differ in their demands, their ability to influence the action process, and their expectations, it can be thought that sustainability strategies should include all actors. For this reason, the sensitivity of businesses to the demands of the actors they interact with is also considered within the scope of the characteristics of sustainability strategies (Biggemann et al. 2014). Sustainability-based value co-creation refers to combining resources, knowledge, and capabilities between products and services and actors in the service ecosystem through sustainability criteria. Thus, this situation contributes to increase market transparency and decrease information asymmetry (Darby and Karni 1973; Arnold 2017). Implementation and management of sustainability strategies may be hampered by erroneous cooperation between actors. As a result of the establishment of vertical cooperation with the actors in the ecosystem (Lacoste 2012) or if some actors in the ecosystem have more than one identity, through both vertical and horizontal cooperation (Balmer and Greyser 2002), value co-creation creates an active and social value. (Scandelius and Cohen 2016).

Sustainability, interpreted as trying to strike a balance between social, economic and environmental imperatives when applied to aviation, the broad concept of sustainability is often limited to "reducing the environmental impact as much as possible" (International Civil Aviation Organization 2011). In improving sustainability in aviation, stakeholder engagement is considered as a possible effective complementary mechanism for both regulatory and market mechanisms (Amaeshi and Crane 2006). In order to make the common value created by the actors in the airline ecosystem more sustainable, they support companies more and they are also affected by the role of environmental and social problems that arise as a result of this support. For this reason, airline companies harmonize themselves by integrating their socially responsible aspects of business practices for sustainable development and competition (Campbell 2007; Abdi et al. 2020). When evaluated based on the assumption that actors in the airline ecosystem are key players and driving forces in making airline businesses effective in terms of sustainability, especially environmentally (Russo and Fouts 1997), value creation processes with the actors in the airline businesses' ecosystem will contribute to gaining the market advantage over their competitors. Sustainability processes are a complex problem that affects multiple businesses, groups, and communities, requiring solutions to be developed in cooperation with the most relevant actors of the industry (Martín-de Castro et al. 2016).

The sustainability-related challenges faced by the aviation industry in today's world, where environmental concerns are becoming more prominent, show that the information required to increase and understand the research efforts on this subject should be focused more directly, but not indirectly (Mauser et al. 2013). If the proactive participation of the actors in the ecosystem in value creation is ensured and the building blocks of common value creation for a favorable experience environment are brought together, it will be possible to create a more shared sense of responsibility and dependence between the main actors (companies and customers) (Sheth and Uslay 2007). The business that develops a value co-creation strategy not only develops its marketing development but also constitutes a source of competitive advantage. Value co-creation strategies that affect all network structures and management processes of companies also contribute to their sustainability focus. The data obtained from the management of DART tools will also provide companies with the opportunity to structure existing policies and procedures in the process of value co-creation (Albinsson et al. 2016). The fact that the passengers, who are in the airline ecosystem and can be defined as the most important actors of the system, are more conscious and more demanding on information (Prahalad and Ramaswamy 2004a, b), increasing the interaction based on cooperation. This trust-based interaction will ensure the elimination of opportunistic behaviors (Schiavone et al. 2014). However, open access brings some problems with it. The emergence of high-volume customer input and the screening of this information can cause a significant burden. Also, it can be stated that there may be problems regarding property rights after the evaluation of ideas (Hoyer et al. 2010). As actors become co-creators of value, they will want to learn more about the potential risks associated with the consumption and production of certain services and goods (Ramaswamy 2005).

The responsibility to inform the passengers in the ecosystem about the potential risks related to the service process belongs to the airline companies. And these notifications play a role in establishing a reliable relationship between passengers and the airline business. However, a successful interaction can be mentioned if the information (presented within the framework) is presented transparently to the actors. Firms increase the willingness of the actors to accept the quality of the service process by ensuring the transparency of the process of creating value with information (Taghizadeh et al. 2016). As a result of this interaction, the transition of sustainability strategies from airline operations to passengers is accelerated, and service innovations are better managed. Transparency, which increases the integrity and performance of the relationship between the actors in the airline ecosystem, will also bring a high level of trust in the relationship networks.

3.4 Conclusion

In the study that deals with sustainability based on value co-creation; the dialogue, access, risk, and transparency components between the actors have been examined specifically for the airline ecosystem. Nowadays, it is important to evaluate sustainability strategies, which are generally considered in terms of cost and service quality components (e.g., better food, timely arrival, lower accident rate) in the airline industry (Veliyath and Fitzgerald 2000), by considering stakeholder engagement.

The interests of mobility, motivation, results, development, and sustainability for each actor in the airline ecosystem can be diverse, even contradictory. These different perspectives cause differentiation of policies regarding sustainability strategies in the aviation field (Payán-Sánchez et al. 2018) and the participation of the actors in the airline ecosystem in the value creation process is important for better management of corporate sustainability. Studies on sustainability management state that actor involvement can increase firm performance (Seuring and Gold 2013) and also, integrating different views to reach solutions for environmental problems can be a global benefit (Tress et al. 2005). When the level of organizational knowledge is low, performance related to environmental sustainability will also be low (Ordieres-Meré et al. 2020). Increasing environmental pressures on the aviation (especially thoughts on the necessity of taxation) industry cause ignored economic and social impacts. It is important for local and global economies to actively support sustainability strategies in the industry, which creates 65 million jobs (direct and indirect) and an economy of 2.47 trillion dollars globally. The fact that more passengers and transporters have access to aviation worldwide than ever before contributes positively to the sustainable growth of the industry (International Air Transport Association (IATA) 2019).

Sustainability of aviation supported by all countries in the world, as the conclusion of the passengers' individual efforts of less carbon footprint for a cleaner and safe world, the sustainability strategies are also included value co-creation processes in the airline ecosystem. Dialogue processes developed with the actors in the airline ecosystem will provide a better understanding of the needs, expectations, and wishes of stakeholders in different interest groups. The dialogue will also play a role in strengthening the interaction within the framework of certain rules by enabling actors to become effective in access and risk assessment. Transparency in the interaction process of the actors contributes to sustainable relations and value co-creation activities by increasing the actors' trust towards each other. The Lufthansa Group, which attributes its economic success largely to the commitment and motivation of its employees. Its responsible and sustainable approach to its stakeholders, environment, and employees ensures financial stability. At the same time, this relationship with the actors leads to the acceptance of the business model inside and outside the company. Careful management of opportunities and risks improves the dialogue between actors and increases their active participation in activities, as well as the competitive advantage of the company within the sector (Lufthansa Group 2019).

As a result, innovative, smart, and environment-friendly sustainable solutions will increase the competitiveness of the airline industry (EASA 2019) and ensure that the actors participating in the value co-creation will play a role in short- and long-term sustainability policies. The value co-created based on mutual trust between actors is effective in increasing sustainability performance as well as providing a competitive advantage in the industry. Ensuring a transparent manner of information flow in the interaction of airline ecosystem actors will ensure that new product/service developments, as well as existing relationships, are built on more solid foundations. The study which deals with the sustainability process based on value co-creation was evaluated within the framework of the DART model. Evaluating the actors in

Table 3.1 The axioms of S-D logic (Vargo and Lusch 2016)

Axiom	Description
Axiom 1/FP1	Service is the fundamental basis of exchange
Axiom 2/FP6	Value is co-created by multiple actors, always including the beneficiary
Axiom 3/FP9	All social and economic actors are resource integrators
Axiom4/FP10	Value is always uniquely and phenomenologically determined by the beneficiary
Axiom 5/FP11	Value co-creation is coordinated through actor-generated institutions and institutional arrangements

the ecosystem separately will provide a clearer understanding of the relationship between sustainability and value co-creation related to the sector (Table 3.1).

References

Abdi Y, Li X, Càmara-Turull X (2020) Impact of sustainability on firm value and financial performance in the air transport industry. Sustainability 12(23):9957

Agarwal N, Soh C, Yeow A (2016) Value co-creation in service ecosystems: a member perspective

Albinsson PA, Perera BY, Sautter PT (2016) DART scale development: diagnosing a firm's readiness for strategic value co-creation. J Market Theory Pract 24(1):42–58

Amaeshi KM, Crane A (2006) Stakeholder engagement: a mechanism for sustainable aviation. Corp Soc Responsib Environ Manag 13(5):245–260

Anderson JC, Kumar N, Narus JA (2007) Value merchants: demonstrating and documenting superior value in business markets. Harvard business press

Aquilani B, Silvestri C, Ruggieri A (2016) Sustainability, TQM and value co-creation processes: the role of critical success factors. Sustainability 8(10):995

Arnold M (2017) Fostering sustainability by linking co-creation and relationship management concepts. J Clean Prod 140:179–188

Assiouras I, Skourtis G, Giannopoulos A, Buhalis D, Koniordos M (2019) Value co-creation and customer citizenship behavior. Ann Tour Res 78:102742

Balmer JM, Greyser SA (2002) Managing the multiple identities of the corporation. Calif Manag Rev 44(3):72–86

Biggemann S, Williams M, Kro G (2014) Building in sustainability, social responsibility and value co-creation. J Bus Ind Mark 29(4):304–312

Björkdahl J, Holmén M (2013) Business model innovation–the challenges ahead. Int J Prod Dev 18(3/4):213–225

Borys B, Jemison DB (1989) Hybrid arrangements as strategic alliances: theoretical issues in organizational combinations. Acad Manag Rev 14(2):234–249

Campbell JL (2007) Why would corporations behave in socially responsible ways? An institutional theory of corporate social responsibility. Acad Manag Rev 32(3):946–967

Clarkson ME (1995) A stakeholder framework for analyzing and evaluating corporate social performance. Acad Manag Rev 20(1):92–117

Cossío-Silva FJ, Revilla-Camacho MÁ, Vega-Vázquez M, Palacios-Florencio B (2016) Value co-creation and customer loyalty. J Bus Res 69(5):1621–1625

Darby MR, Karni E (1973) Free competition and the optimal amount of fraud. J Law Econ 16(1):67–88

EASA (2019) European aviation report 2019

Edvardsson B, Tronvoll B, Gruber T (2011) Expanding understanding of service exchange and value co-creation: a social construction approach. J Acad Mark Sci 39(2):327–339

Fernando Y, Chukai C (2018) Value co-creation, goods and service tax (GST) impacts on sustainable logistic performance. Res Transp Bus Manag 28:92–102

Flint DJ (2006) Innovation, symbolic interaction and customer valuing: thoughts stemming from a service-dominant logic of marketing. Mark Theory 6(3):349–362

Furrer O (2010) A customer relationship typology of product services strategies. In: The Handbook of Innovation and Services. Edward Elgar Publishing

Gebauer H, Gustafsson A, Witell L (2011) Competitive advantage through service differentiation by manufacturing companies. J Bus Res 64(12):1270–1280

Geum Y, Park Y (2011) Designing the sustainable product-service integration: a product-service blueprint approach. J Clean Prod 19(14):1601–1614

Grandy G, Levit T (2015) Value co-creation and stakeholder complexity: what strategy can learn from churches. Qual Res Organ Manag Int J

Grönroos C (2011) Value co-creation in service logic: a critical analysis. Mark Theory 11(3):279–301

Grönroos C, Voima P (2013) Critical service logic: making sense of value creation and co-creation. J Acad Mark Sci 41(2):133–150

Gummesson E, Mele C, Polese F, Galvagno M, Dalli D (2014) Theory of value co-creation: a systematic literature review. Manag Serv Qual

Hatch MJ, Schultz M (2010) Toward a theory of brand co-creation with implications for brand governance. J Brand Manag 17(8):590–604

Hoyer WD, Chandy R, Dorotic M, Krafft M, Singh SS (2010) Consumer cocreation in new product development. J Serv Res 13(3):283–296

International Air Transport Association (IATA) (2019) Annual review 2019. IATA

International Civil Aviation Organization (2011) Aviation & sustainability. Determining the complex environmental, economic and social impacts that are defining aviation's future. ICAO J 66(6):1–36

Karpen IO, Bove LL, Lukas BA (2012) Linking service-dominant logic and strategic business practice: a conceptual model of a service-dominant orientation. J Serv Res 15(1):21–38

Karpen IO, Bove LL, Lukas BA, Zyphur MJ (2015) Service-dominant orientation: measurement and impact on performance outcomes. J Retail 91(1):89–108

Lacoste S (2012) "Vertical coopetition": The key account perspective. Ind Mark Manag, 41(4):649–658

Lacoste S (2016) Sustainable value co-creation in business networks. Ind Mark Manage 52:151–162

Laperche B, Picard F (2013) Environmental constraints, product-service systems development and impacts on innovation management: learning from manufacturing firms in the French context. J Clean Prod 53:118–128

Lindström T, Polsa P (2016) Coopetition close to the customer—a case study of a small business network. Ind Mark Manage 53:207–215

Lufthansa Group (2019) Balance sustainability report 2019, 25th issue. https://www.lufthansagroup.com/media/downloads/en/responsibility/LH-sustainability-report-2019.pdf

Lusch RF, Nambisan S (2015) Service innovation: a service-dominant logic perspective. MIS Q 39(1):155–176

Ma Y, Rong K, Luo Y, Wang Y, Mangalagiu D, Thornton TF (2019) Value co-creation for sustainable consumption and production in the sharing economy in China. J Clean Prod 208:1148–1158

Martín-de Castro G, Amores-Salvadó J, Navas-López JE (2016) Environmental management systems and firm performance: improving firm environmental policy through stakeholder engagement. Corp Soc Responsib Environ Manag 23(4):243–256

Mauser W, Klepper G, Rice M, Schmalzbauer BS, Hackmann H, Leemans R, Moore H (2013) Transdisciplinary global change research: the co-creation of knowledge for sustainability. Curr Opin Environ Sustain 5(3–4):420–431

Mazur J, Zaborek P (2014) Validating DART model. Int J Manag Econ 44(1):106–125

McKelvey M, Holmén M (eds.) (2006) Flexibility and stability in the innovating economy. Oxford University Press

Nidumolu R, Prahalad CK, Rangaswami MR (2009) Why sustainability is now the key driver of innovation. Harv Bus Rev 87(9):56–64

Ordieres-Meré J, Remón TP, Rubio J (2020) Digitalization: an opportunity for contributing to sustainability from knowledge creation. Sustainability 12(4):1460

Payán-Sánchez B, Plaza-Úbeda JA, Pérez-Valls M, Carmona-Moreno E (2018) Social embeddedness for sustainability in the aviation sector. Corp Soc Responsib Environ Manag 25(4):537–553

Payne AF, Storbacka K, Frow P (2008) Managing the co-creation of value. J Acad Mark Sci 36(1):83–96

Porter ME, Van der Linde C (1995) Toward a new conception of the environment-competitiveness relationship. J Econ Perspect 9(4):97–118

Prahalad CK, Ramaswamy V (2001a) The collaboration continuum. Optim Magaz 31–39

Prahalad CK, Ramaswamy V (2004b) The future of competition: co-creating unique value with customers. Harvard Business Press

Prahalad CK, Ramaswamy V (2001b) The value creation dilemma: new building blocks for co-creating experience. Harv Bus Rev 18(3):5–14

Prahalad CK, Ramaswamy V (2004a) Co-creation experiences: the next practice in value creation. J Interact Mark 18(3):5–14

Ramaswamy V (2005) Co-creating experiences with customers: new paradigm of value creation. TMTC J Manag 8:6–14

Ramaswamy V (2008) Co-creating value through customers' experiences: the Nike case. Strat Leadersh

Ramirez R (1999) Value co-production: intellectual origins and implications for practice and research. Strateg Manag J 20(1):49–65

Ranjan KR, Read S (2016) Value co-creation: concept and measurement. J Acad Mark Sci 44(3):290–315

Rantala T, Ukko J, Saunila M, Havukainen J (2018) The effect of sustainability in the adoption of technological, service, and business model innovations. J Clean Prod 172:46–55

Russo MV, Fouts PA (1997) A resource-based perspective on corporate environmental performance and profitability. Acad Manag J 40(3):534–559

Russo-Spena T, Mele C (2012) "Five Co-s" in innovating: a practice-based view. J Serv Manag 23(4):527–553

Saarijärvi H (2012) The mechanisms of value co-creation. J Strateg Mark 20(5):381–391

Salzmann O, Ionescu-Somers A, Steger U (2005) The business case for corporate sustainability: literature review and research options. Eur Manag J 23(1):27–36

Scandelius C, Cohen G (2016) Sustainability program brands: platforms for collaboration and co-creation. Ind Mark Manage 57:166–176

Schiavone F, Metallo C, Agrifoglio R (2014) Extending the DART model for social media. Int J Technol Manag 66(4):271–287

Sesliokuyucu OS, Polat İ (2020) Dialogue and transparency in value co-creation: An empirical analysis of airline passengers. J Aviat Res 2(2):168–181

Seuring S, Gold S (2013) Sustainability management beyond corporate boundaries: from stakeholders to performance. J Clean Prod 56:1–6

Sheth JN, Uslay C (2007) Implications of the revised definition of marketing: from exchange to value creation. J Public Policy Mark 26(2):302–307

Taghizadeh SK, Jayaraman K, Ismail I, Rahman SA (2016) Scale development and validation for DART model of value co-creation process on innovation strategy. J Bus Indus Market

Taneja NK (2016) Airline industry: poised for disruptive innovation? Routledge

Thorelli HB (1986) Networks: between markets and hierarchies. Strateg Manag J 7(1):37–51

Tress G, Tress B, Fry G (2005) Clarifying integrative research concepts in landscape ecology. Landscape Ecol 20(4):479–493

Vargo SL (2008) Customer integration and value creation: paradigmatic traps and perspectives. J Serv Res 11(2):211–215

Vargo SL, Lusch RF (2004) Evolving to a new dominant logic for marketing. J Mark 68(1):1–17

Vargo SL, Lusch RF (2008) Service-dominant logic: continuing the evolution. J Acad Mark Sci 36(1):1–10

Vargo SL, Lusch RF (2016) Institutions and axioms: an extension and update of service-dominant logic. J Acad Mark Sci 44(1):5–23

Vargo SL, Maglio PP, Akaka MA (2008) On value and value co-creation: a service systems and service logic perspective. Eur Manag J 26(3):145–152

Veliyath R, Fitzgerald E (2000) Firm capabilities, business strategies, customer preferences, and hypercompetitive arenas: the sustainability of competitive advantages with implications for firm competitiveness. Compet Rev Int Bus J

Von Hippel E (2005) Open source software projects as user innovation networks. Perspect free Open Source Softw 267–278

Williams J, Aitken R (2011) The service-dominant logic of marketing and marketing ethics. J Bus Ethics 102(3):439–454

Yi Y, Gong T (2013) Customer value co-creation behavior: Scale development and validation. J Bus Res 66(9):1279–1284

Zwass V (2010) Co-creation: Toward a taxonomy and an integrated research perspective. Int J Electron Commer 15(1):11–48

Inci Polat (Ph.D.) is currently an Assistant Professor in the Department of Aviation Management at Suleyman Demirel University. She received her MSc. in International Trade and Logistics at Gaziantep University. She graduated with her Ph.D. in Civil Aviation Management from Anadolu University, Turkey. She published papers about value co-creation/destruction, service innovation and strategic orientation national/international journals and books. Her research field includes S-D logic, value co-creation/destruction, environmental sustainability, service innovation, airline marketing.

Chapter 4
Corporate Social Responsibilities in Air Transport: A Research Agenda on the Effects of the COVID-19 Pandemic

Yeşim Kurt

Abstract The subject of this research is the corporate social responsibilities of aviation organizations. The research aims to reveal the relationship between aviation organizations and CSR initiatives and the effects of the COVID-19 pandemic on this relationship. The research area has been created with a global approach, from airport and airline organizations in different parts of the world. The research utilizes the literature review and secondary sources obtained from the websites of the relevant aviation organizations. First of all, this research, which includes general information about CSR, then reveals the CSR initiatives carried out by aviation organizations before COVID-19. In the following section, why CSR initiatives should be maintained during the COVID-19 pandemic is discussed. Finally, the research includes CSR initiatives undertaken by aviation organizations about the COVID-19 pandemic. According to the results of the research, airport organizations mostly carry out ethical and legal responsibilities related to health, hygiene, and safety measures against the pandemic. In addition to the aforementioned measures, airline organizations carry out voluntary corporate social responsibilities such as donating and transporting medical protective equipment, carrying the COVID-19 vaccine, free transport of pandemic groups, especially healthcare workers, or gifting flight miles to these groups.

Keywords COVID-19 · Air transport · Airport management · Airline management · Corporate social responsibility

4.1 Introduction

The detection of the COVID-19 disease in Wuhan, China on December 31, 2019, and its rapid spread to become a global problem in the future has brought the whole world to the brink of an unprecedented crisis. The rapid spread of this virus has led to

Y. Kurt (✉)
Lüleburgaz Faculty of Aeronautics and Astronautics, Department of Civil Aviation Management, Kırklareli University, Kırklareli, Turkey
e-mail: yeshimkurt@klu.edu.tr

© The Author(s), under exclusive license to Springer Nature Singapore Pte Ltd. 2022
K. Kiracı and K. T. Çalıyurt (eds.), *Corporate Governance, Sustainability, and Information Systems in the Aviation Sector, Volume I*, Accounting, Finance, Sustainability, Governance & Fraud: Theory and Application, https://doi.org/10.1007/978-981-16-9276-5_4

the mass sickness of millions of people and the death of many in a short time.[1] While this situation started to threaten the health systems and economies of the country, states have taken measures to prevent the spread of the disease. In this new era where remote work is widespread, face-to-face education is suspended, home quarantines are experienced, and people cannot socialize, various sectors have come to the brink of the economic crisis. This crisis is so great that it is seen as equivalent to the great depression of the 1930s. The tourism and travel industry are also among the areas most affected by this crisis (Chua et al. 2020, p. 1). States that want to prevent the spread of the virus have implemented flight bans by closing their air borders. These prohibitions brought air transportation to a halt (International Finance Corporation 2020, p. 1; Kurt 2020, p. 195).

The fact that passenger transportation by airline came to a halt, by affecting airline organizations, airport organizations, and other aviation organizations providing service to the airport, brought this constantly growing sector to a standstill. This effect has been so great that travel bans have been imposed on 96% of all flight destinations in the world in a short period from the time the virus appeared until April 6, 2020 (UNWTO 2020). The International Civil Aviation Organization (ICAO) estimates that airlines lost revenues of $ 112–135 billion in the first half of 2020 (Buhusayen et al. 2020, p. 1). Similarly, a 60%–80% reduction in annual international passenger transport is expected in 2020 compared to the previous year (Han et al. 2020, p. 1). It is thought that it will take years to compensate for these great losses in air transportation. According to the authorities, it does not seem possible for this sector to reach 2019 levels even before 2023 (Elias 2020).

In this period when individuals, employees, and societies need support most, it is also an important research topic whether organizations will turn to socially responsible behaviors. To what extent will sectors such as air transport, which is also in a great economic crisis, respond to the expectations of the society in this period? How much ethical behavior will he be able to display? To what extent will it be able to fulfill its economic and ethical responsibilities towards its stakeholders? What responsibilities will it take for the well-being of its employees? The answers to these questions point to a paradox worth discussing. Because, according to some, organizations whose resources are decreasing in this period may show unethical behaviors or behaviors towards reducing long-term CSR investments (He and Harris 2020, p. 177).

There are various studies in the literature that raise the relationship between COVID-19 and CSR (Chua et al. 2020; Goldston 2020; Qiu et al. 2021; Lee 2020; Ebrahim and Buheji 2020; Marom and Lussier 2020; García-Sánchez and García-Sánchez 2020; Crane and Matten 2020; Roxana-Loredana et al. 2020; Aini 2020; Popkova et al. 2020). All these studies show the relationship between COVID-19 and CSR in different sectors; uncertainties brought by this period, the future of CSR, ethical CSR, the impact of CSR on the value of the firm, etc., argues in many ways. However, in the literature, no research directly focusing on the relationship between

---

[1] https://www.who.int/emergencies/diseases/novel-coronavirus-2019 (Access of date: 03 Jan 2021).

CSR and COVID-19 in air transport has been encountered. This book section, it is aimed to understand and explain the approaches of aviation organizations to CSR in the COVID-19 period. For this purpose, it is planned to discuss CSR and the relationship between these organizations comprehensively.

This book section, which is based on literature research and secondary sources on CSR initiatives of aviation organizations, includes various subjects. In the second title of the research, the definition of CSR, motivation sources, and CSR typologies are framed. In the third part, the relationship between air transport and CSR is emphasized. This relationship is exemplified by the CSR typologies that aviation organizations have turned and the CSR initiatives they implemented in the pre-COVID-19 period. In the fourth part of the study, it is discussed why commercial organizations should carry out CSR initiatives during the pandemic period. This requirement is discussed by revealing the need for CSR in society and the long-term benefits of meeting this need for organizations. In the last section, the corporate social responsibilities of aviation organizations related to COVID-19 during the pandemic period are presented with examples from airport and airline organizations.

4.2 Corporate Social Responsibility: Definitions, Motivations, Typologies

4.2.1 Definitions

CSR was first mentioned by Bowen (1953) in the book "Social Responsibilities of the Businessman"; It is expressed as the organizations implementing practices that are compatible with the society they live in and to improve this society. Over time, this concept, which attracted the attention of both academics and various authorities or non-governmental organizations, has been tried to be explained with similar definitions. Kotler and Lee (2008, p. 3) express CSR as the necessity of organizations to work with society to improve society. CSR has a perspective that is based on the fact that organizations should undertake social responsibilities towards society and that associates the sustainability of organizations with the sustainability of society (Marom and Lussier 2020). CSR is related to the policies realized by the organizations considering the interests of the society rather than their interests (Theaker 2004, p. 183). CSR is defined as the improvement of society by integrating social and environmental issues into the organization (UN 2004, p. 4). According to the Organization for Economic Co-operation and Development, this concept is that organizations contribute to environmental and social development through their activities (OECD 2001, p. 13). According to another statement, CSR is related to the decisions and practices of organizations to improve economic, legal, ethical, philanthropic, and environmental factors (Han et al. 2020, p. 2). According to Carroll (1979, p. 499), one of the researchers who carried out important research on CSR, this concept is that organizations exhibit economic, legal, ethical, and philanthropic behaviors related to the society they live in.

4.2.2 Motivations

What are the reasons that motivate business organizations, whose main purpose is to make a profit, to allocate resources for CSR initiatives? Why do these organizations carry out voluntary responsibilities that show that they are more sensitive to the environment and add value to society besides their economic responsibilities? What are the dynamics that cause organizations that aim to make more profit, on the one hand, to devote their resources to aid activities on the other hand? These questions have been discussed and researched by different academics in various fields for years. Answers to these important questions are sought in this title of the book chapter.

CSR was first questioned with the great depression in the 1930s. The great economic crises that people are going through have also paved the way for the existence of commercial organizations to be questioned. In the following years, this issue, which attracted the attention of academicians, has become debated in the world of science as well as in society.

In the following period, one of the main reasons that led organizations to these initiatives is globalization and liberalization developments. These developments, which made the competition much more intense, made it necessary for organizations to do different jobs than others to survive. In this period, the expectations of organizations, their society, their customers, or different stakeholder groups started to change. Organizations have tried to achieve sustainability by responding to these expectations (Vogel 2007, p. 8; McWilliams and Siegel 2001, p. 117). In other words, globalization has put pressure on organizations to act more responsibly (Blowfield and Murray 2014, p. 3–4). Multinational organizations operating in cross-border regions due to globalization have had to meet the expectations of people and governments in new countries.

In this new period, when the damage caused by organizations to the environment is being discussed, the choices of customers have started to favor organizations that are less harmful to the environment. For this reason, organizations that want to be preferred more have started to carry out activities that will create the impression of a good corporate citizen and perform activities that will be deemed respectable by the social, economic, and institutional environment. Non-governmental organizations also emerged in this period and started to expect organizations to compensate for the damage they caused to the environment. In other words, organizations have turned to environmental problems to manage pressures (Kramer and Porter 2006, p. 5).

Also, one of the reasons why organizations carry out voluntary activities to improve society is the belief that these activities will increase brand value and their desire to make advertising (Brammer and Millington 2005, p. 517). In a study conducted by Kurt (2019a), it has been revealed that airline organizations tend to corporate social responsibilities with market logic, even voluntary activities are used as corporate communication and marketing communication, and these activities are seen as a tool to be more preferred by customers. In another research, it was revealed that CSR activities also create advantages in attracting and keeping strategic human resources to the organization (Justice 2002).

In addition to all these developments, the new standards that emerged with globalization, many standards, initiatives, or guidelines set by the country authorities or global professional organizations have also been guiding the CSR initiatives of commercial organizations. Kurt (2019b); In another research, these global organizations and initiatives that direct CSR activities of organizations have been revealed. According to her, there are 16 corporate actors and initiatives that direct business organizations to legal, ethical, and voluntary responsibilities in various fields of activity. These are listed as follows (Kurt 2019b, p. 42):

1. International Labour Organisation—(ILO)
2. Organization for Economic Co-Operation and Development—(OECD)
3. United Nations Global Compact
4. United Nations: Environment Programme Finance Initiative—(UNEP-FI)
5. European Commission: White and Green Paper
6. United Nations: The Universal Declaration of Human Rights—(UDHR)
7. Caux Round Table: Principles of Caux
8. Global Sullivan Principles
9. Coalition For Environmentally Responsible Economies—(CERES)
10. Financial Times Stock Exchange: FTSE4GOOD index
11. Global Reporting Initiative—(GRI)
12. Global Corporate Citizenship Initiative—(GCCI)
13. Social Accountability 8000
14. AccountAbility—AA100
15. British Standards Institution –(BSI): Social Responsibility Code of Conduct; Occupational Health and Safety Assessment Systems—(OHSAS 18,001)
16. International Organization for Standardization—(ISO): ISO 45001, ISO 26000, ISO 14001.

4.2.3 Typologies

In which areas and what type of CSR initiatives commercial organizations have undertaken, researchers have also explained various classifications. Although many researchers reveal models and classifications for CSR, one of the most familiar of them is Carroll's social responsibility pyramid. Carroll (1991) classified the corporate social responsibilities of organizations into 4 levels by likening them to a pyramid and placed the economic responsibility of organizations at the bottom of the pyramid. According to him, the secondary responsibility of the organizations is legal, the third responsibility is ethics and the highest level of responsibility is volunteerism. Economic responsibilities are the most fundamental and indispensable responsibilities of organizations because it is not possible for organizations that cannot fulfill this responsibility and make their unprofitable existence sustainable to fulfill their other responsibilities. Legal responsibilities are related to the compliance of organizations with the laws of the countries and regions in which they operate. Ethical responsibilities are about the ethical behavior of organizations, being transparent and fair,

Table 4.1 Proposal for CSR Typologies

Economic and legal responsibilities	Commercial CSR	CSR ethics	Altruistic CSR	Strategic CSR
Decisions that managers would make to guarantee the creation of value for shareholders, the job security of employees, and the quality of products and services They suppose the fulfilment of laws and informal rules of the game	CSR actions closely related to business activity and aimed at obtaining economic benefits associated with attracting new customers or increasing the confidence of existing ones	Fair and equitable business decisions to avoid damages	Philanthropic actions aimed at preventing potential harm and alleviating negative externalities that affect the welfare state without necessarily entailing economic benefits for the company	Ethical and altruistic actions selected to guarantee the creation of value for shareholders and investors and aimed at generating benefits for the company through their impact on image and reputation

the right thing to do, and the right way of doing things. Responsibilities involving volunteering at the highest level are related to activities that organizations are not responsible for, do not make a profit, and philanthropic to improve society and the world (Carroll 1979, 1991).

On the other hand, García-Sánchez and García-Sánchez (2020, p. 5) classify CSR typologies with a striking proposal in a new study. This recommendation is shown in Table 4.1.

One of the most important issues that draw attention in Table 4.1 is Strategic CSR. Accordingly, the CSR initiatives that are selectively implemented by organizations to manage their image and reputation while fulfilling their economic and commercial responsibilities towards their stakeholders and shareholders are summarized as strategic CSR.

The target group of CSR initiatives of organizations in different typologies, as summarized above, may differ from each other. Organizations; can take such initiatives for different stakeholder groups such as employees, managers, shareholders, environment, government, society, competitors, suppliers, or customers (Kurt and Besler 2019).

4.3 Corporate Social Responsibility in Air Transportation

In the previous sections of this unit, basic information about CSR has been given. The definition of CSR, in which areas, in which types it is carried out, CSR typologies are briefly summarized. In this section, the relationship between air transport and CSR is included. This relationship is explained firstly with examples of the areas or typologies in which aviation organizations tend to engage in CSR initiatives, and secondly with examples from the CSR practices they have realized. According to the results of some studies referring to the relationship between air transport and CSR, such initiatives create advantages that increase the financial performance values of airline organizations (Lee and Park 2010; Lee et al. 2013). Kurt and Besler, (2019), all airlines operating in Turkey Turkey in a study in which they examined the CSR initiatives of the organization; It has revealed that there are CSR initiatives in four basic dimensions: economic, legal, ethical, and voluntary. Under these dimensions, it has been determined that practices aimed at the environment, customers, society, managers, and employees are the majority. Other studies are focusing on the relationship between aviation organizations and CSR. Another study revealed that aviation organizations carry out CSR initiatives for social, environmental, and economic responsibilities (Kemp and Vinke 2012). According to another important study focusing on the relationship between airline organizations and CSR, these organizations focus on environmental issues rather than social and economic dimensions (Cowner-Smith and De Grosbois 2011).

In addition to these researches encountered in the literature, giving examples from their sentences in which typologies or areas aviation organizations have undertaken these practices will contribute to a better understanding of this relationship. These examples are shown in Tables 4.2 and 4.3[2]:

As shown by numerous examples in Table 4.2, aviation organizations have CRS policies in various sizes and areas, just like commercial organizations in other sectors. It will also contribute to a better understanding of this section by revealing what kind of initiatives aviation organizations have taken against CRS policies and which CSR activities they have undertaken. Table 4.3 presents examples of CSR activities carried out by various airline and airport organizations.

As exemplified in Table 4.3, aviation organizations show responsible behaviors that contribute to environmental sustainability, as well as initiatives that contribute to society and children. Both airline and airport organizations take measures for energy management, fuel savings, and emission gases and legitimize these measures through certification.

4.4 Why Should CSR Initiatives Be Continued During COVID-19 Pandemic Period?

In this section, which discusses why CRS initiatives should continue during the pandemic period, firstly the need for CSR initiatives in the society is mentioned, and then the benefits of the initiatives to meet these needs are discussed.

---

[2] The examples given in Tables 4.2 and 4.3 were obtained from the websites of the relevant airport and airline organizations between December 15, 2020, and January 3, 2021. Relevant examples were obtained from aviation organizations' press releases, sustainability or social responsibility reports and included in this research as direct citations.

Table 4.2 Statements of aviation organizations on CRS typologies

Organization type	CRS typologies
Airport organizations	Belfast International Airport understands Corporate Responsibility (CR) as its responsibilities to its employees, customers, suppliers, regulators, community, and the environment. The senior management team supports the objectives and principles of CR and this policy will help to translate these into a common, consistent approach for everyone
	At Liverpool John Lennon Airport, we take our Corporate Social Responsibility (CSR) as just that. A responsibility. As a business, we are aware of the impact on our local Economy, Colleagues, and Customers. But as a member of the community, we work hard to ensure we create a sustainable, responsible business—one that gives back to those living and working within our local area
	Hermes Airports is firmly committed to high standards in Health, Safety, and Environmental Management System for all employees, customers, passengers, partners, tenants, and the communities nearby Larnaka and Pafos International Airports
	Riga Airport's entire range of activities and functions is performed according to the principles of Corporate Social Responsibility (CSR). In addition to the successful implementation of its business strategy, the company follows a consistent corporate policy that fully respects its commitment towards passengers, customers, business partners, employees, shareholders, local communities, as well as all other stakeholders
Airline organizations	Norwegian strives to be a good corporate citizen in every area of operation. The company is committed to operating under responsible, ethical, sustainable, and sound business principles, with respect for people, the environment, and society
	Southwest takes great pride in how we care for our Employees, Customers, Shareholders, and Communities. The One Report helps us bring to life our purpose of connecting People to what's important in their lives through friendly, reliable, and low-cost air travel, and we are excited to share it
	At United, we are committed to building a sustainable future. This report highlights our involvement in the communities we serve, our commitment to the environment, and the investments we're making in our fleet, technology, facilities, and co-workers. We believe these actions play an important role in ensuring that we operate an environmentally and socially responsible business
	American Airlines has produced an annual Corporate Responsibility Report since 2007, but 2020 marked a new phase in the evolution of our reporting on environmental, social, and governance (ESG) issues

4.4.1   The Need for CRS During the COVID-19 Pandemic

The COVID-19 pandemic has caused a global crisis all over the world by creating huge impacts on different segments such as individuals, societies, economies, and

Table 4.3 Aviation organizations' CRS initiatives Pre-COVID-19

Organization type	CRS initiatives
Airport organizations	Taoyuan International Airport Corporation, Ltd (referred to as TIAC) has published the first Corporate Social Responsibility Report (referred to as the report), which is compiled with international standard—GRI standards and is verified under AA1000 and ISAE3000 Assurance Standards
	Hermes Airports has acquired the fourth level of the ACA Certification Level 3 + Neutrality (achieving net-zero carbon emissions) for Larnaka and Pafos airports, which relates to the offset of carbon dioxide emissions. The accreditation confirms the steady and long-term reduction of carbon dioxide emissions while at the same time confirms the occurrence of its energy policy and strategy
	Esenboğa Airport receives a Certificate for "Energy Management System. Operated by TAV Airports, Ankara Esenboğa Airport completed all the work required to control energy consumptions and received the ISO 50001 Energy Management System certificate issued by The International Standardization Organization (ISO)
	London Stansted Airport. This year, through a variety of funds at our airport, we contributed £256,672 in funding to 204 local groups. We are pleased to have seen an 83% increase in donations, and a 104% increase in the number of projects funded compared to last year
Airline organizations	Etihad Airways and Boeing commence test flights for sustainable air travel on 787–10 SAS aims to achieve a 25% reduction in CO2 emissions 5 years earlier than planned China Eastern Airlines aircraft fuel consumption per ton-km down 4% from 2017 to 2019
	China Eastern Airlines aircraft fuel consumption per ton-km down 4% from 2017 to 2019 China Eastern Airlines stated (19 Aug 2020) its aircraft fuel consumption per ton kilometer reduced by 4.3% from 2017 to 2019
	In 2019, Delta was again recognized by Great Place to Work, improved our Net Promoter Score, donated a record 13,064 pints of blood, improved fuel efficiency, and this year, committed to investing $1 billion over the next 10 years to become carbon neutral
	American Airlines is donating 10 million loyalty miles to St. Jude Children's Research Hospital® to support its mission to help patient families as it leads the way the world understands, treats, and defeats childhood cancer and other life-threatening diseases

states. While this crisis threatens human health and social life, on the one hand, it has created a global rupture in which many segments of society need help by affecting and changing business life at an unprecedented level.

The primary impact of COVID-19 pandemic is its threats to human health and social life. According to the data of the World Health Organization, in the period from the day the virus was detected to January 5, 2021, 84,474,195 people were detected to have this virus, and 1,848,704 people died due to this virus. States that want to prevent the rapid spread of the virus have introduced various prohibitions and restrictions. Education and training systems started to be run remotely, physical distance rules were developed; crowd activities such as entertainment, travel, and shopping have been discontinued; and the use of protective physical equipment has been made mandatory to prevent the virus from spreading through droplets (Blake-Beard et al. 2020). The constant presence of people at home has started to cause different health problems while protecting them from the virus. During this period, people's psychology began to deteriorate, alcohol, etc. The use of harmful substances has increased, and sedentary life has started to trigger other physical health problems. (García-Sánchez and García-Sánchez 2020, p. 3; Settersten et al. 2020, p. 3; Venkatesh 2020, p. 3.) The rapid increase in the number of people infected with the virus worldwide is also huge in health systems caused problems. The vast majority of hospitals were reserved to serve COVID-19 patients, hospital occupancy rates increased rapidly, COVID-19 patients could not receive treatment because the number of patients continued to increase, all healthcare professionals were removed from their routines to provide adequate healthcare services to the increasing number of patients was commissioned for. This situation led to both the fatigue of healthcare professionals and working at risk (Halawi et al. 2020, pp. 759–760), causing many of them to become COVID-19 patients and sometimes death.

Other important effects of this pandemic have been on business life and economies. The prohibitions imposed by the authorities of the country on sectors such as entertainment, tourism, shopping, travel, or the curfews imposed to maintain social distances have forced people to work and spend time in their homes. The spending of people who do not go out on the streets has also changed and contracted. This situation has caused the economies of many sectors to suffer. Most of the organizations operating in sectors such as transportation, tourism, food, and beverage have entered into an economic crisis and these contractions have become threatening the global economic markets (Altamirano and Collazo 2020, p. 65). According to a study, this economic contraction in the business world causes global GDP to fall by as much as 5.2% (Mukherjee et al. 2020, p. 179). In the sectors that entered the crisis, many people started to lose their jobs, and those who were not fired were cut their salaries (García - Sánchez and García-Sánchez 2020, p. 3; Venkatesh 2020, p. 2; Altamirano and Collazo 2020, p.). This situation has also forced people living with the threat of the virus to struggle with economic concerns.

People whose social lives, health, and work lives are changed and threatened due to COVID-19 have become in need of support like never before. This support makes itself felt in different ways, both in terms of their economy, health, morale, and motivation. Employees' expectations from organizations in this period have changed in the direction of following ethical policies that prioritize their health, take adequate hygiene and health measures for them, provide adequate training support for remote work, and do not leave them unemployed while experiencing such an economic collapse. People have come to expect these expectations from both commercial organizations and their governments.

In addition to these expectations and needs, the fact that organizations have been affected by the crisis at least as much as individuals put commercial organizations in a difficult situation. Many organizations in different sectors, whose activities have come to a halt, are on the brink of bankruptcy. While the COVID-19 crisis affects all individuals and systems in both micro-, meso-, and macro-terms to this extent, it is also a matter of discussion what will be the approach of organizations that are struggling to exist themselves to corporate social responsibilities. What may be the reasons that will motivate these organizations, which are on the verge of closing down, towards corporate social responsibility initiatives in such a difficult period? How were CSR initiatives of organizations affected during this period? Are the resources allocated by organizations for these initiatives ongoing? To what extent will organizations be able to respond to their responsibilities to the laws of their employees, customers, shareholders, or the countries to which they belong? The answers to these questions are discussed in the following section.

4.4.2 Benefits of CRS Initiatives for Organizations During the COVID-19 Pandemic

During the crisis period organizations are tested in terms of their ethical behavior and commitment to CSR. Some organizations have turned to pursue short-term gains even through fraud and abuse due to the financial difficulties caused by this epidemic, and this crisis has probably triggered them to reduce their long-term CSR investments. Despite these, some organizations also turn to various CSR initiatives related to the pandemic during this period (He and Harris 2020, p. 177). Others experience an uncertainty about CSR strategies. In this section, the advantages of making decisions to realize CSR initiatives during the pandemic period will provide organizations.

In the past, studies have been conducted to reveal the positive effects of CSR on the brand value and reputation of organizations (Yang et al. 2020, p. 3). There are even studies revealing that CSR initiatives carried out in the pre-pandemic period contributed to less damage to the organization during the COVID-19 crisis period (Huang et al. 2020, p. 1). In some studies conducted in the post-COVID-19 period, it has been suggested that undertaking CSR initiatives during this crisis period will have consequences for the advantage of organizations.

During the pandemic period, 78% of employees expect organizations to take action to protect employees and the local community, according to a study conducted simultaneously in ten different countries. Seventy-nine percent of this period; organizations expect adaptations in favor of employees, such as working remotely, canceling non-compulsory work, and stopping business travel. Seventy-three percent of employees expect adaptations such as improvements in human resources policies and paid sick leave.[3] Another study conducted by the same institution with 12,000 people in 12 different countries also reveals remarkable results. Ninety percent of people expect brands to do everything in their power to protect the well-being and financial strength of their employees and suppliers, even if they incur significant financial losses until

---

[3] https://www.edelman.com/research/edelman-trust-covid-19-demonstrates-essential-role-of-pri vate-sector (Access of date: 03 Jan 2021).

the pandemic ends. Besides, 89% of respondents want to produce products that help consumers meet today's challenges, as well as provide free or lower priced products to healthcare workers, high-risk people, and those whose jobs are affected. At the same time, 86% of the participants expect brands to step in where states are lacking. One of the remarkable results of the study is that more than one-third of the participants preferred new brands that displayed responsible behavior during this period.[4] Accordingly, the continuation of responsible behaviors of organizations that are in financial difficulties during such crisis periods will attract the attention of society and customers and will contribute to gaining new customers in the post-crisis period (Huang et al. 2020, p. 4).

According to a study conducted on accommodation organizations during the pandemic period, CSR initiatives aimed at stakeholders such as the society, customers, and employees contribute to the protection and even increase of the share value of the organizations and lead them to recover from this crisis more easily (Qiu et al. 2021, p. one).

Meeting the expectations explained above will bring sustainability to organizations in the long run. Employees in organizations that act ethically and responsibly towards their employees in times of crisis will contribute to this period more strongly and optimistically and employees will be motivated. As a result, the organizational commitment and organizational trust levels of the employees will also increase (Qiu et al. 2021, p. 2). At the same time, these initiatives in favor of the employee will continue to attract a qualified workforce into the organization in the post-crisis period (Huang et al. 2020, p. 1; Kurt 2020, p. 195).

For all these reasons, organizations in the new era; should further clarify CSR strategies for society and vulnerable groups. The way to survive in the long term in the post-COVID-19 era is to develop new visions and policies in line with the current situation (García-Sánchez and García-Sánchez 2020, p. 1).

4.5 CRS Initiatives in Air Transportation Post Covid-19 Era

Flight bans put forward by the country's authoritarian who wanted to prevent the spread of the virus during the pandemic period brought air transportation to a halt. Therefore, airports, airlines, and many other aviation organizations have come to the brink of an economic bottleneck. Since aviation organizations have a serious impact on the economies of their regions or countries, the sector's cessation has caused the global economy to worsen (International Finance Corporation 2020, p. 1; Flight Safety Foundation 2020, p. 4).

Commercial aviation organizations, which were in trouble during this period, have come to expect support from the higher authorities they are affiliated with.

---

[4] https://www.edelman.com/research/covid-19-brand-trust-report (Access of date: 03 Jan 2021).

For example, the Federal Aviation Authority (FAA), training, simulator, etc., during the epidemic. it waived services and certain earnings and sought to alleviate the economic pressures of airlines. In some of the important airline organizations, the sustainability of personnel employment has been compromised, layoffs, referrals, etc., applications have become visible. For example, it has been stated that Delta Airlines notified its 2,500 pilots on leave, directed some to early retirement, and aimed to reduce the workforce (Elias 2020, p. 5). Apart from this, many sector employees also face similar threats. In the light of all these developments, the approaches of aviation organizations towards CSR during the COVID-19 period, whether they have undertaken CSR initiatives in this period, and whom they have attempted in what kind of areas are among the important issues that are sought to answer in this book section. In this part of the unit, sample applications are given in response to related questions.

During the pandemic period, both the country authorities, the health authorities, and the aviation authorities have assumed responsibilities for reliable transportation in the sector under necessary hygienic conditions with various guidelines, regulations, and directives they have published.[5] Apart from the authorities, it is also a matter of curiosity about what kind of CSR initiatives commercial aviation organizations, especially airline and airport organizations have undertaken during the pandemic period. To clarify this issue, examples of initiatives of various airport organizations that can be associated with CSR during the COVID-19 period are given in Table 4.4.[6]

When Table 4.4 is analyzed, it is seen that airport organizations mostly assume responsibilities for health and hygiene measures to reduce the risk of COVID-19 transmission during the COVID-19 period. Airport organizations develop ethical practices such as regulations following physical distance rules and the use of protective equipment for their employees and their customers using airports. One of the important examples of this is that many airports have turned to the AHA accreditation program and provided certification. This program is an ethical responsibility recommended by Airport Council International (ACI). Since the security and hygiene measures taken by airports other than this certification may be mandatory by some state authorities, many of the practices exemplified in the table can also be shown as examples of legal responsibilities.

One of the aims of this chapter is to reveal what kind of social responsibility initiatives airline organizations have undertaken about the pandemic. Table 4.5 contains direct citations illustrating these initiatives.

When Table 4.5 is examined, it can be seen that airline organizations take different initiatives that they associate with social responsibility as well as providing hygiene and health measures for their employees and customers. These initiatives

---

[5] https://www.icao.int/Security/COVID-19/Pages/default.aspx (Access of date: 03 Jan 2021).

[6] The examples given in Tables 4 and 5 were obtained from the websites of the relevant airport and airline organizations between December 15, 2020, and January 3, 2021. Relevant examples were obtained from aviation organizations' press releases, sustainability or social responsibility reports and included in this research as direct citations.

Table 4.4 Airport organizations' post-COVID-19 CRS initiatives

Organization type	CRS initiatives
Airport organizations	At Hermes Airports, we believe it is our social responsibility to give back to our community. As a result of this, we have purchased dozens of face masks from KARAISKAKIO FOUNDATION for all our Hermes staff, actively thus showing our support, care, and loyalty to this commendable initiative
	At Taoyuan International Airport, we have spared no effort to make the aviation industry survive as well as to preserve jobs by considerably reducing rental fees for airlines, duty-free shops, and other commercial facilities. Going forward, we continue to restore the airport ecosystem hard hit by the pandemic and vitalize the local economy in line with our 'Together' spirit
	Being aware of its responsibility to control the spread of COVID-19 and to ensure the safety of passengers, employees, and partners, Riga Airport has implemented a comprehensive epidemiological safety program #ForbidTheVirusFromTravelling. It provides a set of measures for the protection of the Airport and Airport employees and the safe handling of passengers to minimize the risks of the spread of COVID-19
	Istanbul Airport first to achieve ACI Airport Health Accreditation. Accreditation (AHA) recognizes commitment to health and welfare of travelers, and the public
	Velana International Airport: September 1z 2020—After consistent measures in our fight against COVID-19, we are happy to announce that Velana International Airport (VIA) has received the much-awaited "Airport Health Accreditation" by Airport Council International (ACI) Velana International Airport is the second airport in the Asia Pacific Region to receive the ACI Airport Health Accreditation
	Da Nang International Airport receives Airport Health Accreditation (AHA) from Airports Council International. The accreditation is valid for the next 12 months. This is a recognition for DIA in terms of a high standard in upholding health and safety in its operations

are often carried out concerning transport services, which are the core capabilities of airlines. For example, some organizations have mediated the transportation of protective medical equipment needed during the pandemic period, and sometimes even attempted to both donate and carry these materials. At the same time, it is one of the most important of these responsibilities that they serve as a hope for the end of the pandemic and to serve the transportation of COVID-19 vaccines, which have recently been introduced in some regions. In addition to these, it is observed that airline organizations perform responsible behaviors that include volunteering, such as offering free or discounted transportation services to groups working to cure the pandemic, especially to healthcare workers, and giving free flight miles.

Table 4.5 Airline organizations' post-COVID-19 CRS initiatives

Organization type	CRS initiatives
Airline organizations	Since the beginning of the Covid-19 outbreak in Wuhan, AirAsia has leveraged its strong network in the Asia Pacific to transport critical medical and protective supplies such as face masks, gloves, and other protective materials, as well as helping organizations and individuals worldwide who donated humanitarian goods and equipment to frontliners in China
	Alitalia flight AZ675 landed at Rome Fiumicino Airport on March 24 with two million protective face masks from Sao Paulo, Brazil, for hospitals in the Piedmont region in northern Italy, where health workers are operating under very difficult conditions
	Helping in the fight against Coronavirus in February 2020, Qatar Airways Cargo sent five freighters to China to support coronavirus relief efforts. The five flights—carrying approximately 300 tonnes of medical supplies donated by the airline… A total of 2.5 million face masks and 500,000 bottles of hand sanitizer have been donated by Qatar Airways… Qatar Airways to Give Away 100,000 Complimentary Tickets to Frontline Healthcare Professionals
	American Airlines Transports Its First COVID-19 Vaccine Shipment From Chicago To Miami FORT WORTH, Texas—The American Airlines Cargo team carried its first shipment of coronavirus (COVID-19) vaccine last night. In close collabo…
	In the past month, the airline has transported over 50,000,00 0 kg of medical and aid supplies to impacted regions around the globe. This equates to roughly 500 fully loaded Boeing 777 freighters
	United Airlines is partnering with New York City to provide free round-trip flights for medical volunteers who want to help in the frontline fight against the COVID-19 crisis. The airline is working closely with the Mayor's Fund to Advance New York City and a network of medical volunteer organizations, including The Society of Critical Care Medicine, to coordinate travel for doctors, nurses, and other medical professionals from across the country to help treat patients, in this time of unprecedented need
	In response to the COVID-19 pandemic, we've made changes to our operations and procedures to better support the well-being and comfort of our Employees and Customers… We've employed stringent cleaning and physical-distancing practices such as using electrostatic and anti-microbial spray treatments in the cabin, implemented physical-distancing measures, modified boarding procedures, and provided masks for Employees

4.6   Conclusion and Recommendations

The COVID-19 pandemic creates a situation in which society most needs help, with high expectations from both governments and business organizations. On the other hand, this pandemic creates a paradox in which commercial organizations also need economic support, and are difficult to respond to society's expectations. This paradox is also evident in air transport, which has shown a continuous growth trend until the COVID-19 crisis. It is a matter of curiosity to what extent aviation organizations, whose activities have come to a halt and are struggling with major financial crises, turn to CSR initiatives in such a period. Therefore, in this unit, it is aimed to reveal the CSR initiatives of aviation organizations and the effects of the COVID-19 pandemic on these initiatives.

This research, which is based on the literature review and secondary sources obtained from the websites of aviation organizations, focused on the CSR initiatives of airport and airline organizations. On the one hand, the study included the CSR initiatives carried out by these organizations in the pre-COVID-19 period, on the other hand, the types of initiatives these organizations took in the name of social responsibility in the post-COVID-19 period were demonstrated with examples. In the pre-COVID-19 period, it was found that it mostly focused on environmental ethical responsibilities and sometimes voluntary social responsibilities. In the post-COVID-19 period, airport organizations assume responsibilities for improvements in required health, hygiene, and safety issues at airports for their customers and employees. Airline organizations, on the other hand, carried out social responsibilities related to the transportation service with their capabilities. These organizations have undertaken the task of donating and transporting medical protective equipment needed to protect against the pandemic, volunteering for the transport of COVID-19 vaccines that create excitement for the world, and transport groups such as healthcare workers who are at the forefront of the pandemic, free of charge or at a discount, as a gift to them with flight miles. They fulfilled voluntary responsibilities. According to some researches in the literature, such initiatives will provide various advantages to organizations. These initiatives will increase employees' loyalty to their organizations. It will attract new talents to these organizations after the pandemic; customers will prefer these organizations more. It is thought that brand and share values will increase and they will overcome the effects of the crisis more easily than other organizations (Huang 2020, p. 4; Kurt 2020, p. 192; Qiu et al. 2021, p. 1).

This research is limited to literature and document review. To understand and explain the relationship between CSR initiatives and the COVID-19 pandemic in more depth, it is necessary to continue researches both in the aviation industry and in different sectors. What impact has the pandemic era had on the realization or non-realization of such initiatives? What could be the short- and long-term benefits or harms of taking such initiatives for organizations? What awaits the organizations that did not take social responsibility initiatives during the pandemic period? Answering these and similar research questions is considered important to enrich the literature on the subject.

References

Aini EN (2020) Strategies and programs corporate social responsibility (CSR) facing the era new normal. Int J Innov Rev 1(1):45–52

Altamirano MA, Collazo CER (2020) Leading with emotional intelligence: How leaders of a diverse-based urban college in new york successfully respond to the COVID-19 crisis of 2020. J Educ Innov Commun Spec Issue, 65–75

Blake-Beard S, Shapiro M, Ingols C (2020) Feminine? Masculine? Androgynous leadership as a necessity in COVID-19. Gender Manag Int J

Blowfield M, Murray A (2014) Corporate responsibility. Oxford University Press

Bowen HR (1953) Social responsibilities of the businessman. Harper and Row, New York

Brammer S, Millington A (2005) Profit maximization vs. agency: an analysis of charitable giving By Uk Firms. Cambridge J Econ 29(4):517–534

Buhusayen B, Seet PS, Coetzer A (2020) Turnaround management of airport service providers operating during COVID-19 restrictions. Sustainability 12(23):10155

Carroll AB (1979) A Three-dimensional conceptual model of corporate performance. Acad Manag Rev 4(4): 497–50

Carroll AB (1991) The pyramid of corporate social responsibility: toward the moral management of organizational stakeholders. Bus Horizons, 39–48

Chua BL, Al-Ansi A, Lee MJ, Han H (2020) Tourists' outbound travel behavior in the aftermath of the COVID-19: the role of corporate social responsibility, response effort, and health prevention. J Sustain Tourism, 1–28

Cowper-Smith A, De Grosbois D (2011) The adoption of corporate social responsibility practices in the airline industry. J Sustain Tour 19(1):59–77

Crane A, Matten D (2020) COVID-19 and the future of CSR research. J Manag Stud

Ebrahim AH, Buheji M (2020) A pursuit for a 'holistic social responsibility strategic Framework' Addressing COVID-19 pandemic needs. Amer J Econ 10(5):293–304

Elias B (2020) Addressing COVID-19 pandemic impacts on civil aviation operations (No. R46483). https://crsreports.congress.gov/product/pdf/R/R4648.74. Accessed 29 December 2020

Flight Safety Foundation (2020) Non-medical operational safety aspects supplemental materials: pandemic—aviation safety roadmap version. https://flightsafety.org/wp-content/uploads/2020/05/COVID-19-Roadmap-V2.pdf. Accessed 20 December 2020

García-Sánchez IM, García-Sánchez A (2020) Corporate social responsibility during COVID-19 pandemic. J Open Innov Technol Market Complexity 6(4):126

Goldston J (2020) Pandemically speaking: sustaining corporate social responsibility during times of uncertainty. Int Res J Eng Technol (IRJET) 7(6):5514–5518

Halawi MJ, Wang DD, Hunt III TR (2020) What's important: weathering the COVID-19 crisis: time for leadership, vigilance, and Unity. J Bone Joint Surgery. American 102(9):759

Han H, Lee S, Kim JJ, Ryu HB (2020) Coronavirus disease (COVID-19), traveler behaviors, and international tourism businesses: impact of the corporate social responsibility (csr), knowledge, psychological distress, attitude, and ascribed responsibility. Sustainability 12(20):8639

He H, Harris L (2020) The impact of Covid-19 pandemic on corporate social responsibility and marketing philosophy. J Bus Res

Huang W, Chen S, Nguyen LT (2020) Corporate social responsibility and organizational resilience to COVID-19 crisis: an empirical study of chinese firms. Sustainability 12(21):8970

International Finance Corporation (2020) The Impact of COVID-19 on Airports: An Analysis. https://www.ifc.org/wps/wcm/connect/26d83b55-4f7d-47b1-bcf3-01eb996df35a/IFC-Covid19-Airport-FINAL_web3.pdf?MOD=AJPERES&CVID=n8lgpkG. Accessed 2 Jan 2021

Justice DW (2002) Corporate social responsibility: challenges and opportunities for trade unionists. https://library.fes.de/pdf-files/gurn/00091.pdf. Accessed 3 Jan 2021

Kemp LJ, Vinke J (2012) CSR reporting: a review of the Pakistani aviation industry. South Asian J Global Bus Res 1(2):276–292

Kotler P, Lee N (2008) Corporate social responsibility: doing the most good for your company and your cause. Wiley

Kramer MR, Porter ME (2006) Strategy and Society: the link between competitive advantage and corporate social responsibility. Harv Bus Rev 84(12):78–92

Kurt Y (2019a) Corporate social responsibility in turkish airline industry: a "new institutional theory" based approach. Anadolu University, Social Sciences Institute, Civil Aviation Management Department Doctoral Thesis. Eskisehir

Kurt Y (2019b) Choreographers of corporate social responsibility. Int J Bus Manag Inv (IJBMI) 8(10) Series. I. p 34–44. http://www.ijbmi.org/papers/Vol(8)10/Series-1/E0810013444. pdf. Accessed 13 Oct 2020

Kurt Y (2020) Covid-19 crisis in air transportation: passengers and human resources protection measures. Gaziantep Unıv J Soc Sci 2020 Special Issue, 191–211. https://doi.org/10.21547/jss. 758434. Accessed 12 Dec 2020

Kurt Y, Besler S (2019) Corporate Social responsibilities of turkish airline organizations. Çanakkale Onsekiz Mart Üniversitesi Uluslararası Sosyal Bilimler Dergisi 4(2):181–216. https://doi.org/10. 31454/usb.634728. Accessed 15 Oct 2020

Lee S (2020) Corporate social responsibility and COVID-19: research implications. Tourism Econ, 1354816620978136

Lee S, Seo K, Sharma A (2013) Corporate social responsibility and firm performance in the airline industry: the moderating role of oil prices. Tour Manage 38:20–30

Lee S, Park SY (2010) Financial impacts of socially responsible activities on airline companies. J Hosp Ve Tourism Res 34(2):185–203

Manuel T, Herron TL (2020) An ethical perspective of business CSR and the COVID-19 pandemic. Soc Bus Rev

Marom S, Lussier R (2020) Corporate social responsibility during the coronavirus pandemic: an interim overview. Bus Econ Res 10(2):250

Mcwilliams A, Siegel D (2001) Corporate social responsibility: a theory of the firm perspective. Acad Manag Rev 26(1):117–127

Mukherjee A, Babu SS, Ghosh S (2020) Thinking about water and air to attain sustainable development goals during times of COVID-19 Pandemic. J Earth Syst Sci 129(1):1–8

OECD (2001) Corporate social responsibility partners for progress. Organization for Economic Co-Operatıon and Development. https://www.oecd.org/cfe/leed/corporatesocialresponsibilitypartner sforprogress.htm. Accessed 2 Dec 2020

Popkova E, DeLo P, Sergi BS (2020) Corporate social responsibility amid social distancing during the COVID-19 crisis: BRICS vs. OECD countries. Res Int Bus Financ 55:101315

Qiu SC, Jiang J, Liu X, Chen MH, Yuan X (2021) Can corporate social responsibility protect firm value during the COVID-19 pandemic? Int J Hosp Manag 93:102759

Roxana-Loredana A, Ana-Cristina N, Alexandru B (2020) The new perspectives of corporate social responsibility in the post-corona economy. In : Proceedings of international academic conferences (No. 10612999). International Institute of Social and Economic Sciences

Settersten Jr, RA, Bernardi L, Härkönen J, Antonucci TC, Dykstra PA, Heckhausen J, ... Mulder CH (2020) Understanding the effects of Covid-19 through a life-course lens. 45: Advances in Life Course Research., 100360

Theaker A (2004) Halkla İlişkilerin El Kitabı. (Public Relations Handbook). Çev: Murat Yaz., Mediacat Books, İstanbul

UN (2004) Dısclosure of the ımpact of corporatıons on socıety current trends and ıssues" Unıted Natıons Publıcatıon. http://Unctad.Org/En/Docs/iteteb20037_En.Pdf. Accessed 30 Dec 2020

Venkatesh V (2020) Impacts of COVID-19: a research agenda to support people in their fight. Int J Inform Manag, 102197

Vogel D (2007) The market for virtue: the potential and limits of corporate social responsibility. Brookings Institution Press

World Tourism Organization the United Nations (UNWTO) (2020) COVID—19 related travel restrictions a global review for tourism. https://webunwto.s3.eu-west-1.amazonaws.com/s3fs-public/2020-04/TravelRestrictions_0.pdf. Accessed 27 Dec 2020

Yang L, Ngai CS, Lu W (2020) Changing trends of corporate social responsibility reporting in the world-leading airlines. PloS One 15(6).e0234258

Yeşim Kurt She holds bachelor's, master's, and doctorate degrees in Aviation Management at Anadolu University. She is currently working as a lecturer and vice-dean in the Department of Aviation Management at Kırklareli University. Her fields of study are corporate social responsibility, strategic corporate social responsibility, crisis management, strategic management, and organizational theories. Her research area is the aviation industry.

Chapter 5
Sustainability and Financial Performance: Examining the Airline Industry

Veysi Asker◉ and Kasım Kiracı◉

Abstract The air transport industry has numerous social, cultural and economic benefits. The sector plays a major role in the economic and social development of countries, cities and regions. The air transport industry enables increased interaction among cultures. It speeds up access to long distances and guarantees that aid is quickly and safely delivered to areas in natural disasters. In the twenty-first century, where technological changes and transformations are increasing, there are also significant changes in the air transport industry. In addition, with the globalization, the importance of the air transport industry is increasing. Especially after the 1970s, many deregulations were made for the liberalization of the air transport sector. In this way, the air transport industry has become a global sector. Many airlines were established. In addition, competition among airlines has increased significantly. In this process, airlines with sustainable financial indicators gained competitive advantage. In other words, airlines with sustainable cash flow, debt structure and profitability created more value for their stakeholders. In this study, we focus on airlines with sustainable financial performance. The air transport industry is a capital-intensive sector with a very low margin. Demand for airline changes seasonally. In other words, demand is quite high in summer and very low in winter. In addition, the air transport industry is highly sensitive to incidents, wars and outbreaks occurring spherically. For example, the 2001 terrorist attack and the 2008–2009 global financial crisis have caused the sector to experience financial difficulties. Finally, the COVID-19 pandemic, which began towards the end of 2019 and still continues, has also significantly affected the industry. In such situations and processes, airlines with sustainable financial indicators can take the lead in the competition. Therefore, it is important to analyse the financial statements of airlines in the last 10 years. In the first stage of this study, we have reached the last 10 years of financial data of top 20 airlines operating globally.

V. Asker (✉)
Department of Aviation Management, Civil Aviation High School, Dicle University, 21280 Diyarbakır, Turkey

K. Kiracı (✉)
Faculty of Aeronautics and Astronautics, Department of Aviation, Management, Iskenderun Technical University, Central Campus, 31200 İskenderun, Hatay, Turkey
e-mail: kasim.kiraci@iste.edu.tr

© The Author(s), under exclusive license to Springer Nature Singapore Pte Ltd. 2022   73
K. Kiracı and K. T. Çalıyurt (eds.), *Corporate Governance, Sustainability, and Information Systems in the Aviation Sector, Volume I*, Accounting, Finance, Sustainability, Governance & Fraud: Theory and Application, https://doi.org/10.1007/978-981-16-9276-5_5

In the second stage, we determined variables to measure the sustainable financial performance of airlines. We used the studies in the literature for the determination of the variables. Within the scope of the study, we created four different models in order to reveal the financial performance of airlines. We used the data envelope analysis method to compare the financial performance of airlines. Using this analysis, we will determine which airlines are efficient during the period under review. In this way, we will reveal which airline has a sustainable financial performance. We aim to compare the results obtained in the study with the studies in the literature. In addition, the aim of this study is to fill the gap in the literature.

Keywords Airline industry · Sustainability · Financial performance

5.1 Introduction

The airline industry grew due to the deregulation started in the U.S.A. in 1978. Many airlines were established after 1978. The competition among airlines increased. As a result of the deregulation, ticket prices fell and demand for airlines increased. It is surely beyond doubt that strategies developed by airlines contributed to the improvement of the industry. However, global crises and epidemics led some airlines going bankrupt.

There were many moments affecting historical development of airline industry. For example, oil shock in 1979/1980 led to a cost increase and a revenue decrease for airlines. With Iraqi invasion of Kuwait and the Gulf War in the 1990s, many airlines lost substantial revenue. September 11 terrorist attacks not only led to diminishing demand for airline industry but also an increase in airline costs. Due to the financial crisis in 2008, many airlines were in financial difficulty and had risk of bankruptcy. Lastly, the COVID-19 outbreak led to an unprecedented crisis for airlines. All these global events have significant influence on the performance of airlines.

There are many factors affecting financial and operational performance of airlines. Regulations, mergers (Gudiel Pineda et al. 2018, p. 103), wrong market definition, wrong business model, high fixed costs and the control power of management on variable costs affect the performance. Moreover, overcapacity due to the globalization of airline industry (Jenatabadi and Ismail 2014, p. 25) and destructive competition among airlines affect performance of airlines. It is difficult to determine which factor has more influence on the performance of airlines but financial and operational data are used for measuring performance of airlines (Demydyuk 2012; Gudiel Pineda et al. 2018; Jenatabadi and Ismail 2014; Parast and Fini 2010).

Financial performance metrics enable firms to monitor their financial performances. They are also important for monitoring the state of financial performance based on rivals in the industry. Through indicators on financial tables, it is possible to determine strengths and weaknesses of firms (Mahesh and Prasad 2012, p. 362). The evaluation and measurement of performance is one of widely used methods in comparing firms. Despite many changes related to performance evaluation, the basic method is to determine performance of firm by using financial variables (Fenyves et al. 2015). Using the financial variables in evaluating performance of a firm is critical because it provides a standardized approach with evaluating performance (Ozcan and McCue 1996). Therefore, the use of financial indicators in evaluating financial efficiency of airlines enables the evaluation of financial efficiency and sustainability of firms.

The evaluation and monitoring of financial performance gives important information to managers and investors. Financial evaluation and comparison of firms, especially operating in the same industry, gives valuable information to shareholders about the state of a firm.

The evaluation of financial performance is related to profitability, productivity, investment and economic progress (Abdel-Basset et al. 2020). For this reason, using financial indicators is an appropriate method for measuring both performance of firms by years and financial performance in the relevant industry. Thus, it leads to the progress of firms and monitoring their rankings in the industry. Firms may develop appropriate tactics and strategies by considering their position and ranking in the industry.

In the study, we focus on one of topics which is studied much less. We analyse the evaluation of financial performance of airlines and the change occurred within years. Moreover, considering the competition among airlines, we will present the rankings of airlines in the industry. Our main objective is to evaluate financial performance of top global airlines and compare airlines for the last 10 years. We employ four different models to measure financial performance of airlines accurately. Thus, we both analyse the financial performance of airlines for 2010–2019 and determine which airlines entered COVID-19 process with robust financial position.

5.2 Literature

There are numerous studies analysing financial performance of firms (Abban and Hasan 2020; Baah et al. 2021; Cui et al. 2020; Moon and Min 2020; Pokharel et al. 2020; Sun et al. 2020; Ullah et al. 2020). The main point of these studies is to evaluate and compare the financial performance of firms. In the literature, there are some studies examining the performance of airlines from various aspects. As an example, Hooper and Greenall (2005) analysed the environmental performance of 272 international airlines. Backx et al. (2002) analysed performance of airlines considering their ownership structures. According to the results, mixed public–private airlines outperform public sector airlines. However, private sector carriers outperform mixed public–private ownership airlines. Roman Asatryan and Březinová (2014) focused on the relationship between corporate social responsibility (CSR) and financial performance. The financial statements of 20 airlines which were selected randomly were analysed. The results indicate that there is a strong relationship between CSR and financial performance. Demydyuk (2012) analysed optimal Key Performance Indicators (KPI) for airlines. In the study, critic performance indicators were identified. For example, it was determined that operating profit per passenger is an important KPI for measuring airlines' profitability. In the study examining both positive and negative effect of the deregulation, increase in passenger volume, enlargement of network and decrease in ticket prices are attributed to positive effects of the deregulation.

There are many factors affecting financial performance of airlines. Some of these factors are related to airlines themselves. Other factors affecting financial performance are related to the changes occurring on airlines. There are a few studies in this context in the literature. As an example, Goetz and Vowles (2009) analysed the effect of the deregulation in the U.S.A. on financial performance of airlines. Mahesh and Prasad (2012) analysed the relationship between mergers and acquisitions (M&A) and financial performance of airlines. The results indicate that mergers and acquisitions has no significant effect on the financial performance of airlines in India. Kıracı and Bakır (2020) analysed performance of airlines by using financial and operational data for 2015–2017. In the study, integrated CRITIC Combinative Distance-based Assessment (CODAS) method was used. Gudiel Pineda et al. (2018) offered multiple criteria decision-making model (MCDM) for financial and operational performance of airlines. Bakir et al. (2020) analysed the financial performance of airlines operating in emerging markets by using integration of the PIPRECIA (PIvot Pairwise Relative Criteria Importance Assessment) and MAIRCA (Multi-Attributive Ideal-Real Comparative Analysis) methods. Flouris and Walker (2005) analysed the financial performance of low-cost and full service airlines for 2001 when September 11 terrorist attacks took place. Dayi and Esmer (2019) analysed the financial performance of ten airlines operating in Europe. The ratings of Skytrax were used within the scope of analysis. Kıracı (2019) focused on the effect of airlines' membership of strategic alliances on their financial performance. In the study, Trend Analysis and CRITIC-based TOPSIS methods were used. Mellat-Parast et al. (2015) examined the relationship between the business model and competitive strategy employed by American airlines and their financial performance. The results indicate that there is a relationship between service quality, operational strategy and financial performance. Kıracı and Asker (2019) analysed factors affecting efficiency by using operational data of 45 airlines. Teker et al. (2016) analysed the financial performance of the world's top 20 airlines for 2011–2014. According to the results of the study, total revenue and profitability are among important indicators to measure the financial performance. Kıracı and Bakır (2018) analysed performance of airlines by using CRITIC-based EDAS (Evaluation Based on Distance from Average Solution) method. In the study, performance measuring indicators such as Available Seat Kilometres (ASK), Revenue Passenger Kilometres (RPK), Cost of Available Seat Kilometre (CASK) and Revenue per Available Seat Kilometre (RASK) were used. There are numerous studies dealing with airline performance considering several dimensions in the literature. Our study contributes in various ways to the literature. Firstly, we will analyse the financial performance of airlines for 2010–2019. Therefore, we shed light on the state of airlines' financial performance before COVID-19 epidemic. Secondly, we reveal the long-term financial performance of airlines by using four different models.

5.3 Data Envelopment Analysis

Data envelopment analysis is a measurement method that can be used in situations where it is difficult to compare decision-making units with input and output variables in various scales and can perform the relative efficiency measurement of decision-making units. At the same time, data envelopment analysis is a method that can be used in cases where outputs with similar characteristics are produced by using inputs with similar characteristics and the efficiency of many decision-making units is measured simultaneously (Ramanathan 2003: 26; Tütek et al. 2012: 223). In the data envelopment analysis technique, the best decision-making units that produce maximum output composition by using the input composition at the minimum level are determined and these determined decision-making units constitute the efficiency boundary. This established efficiency limit is accepted as a reference point by other inefficient decision-making units. The distance of inefficient decision-making units to the reference point is calculated radially and expressed in proportional type (Cook et al. 1996: 2).

In the data envelopment analysis, it is very important that the goals and objectives of the decision-making units to be included in the analysis are the same and have similar characteristics. In other words, other features of decision-making units except for density and size characteristics are expected to be the same (Karsak and İscan 2000: 3). Data envelopment analysis does not need a specific functional form between inputs and outputs. Therefore, it has a very flexible structure compared to parametric methods (Arnade 1994: 9). In addition, since inputs and outputs can be expressed in many different units in data envelopment analysis, the efficiency of decision-making units can be evaluated in many aspects (Bowlin 1987: 128). In addition, the data envelopment analysis contributes greatly to the development of managerial activities in institutions and organizations by offering options to enable decision-making units that are inefficient as a result of efficiency measurement to become efficient (Rouyendegh 2009: 53).

Data envelopment analysis has entered a period of rapid development both theoretically and methodologically after the use of the CCR model in 1978. The BCC model, which was first developed in 1984 by Banker, Cooper and Charnes and based on the assumption of variable returns to scale, started to be used. The BCC model, which is found as a result of adding the convexity constraint to the CCR model, has been developed in a way that allows the technical and scale efficiency to be calculated separately. In the following years, data envelopment analysis models such as additive models and multiplicative models have been developed. In addition, methods that show how the efficiency values of decision-making units change from period to period and expressed as window analyses have also started to be used with data envelopment analysis (Başkaya and Avcı 2011: 78). Basically, CCR based on the assumption of "fixed return to scale" and BCC model based on the assumption of "variable return to scale" are widely used in DEA (Ertuğrul and Işık 2008: 206).

5.3.1   CCR Model

The CCR model, which was first used by Charnes, Cooper and Rhodes in 1978, is based on the assumption of constant returns to scale. In the CCR model, the total efficiency of decision-making units, either input or output oriented, can be measured. In order for any decision-making unit to be considered efficient in the CCR model, the relevant decision-making unit must be efficient in terms of both technical and scale efficiency (Lorcu 2008: 71). The mathematical representation of the CCR model is as follows (Yolalan 1993: 44):

$$E_0 = \min\left(\sum_{i=1}^{m} V_i X_{io} / \sum_{r=1}^{s} u_r y_{ro}\right) \tag{5.1}$$

Constraints,

$$\sum_{i=1}^{m} V_i X_{ij} / \sum_{r=1}^{s} u_r y_{rj} \geq 1, j = 1, 2, \ldots n \tag{5.2}$$

$$V_i, u_r \geq \varepsilon r = 1, 2, \ldots si = 1, 2, \ldots m$$

In the model,
u_r : The weight ratio given to "r" number of inputs by "o" decision unit,
V_i: The weight ratio given to "I" number of output by the decision unit "o",
Y_{ro}: "r" number of outputs produced by "o" decision unit,
X_{io}: "i" number of inputs used by "0" decision unit,
Y_{rj}: "r" outputs produced by the decision unit "j",
X_{ij}: "i" number of inputs used by the decision unit "j".
ε: A small positive number.

5.3.2   BCC Model

In 1984, the BCC model was formed when Banker, Charnes and Rhodes added the convexity constraint based on variable return assumption to scale to the CCR model. The feature of the BCC model that differs from the CCR model is that it only measures technical efficiency. The mathematical expression of the BCC model is as follows (Cooper et al. 2007: 90):

$$E_0 = \min\left(\sum_{i=1}^{m} V_i X_{io} - V_o / \sum_{r=1}^{s} u_r y_{ro}\right) \tag{5.3}$$

Constraints,

$$\sum_{i=1}^{m} V_i X_{ij} - V_o / \sum_{r=1}^{s} u_r y_{rj} \geq 1, j = 1, 2, \ldots . n$$

$$V_i, u_r \geq \varepsilon r = 1, 2, \ldots . si = 1, 2, \ldots . . m$$

(5.4)

V_o : Free variable owned by decision-making units.

5.3.2.1 Empirical Evidence

The study examines financial sustainability performances of 20 airlines with data envelopment analysis. Airlines included in the study are chosen among the top airlines in terms of revenue passenger kilometres published in Airline Business in 2019. The data of airlines for 2010–2019 were obtained from the Thomson Reuters Datastream database. There are two different views about the relationship between the number of inputs and outputs and the number of decision-making units in studies conducted using the data envelopment analysis. According to the first view, the number of decision-making units should be equal to twice the sum of the number of inputs and outputs (Dyson et al. 2001, p. 248). According to another view, the number of decision-making units should be equal to three times the multiplication of inputs and outputs or their sum (Cooper et al. 2001, p. 219). Since most of the studies in the literature are based on the first opinion, the first opinion is taken as basis in determining the number of decision-making units and the number of inputs and outputs in this study.

In the data envelopment analysis, since the selection of inputs and outputs can directly affect the success of the analysis, inputs and outputs that are related to each other and reflect the performance of decision-making units should be selected at this stage. In this respect, by examining the studies on financial performance in the literature, inputs and outputs, which are very important in terms of financial sustainability and which are related to each other, are used in this study. The inputs and outputs included in the study are listed in the table below.

As seen in Table 5.1, four different models have been established in order to reflect the financial sustainable performance of airlines in the best way. Through these models, the efficiency of airlines in terms of market value, EBIT, net sales and operating income has been examined.

After set of inputs and outputs were determined, the efficiency measurement for 2010–2019 was carried out through the CCR and BCC models of the data envelopment analysis. While the technical efficiency values of the airlines are calculated with the BCC model, the total efficiency values of the airlines are calculated with the CCR model.

Considering the studies conducted in the literature on the efficiency measurement of airlines, it is found that there are two different opinions. According to the first view, it is stated that the input-oriented model should be used because the power of control over inputs is higher than outputs (Sakthidharan and Sivaraman 2018; Saranga and

Table 5.1 Models, inputs and outputs

	Input	Output
Model 1	• Total debt • Total assets • Property, plant & equip • Operating expenses • Net cash flow-operating activities • Current liabilities • Total capital • Cash	• Market value
Model 2	• Total debt • Total assets • Property, plant & equip • Operating expenses • Net cash flow-operating activities • Current liabilities • Total capital • Cash	• EBIT
Model 3	• Total debt • Total assets • Property, plant & equip • Operating expenses • Net cash flow-operating activities • Current liabilities • Total capital • Cash	• Net sales
Model 4	• Total debt • Total assets • Property, plant & equip • Operating expenses • Net cash flow-operating activities • Current liabilities • Total capital • Cash	• Operating income

Nagpal 2016). According to the second view, it is argued that firms operating in an industry with high fixed costs such as the airline industry have a limited power over inputs and that firms can increase their efficiency by making changes on outputs (Bhadra 2009; Assaf and Josiassen 2011). Because the second opinion is widely accepted in the literature, the output-oriented data envelopment analysis is used in the study. Table 5.2 gives efficiency values of 20 airlines based on 4 different models.

In Model 1, total debt, total assets, property, plant & equip, operating expenses, net cash flow-operating activities, current liabilities, total capital and cash are used as inputs. We use market value as an output. Market value is one of the most important

Table 5.2 Airlines market value effectiveness (Model 1)

CODE	2010 CCR	2010 BCC	2011 CCR	2011 BCC	2012 CCR	2012 BCC	2013 CCR	2013 BCC	2014 CCR	2014 BCC	2015 CCR	2015 BCC	2016 CCR	2016 BCC	2017 CCR	2017 BCC	2018 CCR	2018 BCC	2019 CCR	2019 BCC
SU	0.617	1.000	0.629	1.000	0.375	1.000	0.389	1.000	0.310	1.000	0.151	1.000	0.415	1.000	0.808	1.000	1.000	1.000	0.404	1.000
AC	0.042	0.047	1.000	1.000	0.090	1.000	0.133	1.000	0.336	1.000	0.368	1.000	0.184	0.272	0.266	0.537	0.412	0.550	0.590	0.693
CA	0.635	0.882	0.476	0.820	0.449	0.522	0.234	0.335	0.250	0.251	0.974	0.979	0.494	0.496	0.565	0.590	0.529	0.534	0.721	0.722
AF/KL	0.206	0.227	0.199	0.220	0.090	0.094	0.123	0.147	0.205	0.254	0.101	0.115	0.101	0.110	0.150	0.154	0.115	0.123	0.143	0.148
NH	1.000	1.000	0.382	0.541	0.444	0.461	0.599	0.648	1.000	1.000	1.000	1.000	0.940	1.000	0.775	1.000	0.901	0.934	1.000	1.000
AA	1.000	1.000	1.000	1.000	1.000	1.000	1.000	1.000	1.000	1.000	1.000	1.000	0.790	0.942	1.000	1.000	1.000	1.000	1.000	1.000
MU	0.553	0.568	0.312	0.438	0.394	0.396	1.000	1.000	1.000	1.000	0.510	0.551	1.000	1.000	0.510	0.621	1.000	1.000	1.000	1.000
CZ	0.342	0.377	0.420	0.589	0.355	0.378	0.167	0.195	0.128	0.138	0.758	0.765	0.558	0.558	0.345	0.388	0.515	0.521	0.822	0.864
DL	0.278	0.399	0.240	0.352	0.294	0.426	0.563	0.823	1.000	1.000	1.000	1.000	0.696	1.000	0.839	1.000	1.000	1.000	0.992	1.000
U2	0.556	1.000	0.441	1.000	0.731	1.000	1.000	1.000	1.000	1.000	1.000	1.000	1.000	1.000	0.714	1.000	0.846	1.000	0.758	1.000
JL	1.000	1.000	1.000	1.000	1.000	1.000	0.830	0.890	0.868	0.916	1.000	1.000	0.878	0.907	0.650	0.789	0.866	0.987	0.964	1.000
KE	0.341	0.398	0.232	0.349	0.256	0.258	0.160	0.174	0.181	0.210	0.185	0.186	0.123	0.137	0.172	0.250	0.152	0.154	0.265	1.000
LA	1.000	1.000	1.000	1.000	1.000	1.000	0.702	0.959	0.909	1.000	0.342	0.415	0.736	0.860	0.553	1.000	0.733	0.833	0.551	0.835
LH	0.238	0.321	0.305	0.477	0.226	0.280	0.471	0.497	0.505	0.518	0.307	0.321	0.273	0.274	0.292	0.293	0.379	0.379	0.357	0.370
QF	0.333	0.357	0.209	0.298	0.172	0.182	0.304	0.488	0.408	1.000	0.700	1.000	0.689	1.000	0.641	1.000	1.000	1.000	0.972	1.000
RK	1.000	1.000	0.855	1.000	1.000	1.000	1.000	1.000	1.000	1.000	1.000	1.000	1.000	1.000	1.000	1.000	1.000	1.000	1.000	1.000
SQ	1.000	1.000	1.000	1.000	1.000	1.000	1.000	1.000	1.000	1.000	0.841	1.000	0.858	1.000	0.468	0.880	0.843	0.983	0.568	0.618
WN	0.609	0.632	0.528	0.676	0.611	0.612	0.698	0.704	1.000	1.000	1.000	1.000	1.000	1.000	1.000	1.000	1.000	1.000	1.000	1.000
TK	0.647	1.000	0.263	1.000	0.334	0.575	0.716	1.000	0.946	1.000	1.000	1.000	1.000	1.000	0.588	1.000	0.834	1.000	0.789	1.000
UA/CO	1.000	1.000	1.000	1.000	1.000	1.000	1.000	1.000	1.000	1.000	1.000	1.000	1.000	1.000	1.000	1.000	1.000	1.000	1.000	1.000

indicators showing the performance of firms. Because market value is related to macroeconomic factors (such as inflation, GDP, unemployment, interest) within the country. In addition, social responsibility, relationships with customers and suppliers, management capacity and related internal factors also affect market value. Therefore, we believe that market value is one of the important indicators that reveal financial performance in this study.

As a result of the efficiency measurement in terms of market value efficiency according to Table 5.2, it has been determined that the United Continental is efficient throughout the entire period. American Airlines is efficient in all years except 2016. It has been determined that Ryanair is efficient for the entire period in terms of the BBC model and not only in 2012 in terms of the CCR model. Air China, Air-France-KLM, China Southern Airlines and Lufthansa are not efficient for the entire period.

It has been determined that Southwest is efficient in 2014–2019, Singapore Airlines in 2010–2014, All Nippon Airways in 2010, 2014, 2015 and 2019, China Eastern Airlines in 2013–2014, 2016 and 2018–2019, Delta Airlines in 2014,2015 and 2018, Easyjet in 2013–2016, Japan Airlines and Latam Airlines in 2010–2012, Turkish Airlines in 2015–2016, Air Canada in 2011 and Qantas Airways and Aeroflot in 2018 in terms of both models.

While Turkish Airlines is efficient in 2010–2011, 2013–2014 and 2017–2019, Singapore Airlines in 2015–2016, Qantas Airways in 2014–2017 and 2019, Latam Airlines in 2014 and 2017, Korean Airlines in 2019, Easyjet in 2010–2012 and 2017–2019, Delta Airlines in 2016, 2017 and 2019, All Nippon Airways in 2016–2017, Air Canada in 2012–2015 and Aeroflot in 2010–2017 and 2019 in terms of BBC model, they all are not efficient based on CCR model.

In the Model 2, total debt, total assets, property, plant & equip, operating expenses, net cash flow-operating activities, current liabilities, total capital and cash were used as inputs. We used Earnings before Interest & Taxes (EBIT) as output. EBIT is an important indicator of firms 'profitability. It is widely used in studies related to profitability in the finance literature. It is also popular related to airline rankings. Therefore, EBIT contains information on whether or not airlines are profitable, in other words, have sufficient financial resources. The use of EBIT in the study contributes to revealing financial performance and comparing airlines with each other.

According to Table 5.3, based on CCR and BCC models, no airline is efficient throughout the entire period. Air France-KLM and China Southern Airlines are not efficient for the whole period.

United Continental is efficient in and 2013–2019, Turkish Airlines in 2012–2013, 2018–2019 and 2015, Southwest in 2013 and 2015–2019, Singapore Airlines in 2011, Ryanair in 2013 and 2016–2018, Qantas Airways in 2014 and 2018, Lufthansa in 2010, Latam Airlines in 2011, 2012 and 2016, Japan Airlines in 2010–2013 and 2015–2019, Easyjet in 2015–2016, Delta Airlines in 2013 and 2015–2019, China Eastern Airlines in 2013–2014, American Airlines in 2011–2013, All Nippon Airways in 2012 and 2018, Air China in 2012, Air Canada in 2011–2012 and Aeroflot in 2013 and 2016–2019 in terms of both CCR and BCC models.

Table 5.3 Airlines EBIT effectiveness (Model 2)

CODE	2010 CCR	2010 BCC	2011 CCR	2011 BCC	2012 CCR	2012 BCC	2013 CCR	2013 BCC	2014 CCR	2014 BCC	2015 CCR	2015 BCC	2016 CCR	2016 BCC	2017 CCR	2017 BCC	2018 CCR	2018 BCC	2019 CCR	2019 BCC
SU	0.202	1.000	0.679	1.000	0.911	1.000	1.000	1.000	0.282	1.000	0.178	1.000	1.000	1.000	1.000	1.000	1.000	1.000	1.000	1.000
AC	0.088	0.123	1.000	1.000	1.000	1.000	0.644	1.000	0.245	1.000	0.361	1.000	0.553	0.560	0.803	1.000	0.440	0.539	0.994	1.000
CA	0.303	0.325	0.562	0.699	1.000	1.000	0.551	0.553	0.294	0.330	0.468	0.496	0.667	0.716	0.749	0.872	0.653	0.659	0.669	0.756
AF/KL	0.408	0.444	0.114	0.122	0.824	0.850	0.122	0.124	0.119	0.150	0.225	0.227	0.447	0.458	0.124	0.125	0.404	0.406	0.248	0.256
NH	0.615	0.997	0.367	0.464	1.000	1.000	0.846	0.853	0.187	0.211	0.398	0.468	0.725	0.746	0.757	1.000	1.000	1.000	0.969	1.000
AA	0.049	0.072	1.000	1.000	1.000	1.000	1.000	1.000	0.627	1.000	0.971	1.000	1.000	1.000	1.000	1.000	1.000	1.000	1.000	1.000
MU	0.158	0.169	0.455	0.543	0.960	1.000	1.000	1.000	1.000	1.000	0.370	0.385	0.941	1.000	0.700	0.924	0.681	0.812	0.847	0.865
CZ	0.327	0.370	0.492	0.580	0.903	0.915	0.548	0.556	0.186	0.222	0.354	0.364	0.530	0.592	0.521	0.611	0.540	0.560	0.692	0.735
DL	0.101	0.103	0.287	0.301	0.672	0.869	1.000	1.000	0.225	0.428	1.000	1.000	1.000	1.000	1.000	1.000	1.000	1.000	1.000	1.000
U2	0.101	1.000	0.136	1.000	0.831	1.000	0.938	1.000	0.973	1.000	1.000	1.000	1.000	1.000	0.657	1.000	0.644	1.000	0.794	1.000
JL	1.000	1.000	1.000	1.000	1.000	1.000	1.000	1.000	0.996	1.000	1.000	1.000	1.000	1.000	1.000	1.000	1.000	1.000	1.000	1.000
KE	0.165	0.176	0.085	0.092	0.662	0.683	0.093	0.104	0.063	0.069	0.047	0.054	0.132	0.144	0.763	1.000	0.189	0.189	0.668	1.000
LA	0.320	1.000	1.000	1.000	1.000	1.000	0.209	0.231	0.161	0.183	0.011	0.011	1.000	1.000	0.497	1.000	0.661	0.681	0.457	0.640
LH	1.000	1.000	0.272	0.354	0.706	0.851	0.493	0.501	0.100	0.153	0.620	0.645	0.764	0.782	0.843	0.876	0.942	0.942	0.651	0.681
QF	0.086	0.097	0.297	0.391	0.022	0.022	0.428	0.461	1.000	1.000	0.540	1.000	0.934	1.000	0.686	1.000	1.000	1.000	0.892	1.000
RK	0.123	1.000	0.236	1.000	0.950	1.000	1.000	1.000	0.627	1.000	0.947	1.000	1.000	1.000	1.000	1.000	1.000	1.000	0.934	1.000
SQ	0.049	0.072	1.000	1.000	0.535	1.000	0.569	1.000	0.148	1.000	0.229	0.339	0.442	0.454	0.257	0.380	0.633	0.745	0.478	0.540
WN	0.126	0.158	0.220	0.240	0.585	0.637	1.000	1.000	0.667	0.667	1.000	1.000	1.000	1.000	1.000	1.000	1.000	1.000	1.000	1.000
TK	0.356	1.000	0.283	1.000	1.000	1.000	1.000	1.000	0.759	1.000	1.000	1.000	0.412	1.000	0.564	1.000	1.000	1.000	1.000	1.000
UA/CO	0.355	1.000	1.000	1.000	0.276	1.000	1.000	1.000	1.000	1.000	1.000	1.000	1.000	1.000	1.000	1.000	1.000	1.000	1.000	1.000

Aeroflot is efficient in 2010–2012 and 2014–2015, Air Canada in 2013–2015, 2017 and 2019, All Nippon Airways and Korean Airlines in 2017 and 2019, American Airlines in 2014–2015, China Eastern Airlines in 2012 and 2016, Easyjet in 2010–2014 and 2017–2019, Japan Airlines in 2014, Latam Airlines in 2010 and 2017, Qantas Airways in 2015–2017 and 2019, Ryanair in 2010–2012, 2014–2015 and 2019, Singapore Airlines in 2012–2014, Turkish Airlines in 2010, 2011, 2014, 2016 and 2017 and United Continental in 2010 and 2012.

In the Model 3, total debt, total assets, property, plant & equip, operating expenses, net cash flow-operating activities, current liabilities, total capital and cash are used as inputs. We use net sales as the output. Net sales provides important tips on how efficient and productive firms are. In addition, the stability of net sales is related to the financial strength of the firm. In other words, stable net sales make firms stronger financially. Net sales also sheds light on whether firms are superior in the market share and the competition. Therefore, we believe that the net sales is important for firm efficiency and financial sustainability.

According to Table 5.4, in terms of net sales efficiency based on CCR and BCC models, while Aeroflot, American Airlines, Japan Airlines and United Continental are efficient throughout the entire period, China Southern Airlines is not efficient for the whole period. Moreover, Turkish Airlines is efficient throughout the whole period in terms of BBC model, on the other hand, it is efficient in 2012 based on CCR model.

Air Canada is efficient in 2010–2015 and 2018–2019, Air China in 2010, Air France-KLM in 2010–2012, 2014–2016 and 2019, All Nippon Airways in 2010–2012, 2014–2017 and 2019, China Eastern Airlines in 2013–2014 and 2018, Delta Airlines in all years except for 2014, Easyjet in 2014–2016 and 2019, Latam Airlines in 2010–2012, Lufthansa in 2010–2011, 2013–2017 and 2019, Qantas Airways in 2010, 2011, 2013, 2014, 2016, 2018, and 2019, Ryanair in 2010, 2011, 2014 and 2016–2018, Singapore Airlines in 2010–2014 and Southwest Airlines in all years except for 2012–2013 in terms of both models.

While Air Canada is not efficient in 2016, Air China and China Eastern Airlines in 2012, Air France- KLM in 2013 and 2017, All Nippon in 2013 and 2018, China Eastern in 2012, Easyjet in 2010–2013 and 2017–2018, Korean Air in 2019, Lufthansa in 2012 and 2018, Qantas in 2015 and 2017, Ryanair in 2012,2013, 2015 and 2019, Singapore Airlines in 2015 and Southwest Airlines in 2012–2013 in terms of CCR model, they are efficient based on BCC model.

In the Model 4, total debt, total assets, property, plant & equipment, operating expenses, net cash flow-operating activities, current liabilities, total capital and cash are used as inputs. We use operating income as the output. Operating income is one of the most important indicators showing not only financial but also operational performance of firms because with operating income, it is possible to measure operational capacity and efficiency of a firm. In addition, there is a close relationship between the financial efficiency of firms and operating income. In this study, we aim to measure firms 'financial performance more precisely by using operating income variable. In addition, we believe that this indicator is important in terms of comparing firms with each other.

Table 5.4 Airlines net sales effectiveness (Model 3)

CODE	2010		2011		2012		2013		2014		2015		2016		2017		2018		2019	
	CCR	BCC	CCR	BCC	CCR	BCC	CCR	BCC	CCR	BCC	CCR	BCC	CCR	BCC	CCR	BCC	CCR	BCC	CCR	BCC
SU	1.000	1.000	1.000	1.000	1.000	1.000	1.000	1.000	1.000	1.000	1.000	1.000	1.000	1.000	1.000	1.000	1.000	1.000	1.000	1.000
AC	1.000	1.000	1.000	1.000	1.000	1.000	1.000	1.000	1.000	1.000	1.000	1.000	0.952	0.955	0.998	1.000	1.000	1.000	1.000	1.000
CA	1.000	1.000	0.975	0.985	0.996	1.000	0.922	0.922	0.939	0.939	0.908	0.933	0.967	0.987	0.932	0.952	0.941	0.942	0.931	0.953
AF/KL	1.000	1.000	1.000	1.000	1.000	1.000	0.950	1.000	1.000	1.000	1.000	1.000	1.000	1.000	0.940	1.000	0.958	0.981	1.000	1.000
NH	1.000	1.000	1.000	1.000	1.000	1.000	0.994	1.000	1.000	1.000	1.000	1.000	1.000	1.000	1.000	1.000	0.992	1.000	1.000	1.000
AA	1.000	1.000	1.000	1.000	1.000	1.000	1.000	1.000	1.000	1.000	1.000	1.000	1.000	1.000	1.000	1.000	1.000	1.000	1.000	1.000
MU	0.975	0.978	0.962	0.963	0.991	1.000	1.000	1.000	1.000	1.000	0.844	0.860	1.000	1.000	0.874	0.907	1.000	1.000	0.940	0.957
CZ	0.965	0.983	0.930	0.934	0.949	0.952	0.912	0.913	0.896	0.898	0.881	0.914	0.912	0.929	0.881	0.902	0.933	0.934	0.952	0.960
DL	1.000	1.000	1.000	1.000	1.000	1.000	1.000	1.000	0.921	0.973	1.000	1.000	1.000	1.000	1.000	1.000	1.000	1.000	1.000	1.000
U2	0.962	1.000	0.978	1.000	0.982	1.000	0.988	1.000	1.000	1.000	1.000	1.000	1.000	1.000	0.961	1.000	0.985	1.000	1.000	1.000
JL	1.000	1.000	1.000	1.000	1.000	1.000	1.000	1.000	1.000	1.000	1.000	1.000	1.000	1.000	1.000	1.000	1.000	1.000	1.000	1.000
KE	0.989	0.991	0.946	0.949	0.926	0.934	0.899	0.912	0.908	0.943	0.848	0.909	0.897	0.925	0.909	0.959	0.915	0.918	0.913	1.000
LA	1.000	1.000	1.000	1.000	1.000	1.000	0.954	0.955	0.925	0.961	0.900	0.913	0.978	0.983	0.873	1.000	0.924	0.933	0.893	0.949
LH	1.000	1.000	1.000	1.000	0.992	1.000	1.000	1.000	1.000	1.000	1.000	1.000	1.000	1.000	1.000	1.000	0.971	1.000	1.000	1.000
QF	1.000	1.000	1.000	1.000	0.980	0.981	1.000	1.000	1.000	1.000	0.999	1.000	1.000	1.000	0.989	1.000	1.000	1.000	1.000	1.000
RK	1.000	1.000	1.000	1.000	0.968	1.000	0.995	1.000	1.000	1.000	0.951	1.000	1.000	1.000	1.000	1.000	1.000	1.000	0.990	1.000
SQ	1.000	1.000	1.000	1.000	1.000	1.000	1.000	1.000	1.000	1.000	0.949	1.000	0.910	0.910	0.901	0.934	0.960	0.966	0.933	0.952
WN	1.000	1.000	1.000	1.000	0.998	1.000	0.993	1.000	1.000	1.000	1.000	1.000	1.000	1.000	1.000	1.000	1.000	1.000	1.000	1.000
TK	1.000	1.000	1.000	1.000	0.984	1.000	1.000	1.000	1.000	1.000	1.000	1.000	1.000	1.000	1.000	1.000	1.000	1.000	1.000	1.000
UA/CO	1.000	1.000	1.000	1.000	1.000	1.000	1.000	1.000	1.000	1.000	1.000	1.000	1.000	1.000	1.000	1.000	1.000	1.000	1.000	1.000

Table 5.5 Airlines operating income effectiveness (Model 4)

CODE	2010 CCR	2010 BCC	2011 CCR	2011 BCC	2012 CCR	2012 BCC	2013 CCR	2013 BCC	2014 CCR	2014 BCC	2015 CCR	2015 BCC	2016 CCR	2016 BCC	2017 CCR	2017 BCC	2018 CCR	2018 BCC	2019 CCR	2019 BCC
SU	1.000	1.000	1.000	1.000	0.796	1.000	1.000	1.000	0.326	1.000	0.850	1.000	1.000	1.000	1.000	1.000	1.000	1.000	0.791	1.000
AC	0.630	0.647	1.000	1.000	1.000	1.000	1.000	1.000	0.828	1.000	0.719	1.000	0.637	0.658	0.690	1.000	0.733	0.935	0.705	0.909
CA	1.000	1.000	0.727	0.818	0.937	0.818	0.330	0.333	0.514	0.516	0.582	0.663	0.819	0.923	0.567	0.653	0.540	0.543	0.514	0.616
AF/KL	1.000	1.000	0.040	0.043	0.632	0.747	0.078	0.079	0.075	0.078	0.121	0.125	0.085	0.087	0.305	0.309	0.507	0.509	0.348	0.370
NH	1.000	1.000	0.751	0.771	1.000	1.000	0.890	0.910	0.636	0.638	0.372	0.612	0.633	0.668	0.690	0.965	0.830	0.880	0.980	1.000
AA	0.461	0.474	1.000	1.000	0.634	0.749	1.000	1.000	1.000	1.000	1.000	1.000	1.000	1.000	1.000	1.000	1.000	1.000	1.000	1.000
MU	0.575	0.580	0.481	0.488	0.485	0.667	1.000	1.000	1.000	1.000	0.303	0.331	0.648	0.858	0.152	0.201	0.241	0.280	0.272	0.348
CZ	0.509	0.569	0.108	0.114	0.039	0.042	0.064	0.065	0.172	0.175	0.402	0.490	0.439	0.499	0.219	0.260	0.242	0.242	0.423	0.471
DL	0.783	1.000	1.000	1.000	1.000	1.000	1.000	1.000	0.093	0.093	1.000	1.000	1.000	1.000	1.000	1.000	1.000	1.000	1.000	1.000
U2	0.554	1.000	0.773	1.000	0.827	1.000	0.908	1.000	1.000	1.000	0.898	1.000	1.000	1.000	0.621	1.000	0.821	1.000	0.710	1.000
JL	1.000	1.000	1.000	1.000	1.000	1.000	1.000	1.000	1.000	1.000	0.958	1.000	1.000	1.000	1.000	1.000	1.000	1.000	1.000	1.000
KE	0.866	0.884	0.397	0.409	0.240	0.265	0.015	0.018	0.246	0.355	0.289	0.422	0.451	0.537	0.437	0.646	0.346	0.355	0.175	0.175
LA	1.000	1.000	1.000	1.000	0.167	1.000	0.254	0.278	0.102	0.200	0.037	0.058	0.069	0.071	0.098	1.000	0.223	0.245	0.240	0.411
LH	0.037	0.044	0.044	0.056	0.219	0.240	0.119	0.121	0.157	0.173	0.199	0.212	0.423	0.429	0.544	0.563	0.602	0.602	0.193	0.205
QF	0.808	0.811	0.986	1.000	0.418	0.454	0.799	0.849	0.406	1.000	0.522	1.000	0.850	1.000	0.728	1.000	1.000	1.000	0.874	1.000
RK	1.000	1.000	1.000	1.000	0.813	1.000	0.968	1.000	1.000	1.000	0.781	1.000	1.000	1.000	1.000	1.000	1.000	1.000	0.926	1.000
SQ	0.136	0.142	1.000	1.000	0.411	1.000	0.403	1.000	0.216	1.000	0.146	0.212	0.300	0.315	0.252	0.372	0.602	0.663	0.471	0.561
WN	1.000	1.000	1.000	1.000	0.655	1.000	0.913	0.928	1.000	1.000	1.000	1.000	1.000	1.000	1.000	1.000	1.000	1.000	1.000	1.000
TK	0.942	1.000	0.373	1.000	0.765	1.000	1.000	1.000	1.000	1.000	1.000	1.000	1.000	1.000	0.773	1.000	1.000	1.000	0.814	1.000
UA/CO	1.000	1.000	1.000	1.000	1.000	1.000	1.000	1.000	1.000	1.000	1.000	1.000	1.000	1.000	1.000	1.000	1.000	1.000	1.000	1.000

Table 5.6 The average performance of airlines (2010–2019)

Code	Average score (Model 1)	Average score (Model 2)	Average score (Model 3)	Average score (Model 4)
UA/CO	1.000	0.932	1.000	1.000
JL	0.927	1.000	1.000	0.998
RK	0.993	0.891	0.995	0.974
AA	0.987	0.886	1.000	0.916
U2	0.904	0.854	0.993	0.906
TK	0.835	0.869	0.999	0.933
WN	0.854	0.765	1.000	0.959
SU	0.755	0.863	1.000	0.938
DL	0.745	0.749	0.995	0.898
NH	0.831	0.730	0.999	0.811
QF	0.638	0.643	0.997	0.825
AC	0.526	0.718	0.995	0.855
MU	0.743	0.741	0.963	0.546
SQ	0.903	0.544	0.971	0.510
CA	0.573	0.616	0.956	0.680
LA	0.821	0.603	0.957	0.423
LH	0.354	0.659	0.998	0.259
CZ	0.459	0.530	0.927	0.277
KE	0.259	0.319	0.930	0.418
AF/KL	0.151	0.310	0.991	0.327

According to Table 5.5, total operating revenue efficiency based on CCR and BCC models, while United Continental is efficient throughout the entire period, China Southern Airlines and Lufthansa are not efficient for the whole period. While Japan Airlines is efficient throughout the whole period in terms of BBC model, it is efficient in 2015 based on CCR model. Turkish Airlines is efficient in 2013–2016 and 2018, Southwest Airlines in 2010–2011 and 2014–2019, Singapore Airlines in 2011, Ryanair in 2010, 2011, 2014, 2016, 2017, and 2018, Latam Airlines in 2010–2011, Easyjet in 2014, and 2016, Delta Airlines in 2011–2013 and 2015–2019, China Eastern Airlines in 2013–2014, American Airlines in 2011 and 2013–2019, All Nippon Airways in 2010 and 2012, Air France- KLM and Air China in 2010, Air Canada in 2011–2013 and Aeroflot in 2010, 2011, 2013 and 2016–2018 in terms of both CCR and BBC models.

While Turkish Airlines is not efficient in 2010–2012, 2017 and 2019, Singapore Airlines in 2012–2014, Ryanair in 2012, 2013, 2015 and 2019, Qantas Airways in 2011, 2019 and 2014–2017, Latam Airlines in 2012 and 2017, Korean Airlines in 2019, Easyjet in 2015, 2010–2013 and 2017–2019, Delta Airlines in 2010, Air China in 2012, Air Canada in 2014–2015 and 2017 and Aeroflot in 2012, 2014, 2015 and 2019 in terms of CCR model, they are efficient in terms of BBC model.

Table 5.6 gives the average financial performance scores of airlines for four different models. The aim is to determine which airline has higher average performance scores. The results indicate that the airline with the best financial performance for 2010–2019 is United Continental. This airline is followed by Japan Airlines and Ryanair. Therefore, the table above presents valuable information on financial performance and sustainability.

5.4   Conclusions

In this study, the financial performance and sustainability of 20 airlines for 2010–2019 are analysed by Data Envelopment Analysis. Different models are established in order to present the financial sustainable efficiency of airlines in detail and to measure their financial performance more precisely. While the same inputs are used in all of these models, market value in the first model, EBIT in the second model, net sales in the third model, and total operating income in the fourth model are used as outputs.

CCR and BCC models, which are accepted as basic models of data envelopment analysis, are used in the research. The total efficiency values of the airlines are calculated with the CCR model and the technical efficiency values of the airlines with the BCC model.

According to the results, United Continental is efficient in terms of market value, net sales and total operating income throughout the entire period, and in terms of EBIT in all years except 2010 and 2012, both in terms of CCR and BBC models. From this point of view, it can be said that United Continental outperforms other airlines on all models. China Southern Airlines is efficient in all four models during the entire period. In addition to this, Air France-KLM is not efficient in terms of market value and EBIT and Lufthansa in terms of market value and total operating income throughout the period. It has been determined that Air China is not only efficient in terms of market value throughout the entire period.

Aeroflot, American Airlines and Japan Airlines businesses, which are efficient in terms of net sales throughout the period, achieve lower performance in terms of market value, EBIT and total operating income. Likewise, it has been determined that most of the other airlines have higher efficiency values in terms of net sales, but lower efficiency values in terms of market value, EBIT and total operating income.

With the COVID-19 outbreak, many airlines have experienced financial difficulty and risk of bankruptcy. In this process, we think that airlines that perform well in terms of financial efficiency, in other words, airlines with high financial performance scores will be able to overcome the pandemic relatively easily. Financial efficiency scores are important in terms of comparing and measuring financial performance, but it should not be forgotten that the subsidies to airlines is determinant. As a result, how airlines will perform after COVID-19 and whether they will be ahead of the competition depends on several indicators. Perhaps the most important of these is how well airlines performed before COVID-19. The second is the decisions made by the airlines during the COVID-19 pandemic and the public subsidy rate.

It is thought that this study, in which the financial sustainable efficiency of airlines are evaluated in terms of four different models, will contribute to both the literature on air transport and airlines. In this study, the efficiency of airlines is measured only in terms of financial sustainability. In this respect, both operational and financial sustainable efficiency of airlines can be compared in further studies.

Appendix: Airline List in Study

Airline	IATA code	Airline	IATA code
Aeroflot	SU	Japan Airlines	JL
Air Canada	AC	Korean Air	KE
Air China	CA	Latam Airlines	LA
Air France- KLM	AF/KL	Lufthansa	LH
All Nippon Airways	NH	Qantas Airways	QF
American Airlines	AA	Ryanair	RK
China Eastern Airlines	MU	Singapore Airlines	SQ
China Southern Airlines	CZ	Southwest Airlines	WN
Delta Airlines	DL	Turkish Airlines	TK
Easyjet	U2	United Continental	UA/CO

References

Abban AR, Hasan MZ (2020) The causality direction between environmental performance and financial performance in Australian mining companies—a panel data analysis. Resour Policy 101894. https://doi.org/10.1016/j.resourpol.2020.101894

Abdel-Basset M, Ding W, Mohamed R, Metawa N (2020) An integrated plithogenic MCDM approach for financial performance evaluation of manufacturing industries. Risk Manag 22:192–218. https://doi.org/10.1057/s41283-020-00061-4

Arnade CA (1994) Using data envelopment analysis to measure international agricultural efficeincy and productivity. Economic Research Service, Washington

Asatryan R, Březinová O (2014) Corporate social responsibility and financial performance in the airline industry in Central and Eastern Europe. Acta Univ. Agric. Silvic. Mendelianae Brun. 62:633–639. https://doi.org/10.11118/actaun201462040633

Assaf AG, Josiassen A (2011) The operational performance of UK airlines: 2002–2007. J Econ Stud 38(1):5–16

Baah C, Opoku-Agyeman D, Acquah ISK, Agyabeng-Mensah Y, Afum E, Faibil D, Abdoulaye FAM (2021) Examining the correlations between stakeholder pressures, green production practices, firm reputation, environmental and financial performance: evidence from manufacturing SMEs. Sustain Prod Consum 27:100–114. https://doi.org/10.1016/j.spc.2020.10.015

Backx M, Carney M, Gedajlovic E (2002) Public, private and mixed ownership and the performance of international airlines. J Air Transp Manag 8:213–220. https://doi.org/10.1016/S0969-6997(01)00053-9

Bakir M, Akan Ş, Kiraci K, Karabasevic D, Stanujkic D, Popovic G (2020) Multiple-criteria approach of the operational performance evaluation in the airline industry: Evidence from the emerging markets. Rom J Econ Forecast 23(2):149–172

Başkaya Z, Avcı B (2011) Veri Zarflama Analizi. Bursa: Dora Yayınları

Bhadra D (2009) Race to the bottom or swimming upstream: performance analysis of US airlines. J Air Transp Manag 227–235

Bowlin WF (1987) Eveluating the efficieny of US air force real-property maintanence activities. J Oper Res Soc 38(2):127–135

Cook WD, Kress M, Lawrence MS (1996) Data envelopment analysis in the presence of both quantitative and qualitative factors. J Oper Res Soc 945–953

Cooper WW, Li S, Seiford LM, Tone K, Thrall RM, Zhu J (2001) Sensitivity and stability analysis in dea: some recent developments. J Prod Anal 217–246

Cooper W, Seiford L, Tone K (2007) Data envelopment analysis a comprehensive text with models, applications references and dea-solver software. Springer, Newyork

Cui Y, Khan SU, Li Z, Zhao M (2020) Environmental effect, price subsidy and financial performance: evidence from Chinese new energy enterprises. Energy Policy 112050. https://doi.org/10.1016/j.enpol.2020.112050

Dayi F, Esmer Y (2019) Measuring financial performance of airline passenger transport company in European. In: 33rd international academic conference, pp 60–71. Vienna. https://doi.org/10.20472/iac.2017.33.008

Demydyuk G (2012) Optimal financial key performance indicators: evidence from the airline industry. Account Tax 3:39–52

Dyson RG, Allen R, Camanho AS, Padinovski VV, Sarrico CS, Shale EA (2001) Pitfalls and protocols in DEA. Eur Jounal Oper Res 132:245–259

Ertuğrul İ, Işık AT (2008) İşletmelerin Vza İle Mali Tablolarına Dayalı Etkinlik Ölçümü: Metal Ana Sanayiinde Bir Uygulama. Afyon Kocatepe Üniversitesi İktisadi ve İdari Bilimler Fakültesi Dergisi, 201–217

Fenyves V, Tarnóczi T, Zsidó K (2015) Financial performance evaluation of agricultural enterprises with DEA method. Procedia Econ Financ 32:423–431. https://doi.org/10.1016/s2212-5671(15)01413-6

Flouris T, Walker TJ (2005) The financial performance of low-cost and full-service airlines in times of crisis. Can J Adm Sci 22:3–20. https://doi.org/10.1111/j.1936-4490.2005.tb00357.x

Goetz AR, Vowles TM (2009) The good, the bad, and the ugly: 30 years of US airline deregulation. J Transp Geogr 17:251–263. https://doi.org/10.1016/j.jtrangeo.2009.02.012

Gudiel Pineda PJ, Liou JJH, Hsu CC, Chuang YC (2018) An integrated MCDM model for improving airline operational and financial performance. J Air Transp Manag 68:103–117. https://doi.org/10.1016/j.jairtraman.2017.06.003

Hooper PD, Greenall A (2005) Exploring the potential for environmental performance benchmarking in the airline sector. Benchmarking 12:151–165. https://doi.org/10.1108/14635770510593095

Jenatabadi HS, Ismail NA (2014) Application of structural equation modelling for estimating airline performance. J Air Transp Manag 40:25–33. https://doi.org/10.1016/j.jairtraman.2014.05.005

Karsak E, İşcan F (2000) Çimento Sektöründe Göreli Faaliyet Performanslarının Ağırlık Kısıtlamaları ve Çapraz Etkinlik Kullanılarak Veri Zarflama Analizi İle Değerlendirilmesi. Endüstri Mühendisliği Dergisi 2(3):2–10

Kasım MK, Bakır (2020) Evaluation of airlines performance using an integrated critic and codas methodology: the case of star alliance member airlines. Stud Bus Econ 15:83–99. https://doi.org/10.2478/sbe-2020-0008

Kasım VK, Asker (2019) Etkinlik ve Etkinliği Belirleyen Faktörler: Havayolu Şirketleri Üzerine Ampirik Bir İnceleme. Eskişehir Osmangazi Üniversitesi İİBF Derg 14:25–50

Kiracı K (2019) Does joining global alliances affect airlines' financial performance? In: Kapucu H (ed) Akar, Cüneyt. Contemporary challenges in business and life sciences. IJOPEC Publication Limited, London, pp 39–59

Kiracı K, Bakır M (2018) Critic Temelli Edas Yöntemi İle Havayolu İşletmelerinde Performans Ölçümü Uygulamasi. Pamukkale Üniversitesi Sosyal Bilimler Enstitüsü Dergisi (35):157–174

Lorcu F (2008) Veri Zarflama Analizi (DEA) İle Türkiye ve Avrupa Birliği Ülkelerinin Sağlık Alanındaki Etkinliklerinin Değerlendirilmesi. Yayınlanmamış Doktora Tezi

Mahesh R, Prasad D (2012) Post merger and acquisition of financial performance analysis: a case study of select indian airline companies. Int J Eng Manag Sci 3:362–369

Mellat-Parast M, Golmohammadi D, McFadden KL, Miller JW (2015) Linking business strategy to service failures and financial performance: empirical evidence from the U.S. domestic airline industry. J Oper Manag 38:14–24. https://doi.org/10.1016/j.jom.2015.06.003

Moon H, Min D (2020) A DEA approach for evaluating the relationship between energy efficiency and financial performance for energy-intensive firms in Korea. J Clean Prod 255:120283. https://doi.org/10.1016/j.jclepro.2020.120283

Ozcan YA, McCue MJ (1996) Development of a financial performance index for hospitals: dea approach. J Oper Res Soc 47:18–26. https://doi.org/10.1057/jors.1996.2

Parast MM, Fini EH (2010) The effect of productivity and quality on profitability in US airline industry: an empirical investigation. Manag Serv Qual 20:458–474. https://doi.org/10.1108/09604521011073740

Pokharel KP, Archer DW, Featherstone AM (2020) The impact of size and specialization on the financial performance of agricultural cooperatives. J Co-op Organ Manag 8:100108. https://doi.org/10.1016/j.jcom.2020.100108

Ramanathan R (2003) An introduction to data envelopment analysis a tool for performance measurement. Sage Publications, New Delhi

Rouyendegh BD (2009) Çok Ölçütlü Karar Verme süreci İçin VZA-AHP Sıralı Hibrit Algoritması ve Bir Uygulama. Ankara: Gazi Üniversitesi (Yayınlanmaış Doktora Tezi)

Sakthidhran V, Sivaraman S (2018) Impact of operating cost components on airline efficiency in India: a DEA approach. Asia Pac Manag Rev 23:258–267

Saranga H, Nagpal R (2016) Drivers of operational efficiency and its impact on market performance in the Indian Airline industry. J Air Transp Manag 165–176

Sun Y, Yang Y, Huang N, Zou X (2020) The impacts of climate change risks on financial performance of mining industry: evidence from listed companies in China. Resour Policy 69:101828. https://doi.org/10.1016/j.resourpol.2020.101828

Teker S, Teker D, Güner A (2016) Financial performance of top 20 airlines. Procedia—Soc Behav Sci 235:603–610. https://doi.org/10.1016/j.sbspro.2016.11.035

Tütek H, Gümüşoğlu Ş, Özdemir A (2012) Sayısal yöntemler: yönetsel yaklaşım. İstanbul: Beta

Ullah A, Pinglu C, Ullah S, Zaman M, Hashmi SH (2020) The nexus between capital structure, firm-specific factors, macroeconomic factors and financial performance in the textile sector of Pakistan. Heliyon 6:e04741. https://doi.org/10.1016/j.heliyon.2020.e04741

Yolalan R (1993) İşletmelerarası göreli etkinlik ölçümü. Ankara: Milli Prodüktive Merkezi Yayınları

Dr. Veysi Asker received his first BSc degree in Aviation Management in 2010 from Kocaeli University. He worked as a research assistant Faculty of Aeronautics and Astronautics, Anadolu University between 2014–2020. He obtained his first MSc degree in Aviation Management in 2016 from Anadolu University. He completed his PhD in Aviation Management with the dissertation titled "Operational and Financial Efficiency Measurement in Airlines and Determining Factors of Efficiency" at the Anadolu University, Eskişehir in 2020. Dr. Asker joined Civil Aviation High School, Dicle University in 2020. He has been working as an assistant professor at the Department of Aviation Management at Dicle University since 2020. Veysi Asker has many articles published in international scientific refereed journals on aviation management and airline financing. In addition, he has written many papers presented in congress.

Dr. Kasım Kiracı received his first BSc degree in Aviation Management in 2010 from Kocaeli University and second BSc degree in Economics in 2015 from Anadolu University. He obtained his first MSc degree in Economics from Gebze Technical University and second MSc degree in Aviation Management from Anadolu University. He completed his PhD in Aviation Management with the dissertation titled "Determinants of The Capital Structure in Different Business Models: A Panel Data Analysis On Low Cost and Traditional Airlines" at the Anadolu University, Eskişehir in 2017. Dr. Kiracı joined Faculty of Aeronautics and Astronautics, Iskenderun Technical University in 2018. He has been working as a faculty member at the Department of Aviation Management at İskenderun Technical University since 2018. He received his associate professorship in finance in 2021. Kasım Kiracı has many articles published in web of science journals on aviation economy and airline financing. In addition, there are many book chapters published on different topics of aviation.

Chapter 6
Sustainable Aviation Based on Innovation

Nisa Seçilmiş

Abstract Focusing on sustainable development all over the world has made the issues of sustainability of global economic components and sustainable innovation an international concern. The "sustainable aviation" leg is added to these concerns, with the realization that the air transport industry has serious negative impacts on the environment. The aim of this study is to reveal the importance of the technological steps taken in the fight against climate change within the framework of sustainable aviation. Patent data was used as a technology development indicator to measure the success of political action. According to technology development indicators, a total of 40,523 technological patents have been obtained by the air transport industry for climate change mitigation since 1987, the most patents have been obtained in the efficient propulsion technologies category, and the country with the highest number of patents is the United States.

Keywords Sustainable aviation · Sustainable development · Patent · Innovation · Technology · Sustainable economic development

6.1 Introduction

Ever since Joseph Alois Schumpeter (1883–1950), who first defended "innovation" as an important factor in economic growth and development, innovation has been the subject of many academic studies which investigate the impact of innovation on the economic performance of firms, industries, regions, or countries. The general result obtained from these studies is that innovation is the driving force of economic growth and development. All kinds of innovations leading to increased productivity; provides an increase in social welfare with the positive effects it creates on many socioeconomic factors such as employment, economic growth, and sustainable development. In other words, innovation constitutes the key point of sustainable economic growth, development, and social welfare (Akbey 2014: 1; Seçilmiş 2020:138, 139).

N. Seçilmiş (✉)
Faculty of Aeronautics and Aerospace, Aviation Management Department, Gaziantep University, Gaziantep, Turkey

© The Author(s), under exclusive license to Springer Nature Singapore Pte Ltd. 2022
K. Kiracı and K. T. Çalıyurt (eds.), *Corporate Governance, Sustainability, and Information Systems in the Aviation Sector, Volume I*, Accounting, Finance, Sustainability, Governance & Fraud: Theory and Application, https://doi.org/10.1007/978-981-16-9276-5_6

The importance of innovation is increasing in almost every sector, but it has a special importance in the aviation industry. Aviation sector is a specific sector that developing and growing with innovative activities. Thanks to the rapid advances in information technologies, the airline industry has developed rapidly and increased its weight in today's global economy. The aviation industry, which has grown rapidly in the last hundred years, creates significant economic and social impacts (Daley 2008: 210). In addition to these economic and social benefits it provides, negative environmental effects also occur on climate change, noise, air quality, and local environment (Upham 2003: 4; Kaszewski and Sheate 2004: 186). Due to the rapid development of the sector and the economic, social and environmental impacts it has created, the concept of "sustainable aviation" has been brought to the agenda in the late 1990s. Sustainable aviation policies (air quality, climate change, natural resources, noise, social&economic, surface access) are determined by the close collaboration of different public and private institutions and organizations such as airlines, airports, aircraft and engine manufacturers, local councils, environmental Non-Governmental Organizations (NGOs), academic researchers (Pastowski 2003: 185). Research & development and innovation have a great importance in sustainable aviation policies. Innovative activities are carried out both to increase economic and social benefits and to reduce the negative effects caused by the sector. Innovation thereby supporting sustainable development while serving sectoral sustainability.

In today's information age, the concept of sustainable development has become the main target of all economic policies. Therefore, micro and macro all economic units are restructured in accordance with these sustainable development goals. The purpose of this study is to investigate the impact of innovative activities in the aviation industry on the development and sustainability of the industry. For this purpose firstly, the concept of sustainability will be explained in terms of economic development and aviation sector. Then, the institutions involved in the creation and implementation of sustainable aviation policies and the policies developed by these institutions will be explained. Finally, patents related to the aviation industry will be used as an indicator of innovation to evaluate the effectiveness of the policies. Patents will be categorized in terms of content and associated with sustainable aviation. In the study, it will be tried to shed light on the institutional and political process of sustainable aviation and the impact of innovation on sustainable aviation.

6.2   The Emergence of Sustainability in the Economy

In order to understand the concept of sustainable aviation, first of all, it is necessary to understand the foundations of this concept. For this reason, it is thought that it is useful to first touch on how the concept of sustainability emerged and how it spread to other areas, and what kind of a process it went through globally. Until the last quarter of the twentieth century, economic development plans around the world were focused only on increasing production. The massive increase in production, especially after the Industrial Revolution, significantly increased the use of raw materials, and the balanced functioning of the ecosystem began to deteriorate with these changes. Not only the excessive use of natural resources, but also the pollutants released to the environment during production have created a significant threat to the ecosystem (Sipahi 2010: 333). However, at that time, the deterioration of balances against nature was seen as a cost of economic development and a "reaction and treatment" strategy was applied to solve environmental problems (Kaypak 2011: 23).

By the 1970s, with the increasing number of environmental problems and their transformation from being a regional or national problem to a global dimension, the need to develop a more serious strategy for the solution of environmental problems arose. In 1972, the issues of interdependence between economic development and the environment came to the fore, and in the 1980s, the necessity of a "sustainable development" policy was put forward in order to carry out the development-environment relationship in a balanced way. In Table 6.1, the development of policies containing critical decisions in terms of the concept of sustainable development is summarized chronologically (UN 1987; UN 2021; UNDP 2021; Bozoğlan 2005: 1015–1025; Sipahi 2010: 334; Tıraş 2012: 62, 63).

As can be seen from the table, the concept of "sustainable development" was first included in the 1987 "Our Common Future" report of the World Commission

Table 6.1 Milestones of sustainable development

1972	The limits to growth	Attention was drawn to the devastation caused by the development on the natural environment – Emphasis was placed on the interdependence between the economy and the environment
1972	UN Conference on Human Environment, Stockholm	The United Nations Environment Program (UNEP) was established – Principles emphasizing the relationship between economic and social development and the environment are included – June 5 was declared as UN environment day
1976	"UN Conference on Human Settlements—Habitat I	Urbanization problems of developing countries have been discussed on the international platform
1980	World Conservation Strategy	Emphasis is placed on conservation and development oriented thinking for a sustainable society
1983	Establishment of the United Nations World Commission on Environment and Development	The terms environment and development have started to be used together
1987	Our Common Future (Brutland) Report	The definition of sustainable development concept has been made

(continued)

Table 6.1 (continued)

1972	The limits to growth	Attention was drawn to the devastation caused by the development on the natural environment – Emphasis was placed on the interdependence between the economy and the environment
1988	G7 Summit (Toronto)	Discussions were held on anthropogenic climate change and carbon dioxide emissions
1992	United Nations Conference on Environment and Development (Rio)	– Agenda 21 has been adopted – Sustainable development has turned into an active policy on a global scale – The content of the concept of sustainable development has been expanded and it has started to be used for different disciplines
1993	Establishment of the UN Commission on Sustainable Development	The implementation and monitoring of sustainable development principles and provisions at regional and international level has come to the fore
1997	Kyoto Protocol	In order to prevent anthropogenic climate change, the target was to reduce greenhouse gas emissions by 5% below 1990 levels between 2008 and 2012
1997	UN Follow-Up	– The action plan (Agenda 21) was reviewed in Rio
2000	Millennium Development Summit of the United Nations, New York	A new global partnership has been created to reduce extreme poverty formulated in The eight Millennium Development Goals (MDGs), a set of time-bound targets by the end of 2015
2002	World Sustainable Development Summit (WSDS), Johannesburg	Solutions were sought for the problems encountered in the implementation and realization of the decisions taken in Rio
2012	UN Sustainable Development Conference (Rio + 20)	"The Future We Want" result document has been accepted
2015	UN Sustainable Development Summit, New York	With Agenda 2030 (Sustainable Development Goals (SDGs)), 242 international indicators were determined for 17 goals and 169 targets. New areas such as climate change, economic inequality, innovation, sustainable consumption, peace, and justice were added to the priorities
2015	21st Conference of the Parties (COP 21″), Paris	The Paris Agreement, which regulates climate change mitigation measures, has been adopted

Source Created by the author

on Environment and Development. This concept in the report is defined as "Sustainable development is the development that meets the needs of the present without compromising the ability of future generations to meet their own needs" (Rogner and Popescu 2000: 31, Dinçer 2000: 159). According to this definition; it is understood that a society in which sustainable development and sustainable environment are carried out together, and resources are managed in a way that not only meets current social needs but can continue to meet in the future, is sustainable (Gümüş Akar and Seçilmiş 2019: 31).

In sustainable development, there are economic, social, spatial, cultural, and environmental dimensions, all of which affect each other (Kaypak 2011: 22). However, it should be accepted that the problems arising from the compulsory relationship of economy with nature played a greater role in the emergence of the concept. Sustainability actually resulted from a necessary change and transformation upon understanding that the hegemonic policies established by economic policies on the environment are unsustainable. Because the survival of the economic system or systems depends on the establishment of the relationship between the environment and the economy at an acceptable point (Kılıç 2012: 207). The "reaction and treatment" strategy for the solution of environmental problems, together with the concept and policies of sustainable development, has left its place to the "prediction and prevention" strategy (Kaypak 2011: 23). Developments, policies, and practices regarding sustainable development constitute a multifaceted whole that still continues today.

Sustainable development goals, objectives, and indicators are addressed as a whole in the 2015 SDGs, which are the continuation of the 2000 MDGs. By drawing a new global development framework with the Sustainable Development Goals of 2030, environmental issues such as sustainable cities, climate change, combating drought, and preserving biodiversity have been included in the sustainable development agenda (MFA 2021). Government, civil society, business people, science and technology people (Eşkinat 2016: 278) and every individual in the society are responsible for the SDGs that are still carried out today. Because the SDGs are a future investment project that a global community should work on with a sense of responsibility.

6.3 From Sustainable Economy to Sustainable Aviation

Although the year 1987, when the concept of sustainable development was defined, was a turning point, the fact that the concept has been used for different disciplines since 1992 is proof that sustainability is an important global target that should be focused on with all its dimensions.

Travel industry, which has an important place in the globalizing economic wheel, is gaining importance due to its impact on increasing environmental problems. Impacts on the environment as a result of transportation are defined as impacts that are intrinsically related to the transportation strategies, policies, and programs put in place. These effects occur over a period of time and their results are generally irreversible (Fenly et al. 2007: 64).

The sustainable development strategy aims to achieve social justice, economic growth, and environmental sustainability. "Sustainable transportation" is also a very important variable in achieving these goals. The issue of sustainable mobility has come to the fore with the green paper "The Impact of Transport on Environment" published by the Commission of European Communities (CEC) in 1992 regarding the effects of direct transportation on the environment. Sustainable mobility is defined as controlling the negative effects of transportation on the environment while fulfilling its economic and social duties (Longshurst et al. 1996: 199). Sustainable transport is the capacity to meet the current transport needs without compromising the ability of future generations to meet the same needs. Sustainable transport must meet three basic conditions: the rate of use of renewable resources should not exceed their natural rate of regeneration; the rate of use of non-renewable resources should not exceed the rate at which sustainable renewable substitutes are developed; and the rate of pollution emission should not exceed the rate of assimilation of the environment (Fenley et al. 2007: 64).

In Fig. 6.1, Greenhouse Gases Emissions (GHG) are included to express the environmental pollution caused by transport. The approximate increases for Transport GHG emission values for the years 1990–2018 for selected regions are as follows: OECD 23%, US 20%, EU 16%. Countries with an increase of more than 50% in OECD are Poland (68.1%), Turkey (68.0%), Ireland (58%), Luxemburg (57%), and Slovenia (53%). The countries that cause the highest amount of emissions are Japan, Canada, and Germany (OECD 2021).

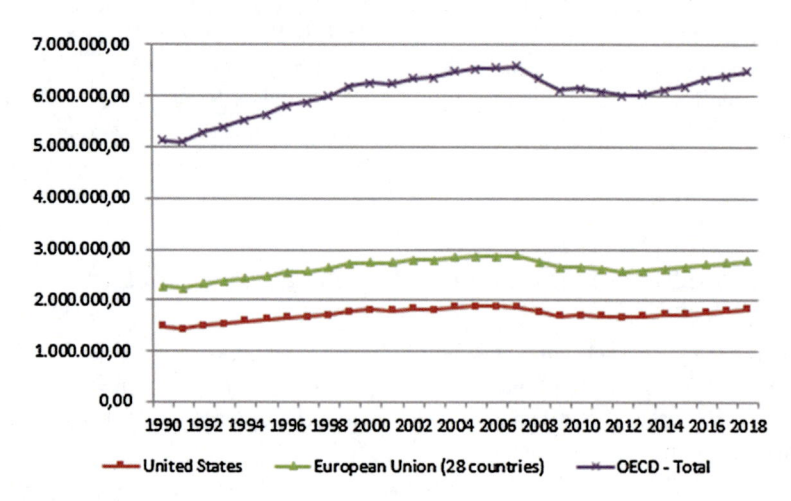

Fig. 6.1 Transport sector greenhouse gases emissions (Tonnes of CO_2 equivalent, Thousands) (*Source* OECD 2021)

Air transport, a mode of transport, is a constantly growing industry. Looking at the historical process, the aviation industry has been affected by global (The Asian crisis in 1998, the U.S. terrorist attack on 11 September 2001, the severe acute respiratory syndrome (SARS) outbreak in 2003, and the 2008–2009 world financial crises), but it has recovered and recaptured the continuous growth trend (ICAO 2021a). The Covid 19 pandemic, which has affected the whole world today, has also seriously affected the aviation industry. When the pandemic is brought under control, as in other global crises, the sector will recover and continue to grow, although it is not clear when it will happen. Air transportation, which produces 4.1% (pre-covid 19) of the global GDP (ATAG 2020a, b: 10), plays an important role in economic and social development by providing connectivity between countries. While it provides economic (direct, indirect, induced, and catalytic) and social benefits to the global economy, it also creates significant negative effects on the environment (Walker and Cook 2009: 378; ATAG 2020b, 34).

Both global and local environmental problems arising from air transportation are very important. It causes climate change with the use of fossil fuels and greenhouse gas emissions on a global scale. Negative environmental problems such as noise, pollution of land-earth-surface waters, decrease in air quality arise at local scale. Noise generated from aircraft and airport operations takeoffs and landings, engine testing, surface transport, and construction is widely considered to be one of aviation's most serious environmental problems. Polluted land, ground and surface waters, and habitat loss at airports resulting from jet fuels, aircraft de-icing operations, waste generation, and land acquisition are other environmental problems. Considering all these effects, the aviation industry, together with the increasing trend of passenger and freight transportation, creates significant negative effects on the environment and human health (Upham 2003: 5; Kılıç et al. 2019: 54; Kaszewski and Sheate 2004: 186).

Global and local environmental effects of aviation on the global and local environment are summarized in Table 6.2.

Combating the aforementioned local and global environmental impacts requires a system that includes all actors on a sectoral basis. Table 6.3 shows the actors and their roles in the fight against climate change. The actors that are directly related to reducing GHG emissions related to air traffic are the aircraft industry (via new technology) and airlines (via operational practices). Airports and air traffic management and control is also the industry component that serves this purpose with its operational practices. Another factor is the demand volume and structure determined by travelers, tourism industry, and shippers (Pastowski 2003: 186).

In the early 1990s, it was determined that 12% of transport-related CO_2 emissions and 2.3% of global CO_2 emissions were caused by aviation. In the very short period between 1990 and 1995, aviation CO_2 emissions increased by 30%. These data show that it produces more CO_2 emissions than other modes of transport (T&E 1998: 2).

Table 6.4, summarizes some aviation pollutants included in the European Aviation Environmental Report 2019. Relatively optimistic picture in HC, CO and particulate matter (PM) values depends on fleet renewal. According to the same report; 3.6% of the total greenhouse gas emissions (in 1990, this rate was 1.4%) and 13.4% of

Table 6.2 Global and local environmental effects of aviation

Causes	Effects
Global	
CO$_2$, CO, NOx, VOCs, and HC emissions from aircraft SO2, aerosol particulates emissions from aircraft NOx and aerosol particulates Land-take due to airport construction and expansion	Climate change, global warming (greenhouse effect) Global cooling (greenhouse effect) Acidification of water (rain&rivers) Decline in biodiversity
Local	
Aircraft liftoffs and takeoffs and operations on the ground Surface access transport Surface run-off from runway and aircraft (de-iceing fluids) and other airport operations Land-take from airport construction and expansion Location of airport	Noise pollution Noise, air pollution, and congestion Water pollution (aquifers, steams, and rivers) Decline in biodiversity and habitat loss, and land-take itself Visual impact/pollution Energy and water consumption Regional and local economic development, increased employment, and level of new industry attracted

Source Kaszewski and Sheate (2004: 187)

Table 6.3 Objectives, options, direct actors and climate policy in civil aviation

Overall objective	Stabilizing the global climate				
Sectoral objective	Limiting GHG emissions from civil aviation				
Determinants/options	Emissions per seat-kilometer	Aircraft operation		Demand for aviation services	
Intermediate determinants	Aircraft technology	Modernity of fleets	Efficiency of air traffic operation	Frequency	Distance
Direct actors Involved	Aircraft industry	Airlines	Airports	Air traffic control	Traveler, tourism industry, shipper
Policies: governmental bodies at various levels; international organizations;	Actor-oriented policies				
NGOs	General sector-oriented policies				

Source Pastowski (2003: 186)

Table 6.4 Some of full-flight emission indicators

	Units	2005	2014	2017	% change (2005–2017)
CO_2	Million tonnes	141	148	163	16
NO_x	Thousand tonnes	669	749	839	25
HC	Thousand tonnes	55	53	57	4
CO	Thousand tonnes	110	102	108	−2
Volatile PM	Thousand tonnes	126	123	136	8
Non-volatile PM	Thousand tonnes	76	55	53	−30

Source EASA, EEA, EUROCONTROL (2019, p. 24)

the emissions from transportation in the EU28 in 2016 originated from the aviation sector, and CO_2 is expected to increase by 40% and NO_x by 45% in 2040. The report underlined that while the amount of carbon monoxide (CO), nitrogen oxide (NO_x), and sulfur oxide (SO_x) produced by other transport modes and other sectors decreased in general, those originating from aviation increased continuously from 1990 to 2015.

The realization of the significant share of aviation in global CO_2 emissions in the 1990s, and the fact that the aviation industry has an ever-increasing growth potential in the global economy, brought along the concern that the environmental problems it would create would increase. For this reason, the need to develop policies for the aviation industry as a sub-branch of sustainable development efforts has emerged as a necessity (T&E 1998: 2). Since these years, policies for sustainable aviation have been started to be developed by local and global aviation authorities in line with this need. Air transport serves to achieve at least 15 targets in the 17 SDGs, and its contribution to the environment-oriented SDGs, with the clearest and wide-ranging action plans it has prepared, is also noteworthy. Aviation is the first sector to set targets for reducing CO_2 gas on a global scale (ATAG 2020a: 1,2).

6.4 Institutional Structures and Policies for Reducing the Environmental Impacts of Aviation

In 1999, at the request of ICAO, the IPCC published for the first time a report specific to an industrial sub-sector, assessing the consequences of greenhouse gas emissions from aircraft engines. There are some options to reduce the impact of aviation emissions, including changes in aircraft and engine technology, fuel options, operational options, and regulatory and economic measures (IPCC 1999: 10–12). This report statements have been a pioneer of institutional and political formations for aviation-related environmental impacts, and steps have been taken for sustainable aviation.

A sustainable air transport system is based on the identification, analysis, and evaluation of the economic, social, and environmental dimension, all of which are interconnected. The contents of these three dimensions are as follows; *economic dimension* operating revenues, costs, and productivity; *social dimension*: system's direct and indirect contribution to employment and GDP; *environmental dimension*: local (airport) and global (airspace) air pollution, airport noise, aircraft accidents, traffic congestion, generation of waste and land use (Janic 2004: 44; Walker and Cook 2009: 378).

Environmental issues are generally more involved in discourses about sustainable aviation than social or economic ones (Walker and Cook 2009: 381). In addition, the operational dimension (such as demand and capacity, quality of service, and safety and security) is considered as a fourth party in sustainable air transport (Janic 2002: 117; Janic 2004: 44).

In this section, important institutions and organizations operating in economic, social, and environmental terms and policies developed in this regard will be included in order to reduce and prevent aviation-related environmental impacts.

6.4.1 Institutions Struggling with Environmental Protection in Aviation

The institutions responsible for the development of environmental policy and standards at the international level are UN and ICAO. The responsibility of managing GHG emissions from aviation has been given to ICAO with the Kyoto Protocol. Therefore, ICAO is the most authoritative policy-making institution on environmental protection in the aviation industry. However, applications that go beyond ICAO standards at the national and local level can also be seen (NSTC 1999: 51), and there are many institutions that develop these applications and operate for the protection of the environment at the national and international level. Some important institutions were established specifically for this purpose and information about these institutions can be summarized as follows:

Committee on Aviation Environmental Protection (CAEP): Established on 5 December 1983 by the ICAO Council, replacing the Committee on Aircraft Noise (CAN) and the Committee on Aircraft Engine Emissions (CAEE), CAEP is a committee that provides technical support to the console. These technical supports consist of; noise, local air quality (LAQ), and reducing international aviation CO_2 emissions policies (aircraft technology, operational improvement, sustainable aviation fuels and market-based measures). International Standards and Recommended Practices (SARPs), determined by stakeholders, including CAEP, are reflected in Convention on International Civil Aviation-Annex 16. Annex 16 covers: aircraft noise-Volume I, aircraft engine emissions-Volume II, aeroplane CO_2 emissions-Volume III, and Carbon Offsetting and Reduction Scheme for International Aviation (CORSIA)-Volume IV (ICAO 2019: 13).

CAEP has held 11 meetings at regular intervals since its establishment and the last one was CAEP/11 in 2019. These meetings and the important decisions taken at these meetings are summarized in Fig. 6.2.

As of February 2021, As of 2021, CAEP, which has 30 Members, 22 Observers and more than 600 experts, carries out its activities with the following working groups divided according to their specific fields (ICAO 2021c):

- Working group 1 (WG1): Aircraft Noise Technical Issues
- Working group 2 (WG2): Airports and Operations
- Working group 3 (WG3): Emissions Technical Issues
- Working group 4 (WG4): CORSIA

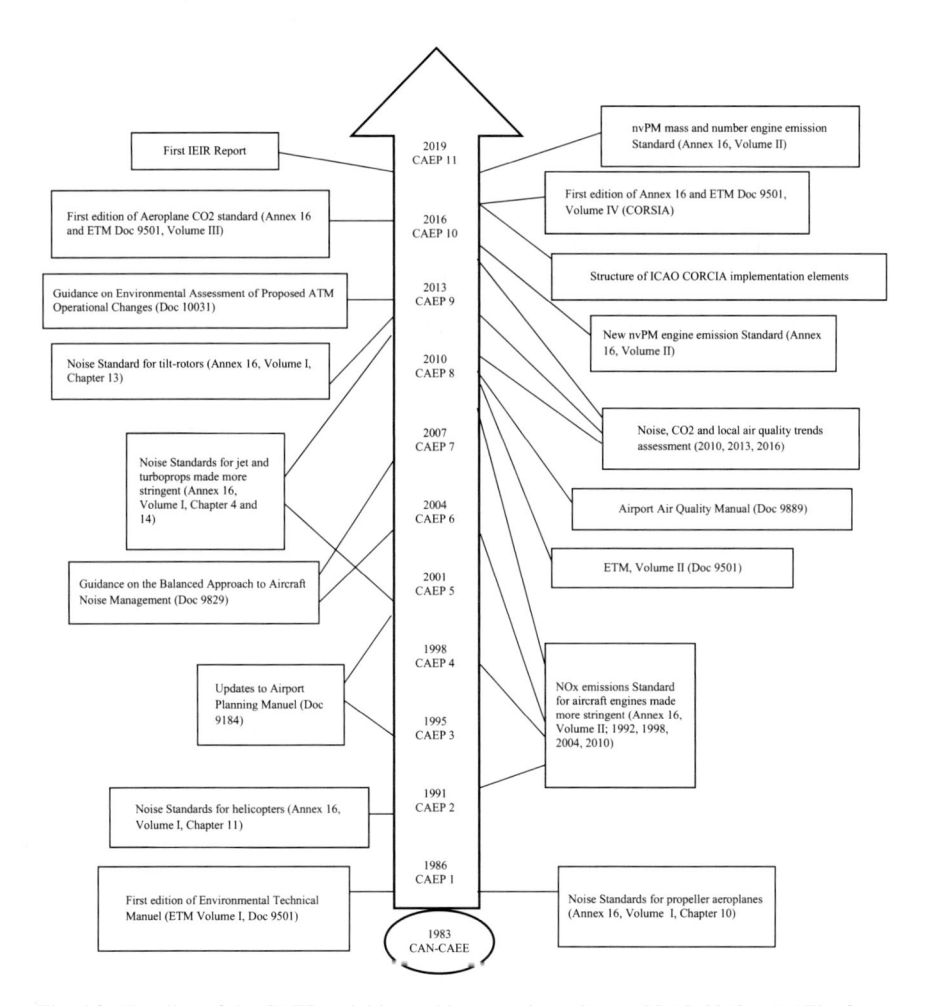

Fig. 6.2 Timeline of the CAEP activities and key practices (*Source* ICAO 2019: 14) *The figure has been rearranged in vertical form by the author in accordance with the content in the source

- Fuels Task Group (FTG)
- Modeling and Databases Group (MDG)
- Forecasting and Economic Analysis Support Group (FESG)
- Aviation Carbon Calculator Support Group (ACCS)
- Impacts and Science Group (ISG)
- Sustainability Certification Schemes Evaluation Group (SCSEG)
- Long-term Global Aspirational Goal Task Group (LTAGTG)

International Coalition for Sustainable Aviation (ICSA): ICSA was established in 1998 to make official observations on ICAO's environmental work. The organization contributes to CAEP's studies on emissions and noise, marked-based solutions, economic and environmental analysis-modeling-forecasting studies (ICSA 2021).

Center for Aviation Transport and Environment (CATE): CATE, affiliated with Manchester Metropolitan University, is an international research center established to study the environmental capacity of the aviation industry and sustainable aviation. With its research, it provides support to all stakeholders in the aviation industry, including ICAO, EU Commission, UK Government, and US FAA, in developing long-term sustainable growth and sustainable aviation policies. The organization's main areas of expertise are Sustainability and the air transport industry, Global climate change, Noise and impacts on society, Local air quality (CATE 2021).

MIT Laboratory for Aviation and Environment (MIT LAE): MIT LAE is an interdisciplinary (Aviation, Mechanical and Chemical Engineering, Atmospheric Science and Economics) research laboratory developing methods to understand and measure the environmental impact of aviation and similar industries. The organization also evaluates and develops new technologies that help reduce aviation's environmental footprint and develops techniques for conducting cost–benefit analyzes of options for reducing environmental impacts. The main areas of expertise are Climate, Air Quality, Fuels, and Technology (MIT LAE 2020).

Aviation Environment Federation (AEF): AEF is a UK NGO and was established in 1975 with the aim of protecting the environment, public health, and quality of life by securing policies and measures that provide effective limits on noise, emissions, and other environmental impacts from the aviation sector. Areas of study include aircraft noise, climate change, air pollution, biodiversity, and public health (AEF 2020).

Nordic Initiative for Sustainable Aviation (NISA): The purpose of the funds is to promote and participate in the development and commercialization of sustainable aviation fuels. The association facilitates and strengthens the conditions for creating a sustainable aviation sector, taking into account national legislation, EU Sustainability Criteria, and international sustainability guidelines. The organization also coordinates initiatives and activities created by nordic and national industry organizations, authorities, UN agencies such as IATA, ATAG, ICAO, EU Flight Path initiatives, and other related activities (NISA 2021).

The Canadian Green Aviation Research and Development Network (GARDN Environment, Aviation, and Innovation): GARDN is a non-profit organization working to improve Canada's aerospace competitiveness. GARDN funds and helps implement joint projects that can reduce the environmental footprint of next-generation aircraft, engines, and avionics systems in Canada (GARDN 2021).

The Concerted Approach of French National Council for Civil Aviation (CORAC): CORAC is a national forum that supports aviation research and development efforts to promote innovations in aviation technology towards the European ACARE ambitious environmental objectives set for 2020. CORAC supports studies aimed at understanding and modeling the impact of air transport on the environment and climate change mechanisms. Among the priority issues identified are combustion products, climate impact, noise emission, and pollution chemistry modeling in the airport environment (Desvallées 2012: 416).

6.4.2 Policies for Environmental Impacts

The International Civil Aviation Organization (ICAO), which provides diplomacy and cooperation in the international arena, is the most authoritative institution that carries out activities to reduce and prevent the impact of aviation on the global environment and carries out its environmental studies through CAEP. ICAO first started working on environmental problems in the 1960s by developing policies and standards on aircraft noise. These studies expanded gradually and reached three targets in 2004: "to limit or reduce the number of people affected by significant aircraft noise", "to limit or reduce the impact of aviation emissions on local air quality (LAQ)", "to limit or reduce the impact of aviation greenhouse gas emissions on the global climate" (ICAO 2016a: 5). In the 35 Assembly, Appendix I of Resolution A35-5 (PDF) in 2004, it was decided to continue to develop the Council's guidelines on measures for aircraft engine emissions.

In connection with this, in 2007 ICAO issued in 2007 Guidance on Aircraft Emissions Charges Related to Local Air Quality (Doc 9884) and a paragraph (Doc 9082) on emissions-related airfares for local air quality in and around airports (ICAO 2021b). The emphasis on "Sustainable Aviation Fuel (SAF) contributes to lower environmental impact and therefore the importance of fuel efficiency and alternative fuels research and development activities" in 2007, 36th Session is another important development this year (ICAO 2019: 171).

The aviation industry's global stakeholder associations (Airports Council International-ACI, Civil Air Navigation Services Organization-CANSO, International Air Transport Association-IATA and International Coordinating Council of Acrospace Industries Associations-ICCAIA) adopted three key goals under the umbrella of the Air Transport Action Group (ATAG) to reduce the climate change problem and CO_2 emissions from air transport in 2008.

These goals are listed as follows (IATA 2019; EASA, EEA, EUROCONTROL 2019: 27):

- Short-term goal: An average improvement in fuel efficiency (CO_2 per Revenue Tonne Kilometer) of 1.5% per year from 2009 to 2020.
- Medium-term goal: A cap on net aviation CO_2 emissions from 2020 (carbon–neutral growth).
- Long-term goal: A reduction in net aviation CO_2 emissions of 50% by 2050, relative to 2005 levels.
- To achieve these goals, four strategies were adopted by all stakeholders:
- Improved technology, including the deployment of sustainable low-carbon fuels.
- More efficient aircraft operations.
- Infrastructure improvements, including modernized air traffic management systems.
- A single global market-based measure, to fill the remaining emissions gap.

In 2009, the First ICAO Conference on Aviation and Alternative Fuels (CAAF/1) was held and the use of SAF was approved. In CAAF/1, it was also recommended to create the ICAO Global Framework for Aviation Alternative Fuels (GFAAF) and to share developments related to SAF through this (ICAO 2019: 171).

In 2010, ICAO's Member States agreed to the global goal to stabilize international civil aviation GHG emissions at 2020 levels and defined a basket of measures designed to help achieve this goal (ICAO 37th Assembly). These measures include aircraft technologies such as lighter airframes, higher engine performance and new certification standards, operational, sustainable alternative fuels, and market-based measures (MBMs). These measures were further expanded with the ICAO 39th Assembly in 2016, and the Carbon Offsetting and Reduction Scheme for International Aviation (CORSIA) was accepted as an international benchmark (ICAO 2016a: 5).

With the CORCIA application, it is expected that the carbon emissions will decrease by an average of 2.5 billion tons between the years 2021–2035. CORSIA is a policy tool used to achieve the medium-term target, which is expected to reach the aviation carbon–neutral growth target from 2020 (ATAG 2021a).

Different carbon pricing policies have also been developed, such as an emissions trading scheme (ETS) and Green taxes and fees, which are implemented at the national, regional or local level, but not adopted by ICAO. The EU ETS was established in 2005 to limit GHG emissions from the energy, manufacturing, and airlines sectors. Working with the principle of "cap and trade", the system is the world's first emissions trading system and will only be valid for flights between airports in the European Economic Area until 2024 (EU 2021). Green taxes are actually aimed at reducing aviation demand and their environmental benefits cannot be measured. In addition, Green taxes are not seen as an appropriate policy tool to be applied against climate change, as the aviation industry consumes its financial resources, delays investments in fleet renewal or research and development, and also rarely transfers tax revenues to technologies that will help reduce emissions (IATA 2019; ATAG 2021a).

In 2017, environmental standards were determined on a global scale by ICAO. These standards include aircraft CO_2 emissions and aircraft engine non-volatile Particulate Matter (nvPM) mass concentration. These standards have been integrated into European legislation by EASA and have been implemented as of January 1, 2020. CO_2 standards provide the design process with an additional requirement that increases fuel efficiency priority in overall aircraft design (EASA, EEA, EUROCONTROL 2019: 35).

The above-mentioned objectives and strategies to reduce or prevent environmental impacts cover all civil aviation activities in general, and it is obvious that more effective results will be achieved when all of them are carried out together. However, it should be noted that the development of new aircraft technology and its inclusion in cleaner and quieter advanced designs is one of the most important ways to reduce the environmental impacts from aviation (EASA, EEA, EUROCONTROL 2019: 29).

6.5 Technological Innovation in Aviation for a Sustainable Environment

Technological innovation is seen as an accelerating and developing tool in achieving sustainable environmental policies. At COP21 in 2015 and the 2030 Agenda for Sustainable Development, the impact of technological innovations integrated with economic and environmental policies on sustainable development was mentioned.

The United Nations Framework Convention on Climate Change-Technology Executive Committee (UNFCCC-TEC) published a report in 2017 in which they explain the important role of technological innovation for low-emission, climate-resilient, and prosperous future and ten important elements required for successful technological innovation (UNFCCC 2017: 1.6; Anser et al. 2021: 6). In the same report, it is underlined that it is difficult or even impossible to achieve the SDGs goals in the Paris Agreement, especially without technological innovation (UNFCCC 2017: 26).

This section is reserved for the evaluation of technological innovations, which are so important for the sustainable environment, in terms of the aviation sector. For this purpose, in this section, aviation-sustainable environment related patent data will be included in order to see the importance of technology development in aviation and the success of innovation efforts.

6.5.1 The Link Between Environment, Innovation, and Aviation

The aviation sector is a sector that uses high technology intensively and grows remarkably depending on technological developments (Fenley 2007: 70; Reynolds-Feighan

2005: 154). R&D activities form the basis of technology development and innovation (Secilmis and Konu 2019: 692). Innovative technological advances in the aviation industry are achieved thanks to the importance and support given by industry stakeholders to R&D. Similarly, R&D activities are given importance in order to reduce the damage caused by the aviation industry to the environment. Every year, Aerospace companies spend $20 billion on research and development, and airports spend $35 billion on new infrastructure (ATAG 2017: 22). The first attention was drawn to the existence of environmental problems for aviation originating in the 90 s. It was also realized in the same years that research and development in aviation should be given importance for environmental benefits. In the report published by the National Science and Technology Council-NSTC, Subcommittee on Transportation Research and Development (1995), it was reported that the growth of aviation will be limited in the twenty-first century due to its environmental impacts, and therefore, technology development is important for the growth and sustainability of aviation. In the same report, the issue of "the systematic development and validation of additional technologies to reduce engine and airframe noise and the development of flight procedures to reduce community noise exposure, and also development of emission reduction technology" is highlighted (NTSC 1995).

The National Research and Development Plan for Aviation Safety, Security, Efficiency and Environmental Compatibility (1999) focused on three objectives for Environmental Compatibility: Local air quality, Global Change, Noise Reduction. These three aims are summarized in the report as follows (NSTC 1999: 9):

Local Air Quality: Through research and development of technology, reduce emissions of NOx and other pollutants that endanger public health and the environment.

Global Change: Through research and development of technology, reduce emissions that affect climate or stratospheric ozone.

Noise Reduction: Through research and development of technology, reduce noise levels in the vicinity of airports and in other places where aircraft noise is perceived as disruptive to the environment.

If attention is paid, research-development is stated as the key point for technology development in these three purposes for environmental protection. The same report also envisions a strong link between coordinated multi-agency, core, and focused R&D and operational activities under the FAA's mandate.

"Invest in new technology" took the first place among the four-pillar strategy agreed upon by 240 IATA member airlines to combat climate change (IATA 2007). Here, new technology means aircraft and engine technology, and the development of biofuels (Grimme 2008: 3).

Continual research and development of safe and cost-effective technology options to reduce emissions is essential to protecting the environment and sustaining the growth of aviation (NSTC 1999, 51). On the other hand, adequate levels of R&D and innovation efforts increase the chance of achieving sustainable development goals in an economy (Seçilmiş 2020: 145). In this context, the development of technology for environmental problems can be considered as an important step towards environmental protection, sustainability of the aviation industry, and sustainable development (Fenley 2007: 75). There are also studies showing that the societal benefits of R&D expenditures made by the aerospace industry are much more than the benefits obtained from manufacturing, and that an R&D expenditure of 100 million dollars contributes an additional 70 million dollars to GDP every year (ATAG 2020a, b: 15).

In Europe, ACARE Vision for 2020 was adopted due to the importance of sustainable innovation in aviation. ACARE includes goals to reduce aviation-related environmental pollution, and in addition the production of environmentally friendly technologies and the improvement of further aircraft aerodynamic performance (technology aimed at reducing aircraft eddy and technologies to reduce viscous friction). To support ACARE's global strategies, the European Commission has funded European Coordination Action KATnet I and II (Key Aerodynamic Technologies for Aircraft Performance Improvement) (Abbas et al. 2011: 83). There are also many examples of countries that produce and implement projects similar to those in Europe.

6.5.1.1 Climate Change and Aviation-Related Sustainable Innovation

Sustainable innovation is basically defined as innovations that provide benefits on the environment and society while producing economic benefits (Štreimikiene, et al. 2020: 231). "aircraft technology and sustainable aviation fuel" ranks first among the stakeholders' strategies in the fight against global climate change in the aviation industry. These two technology development activities, which actually serve for sustainable innovation, will be explained without going into technical details.

New Aircraft Technology

Aircraft technologies developed to reduce the local and global environmental impacts of aviation are based on noise reduction and increasing fuel efficiency. More precisely, in order to reduce GHG emissions, which is a global environmental problem, there is a need to develop aircraft and engine technology in the aviation industry and to develop alternative jet fuels, that is, to allocate resources for research and development (Hodgkinson et al. 2007: 33; Grimme 2008: 3).

Efforts to improve aircraft fuel efficiency are not a goal of aircraft design. The main starting point for improving aircraft fuel efficiency is that fuel prices make up a large part of airline costs. With the emergence of aviation-related environmental problems, it has become more important to develop fuel efficient aircraft technologies (Lee et al. 2001: 174; Lee and Mo 2011: 3778). On the other hand, fuel efficiency is highly influential on speed, payload-range performance, and landing takeoff performance of an aircraft, and these are other reasons to develop fuel-saving technologies (Lee and Mo 2011: 3781). In addition to these economic and environmental factors, there is also social pressure to increase fuel efficiency. With the awareness of the effects of aviation on climate change, human health and the environment, increasing social awareness puts pressure on governments in this regard and accelerates the search for solutions (Lee and Mo 2011: 3786).

Aviation produces 3.16 kg of CO_2 emissions per kg of fuel burned, and therefore it is important to increase the fuel efficiency of the aircraft (EASA, EEA, EURO-CONTROL 2019: 107). Fuel efficiency can be achieved through the development of engine technologies, the development of high-lift wing designs, and the development of lighter airframe materials (Palmer 2015: 31). Not all of these listed have the same effect in reducing fuel consumption. For example, aerodynamic devices such as winglets or sharklets can be strengthened, but these improvements have limited impact on fuel efficiency. One of the most effective improvements is to reduce weight (especially long-distance missions). According to a calculation, a 1% reduction in landing weight can provide fuel savings between 0.75% and 1%, depending on the engine type. New materials such as carbon fiber composites are being designed to reduce weight (Palmer 2015: 32).

New generation aircraft equipped with features such as new engines, increased use of light weight composite materials, more-efficient systems applications and modern aerodynamics, winglets, fuselage airflow control devices, and weight reductions (e.g., A350 XWB, A220, Boeing 787 Dreamliner, EmraerE2, 737MAX, A320neo, A330neo, Boeing 747–8), thanks to these features, contribute to both economically saving fuel and reducing environmental impacts (Hodgkinson et al. 2007: 33,34; Capoccitti et al. 2010: 69,70; ATAG 2020b: 35).

ICAO introduces standards that must be followed while developing aircraft technologies (Annex 16: Volume 1-Aircraft Noise, Volume II-Aircraft Engine Emissions, Volume III-Aeroplane CO_2 Emissions). ICAO emission certification standards regulate smoke and various gas emissions from aircraft engines. These can be listed as unburned hydrocarbons (HC), carbon monoxide (CO), nitrogen oxides (NOX), and non-volatile particulate matter (nvPM) (EASA, EEA, EUROCONTROL 2019: 29,107). ICAO also sets technology goals to encourage the development of new technologies. Latest technology goals are included in ICAO Doc 10,127 (Independent Expert Integrated Technology Goals Assessment and Review for Engines and Aircraft) (ICAO 2021d).

Alternative Fuel

Another issue related to sustainable innovation is the production of alternative fuels. In addition to the environmental effects of aircraft emissions, increases in oil prices and jet fuel prices have led to a tendency towards alternative jet fuels. These alternative fuels are biodiesel (made from soybeans, corn, and other products), hydrogen (a long-considered alternative), and synthetics (made by turning coal, oil shale, or natural gas into a liquid that can act like traditional jet fuel). Although ethanol is also an option, it requires more energy consumption due to the need for a larger engine (Hodgkinson et al. 2007: 35). All-electric and hybrid-electric aircraft with ongoing R&D processes will also be able to offer alternative fuels for fully commercial flights in the future. Even between 2023 and 2025, it is considered possible to provide air taxi services in some cities. Today, fossil-based non-fuel fuels, which are called sustainable fuels, currently constitute 1% of global jet fuels (ATAG 2020b: 35, 36).

6.5.2 Evaluation of Environmental Innovation in Aviation with Patent Data

Patent data focusing on the outputs of the invention process are widely used to measure technological innovations. Panel data are widely used because; patent data are commensurable, measures intermediate outputs, quantitative and widely available, and divided into specific technological fields (Haščič and Migotto 2015: 16). Especially the last feature is extremely important for academic studies in specific fields. Panel data will be used in this study, as it will be an effective indicator in terms of measuring how much aviation contributes to the issue of the sustainable environment with technological innovations.

Patent data were obtained from the European Patent Office (EPO) and The Cooperative Patent Classification (CPC) was used for data scanning (EPO 2021). In the CPC, patents associated with the development of anti-climate change technology are in the main group "Y02-Technologies or Applications for Mitigation or Adaptation Against Climate Change". Under this main category, those related to transport are categorized as "Y02T-Climate Change Mitigation Technologies Related to Transportation", and those related to aviation in Y02T are categorized as "Y02T50-Aeronautics or air transport". The Aeronautics or air transport group is divided into five subgroups that show what they are taking against climate change. These subgroup codes and topics are as follows: Y02T50/10-Drag reduction, Y02T50/30-Wing lift efficiency, Y02T50/40-Weight reduction, Y02T50/50-On board measures aiming to increase energy efficiency, Y02T50/60-Efficient propulsion technologies, e.g., for aircraft, Y02T50/80-Energy efficient operational measures, e.g., ground operations or mission management.

Within the Aeronautics or air transport (Y02T50) main group, the total number of patents obtained since 1897 is 40,523. Figure 6.3 represents ten years cumu-

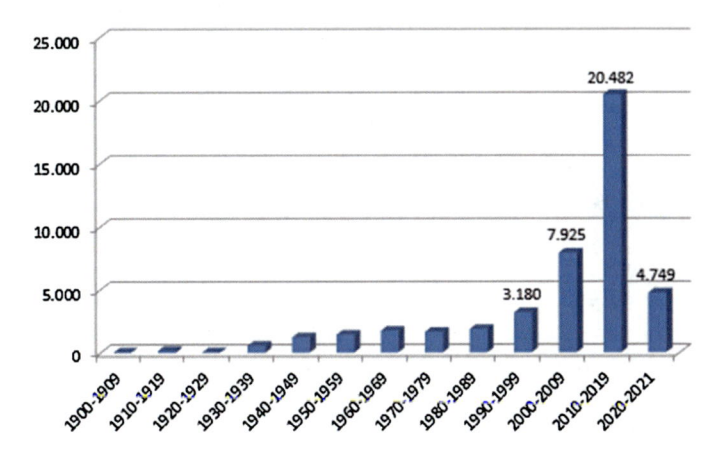

Fig. 6.3 Climate change mitigation technologies related to aeronautics or air transport (10 years cumulative data)

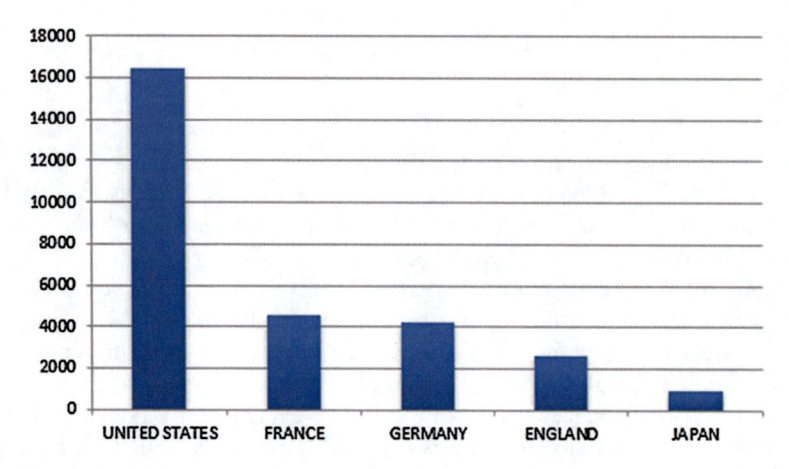

Fig. 6.4 Top five countries in patent ranking

lative patents developed against climate change on Aeronautics or air transport (Y02T50). Aviation technologies-related patent applications made between 1990 and 1999 increased by 70% compared to the previous decade. After the 1999 IPCC report, the number of patents increased with the effect of sustainable aviation policies in the 2000s. Between 2000 and 2009, patents increased by 149% over the previous decade. There were a total of 20,482 patents in the 2019–2019 range, an increase of 158% over the previous decade. The number of patents in a short period of 17 months between 01.01.2020 and 31.05.2021 is 4,749.

Figure 6.4 shows the top five countries that obtained patents under the main group Y02T50. Of the total 40,523 patents, 16,465 are owned by the United States (US). This means that the USA produces 41% of the technologies produced in the field of combating against climate in the field of aviation and is the most successful country in this regard. With a similar logic, 11% of the total patents on this subject were obtained by France, 10% by Germany, 7% by England, and 2% by Japan. the total of patents of these five countries is 28,886 and 71% of the patents within the Y02T50 belong to these countries.

Figure 6.5 shows the shares of subcategories in total patents in the field of climate change mitigation technologies related to air transportation.

According to the pie chart, the results are in order: Efficient propulsion Technologies %62, Weight reduction %17, Drag reduction %11, On board measures aiming to increase energy efficiency %6, Wing lift efficiency %3, and Energy efficient operational measures %2. The chart is important in terms of understanding in which subjects more technology is developed. As can be seen from the figure, the most patented subgroup is Y02T50/60-Efficient Propulsion Technologies (26.145 patents). The category with the least number of patents is Y02T50/80-Energy efficient operational measures (715 patents).

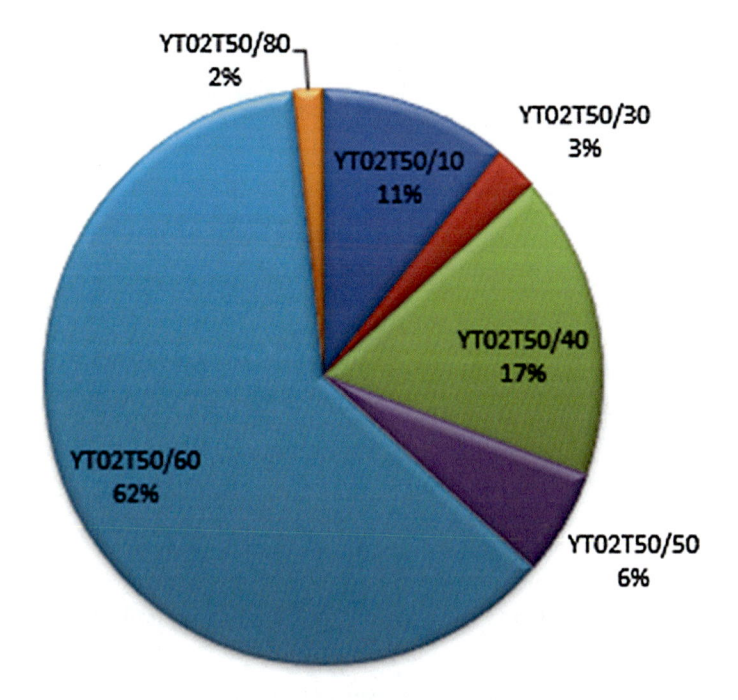

Fig. 6.5 Shares of subcategories

In Fig. 6.6, the top five countries with the most patents and their patent numbers are given in each subcategory. The patent numbers of the top five countries constitute approximately 72% of the total. In general, the US is in the first place in every sub-title. America alone produced approximately 57% of the total patents produced by the first five countries. France, Germany, England, and Japan are the countries that always exist, although their ranking changes in the top five. In the Y02T50/80 category, Israel is among the top five countries with 15 patents. As a different country, South Korea takes its place in the ranking with 131 patents in the Y02T50/50 category. Although not included in the figure, there is a lower category "Y02T50/678-Aviation using fuels of non-fossil origin" and a total of 482 patents have been approved in this category.

Since 1987, within the scope of Climate change mitigation technologies related to transportation, it has been determined that there have been 6,825 patent applications related to road transport (YT02T 10) and 11,402 patent applications related to maritime or waterways transport (YT02T 70). It can be said that Air Transport, with 40,523 patents, puts much more effort into climate change than the other two modes of transport.

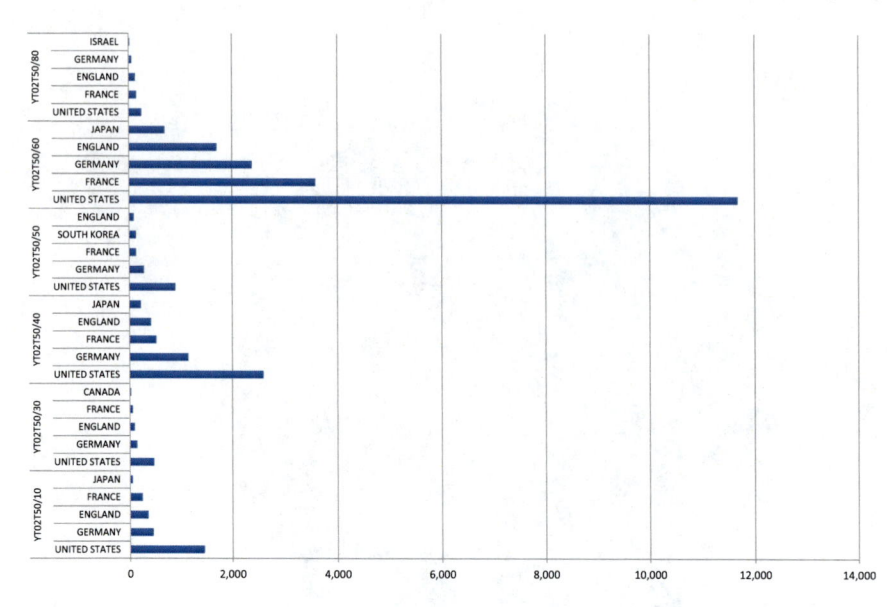

Fig. 6.6 Subgroup patents against climate change related to aeronautics or air transport

6.6 Do Aviation Technology Innovation Contribute to a Sustainable Environment?

There are some doubts about the environmental sustainability of the air transport industry (Goetzh and Graham 2004: 265; Upham 2003: 15). Namely, Developing new technologies in aircraft manufacturing can be effective, but involves uncertainties in terms of time and cost. In addition, fleets of aircraft have a long life and it takes time to replace these fleets. Another technology-based solution for reducing CO_2 emissions is the substitution of petroleum-based kerosene with biofuels. However, there are some economic, ethical, and technical limitations in the production of biofuels (Grimme 2008: 5,6). In addition, although alternative fuels are developed, emissions from biofuels also occur, and the emissions from many biofuels could be greater than that from conventional fuels (Forsyth 2011: 29). Even if benefits such as quieter engines and fuel savings are achieved, there are concerns that these gains will become ineffective due to the continuous increase in air transport (Kaszewski and Sheate 2004: 186).

There is an opinion that air traffic control (Forsyth 2011: 29) and air operational efficiency, which are other alternative ways to reduce emissions by saving fuel, have a limited effect (Grimme 2008: 6, 7). It is also argued that the effectiveness of taxes used as a tool to promote cleaner / quieter aircraft production is questionable and there is an experience that they do not contribute to aviation decarbonization (IATA 2019).

The IPCC concluded that "the increase in aviation emissions attributable to increased air travel demand will not be fully offset by reductions in emissions achieved through technological improvements alone" (Hodkingson et al. 2007: 16). IATA argues that taxing passengers and airlines is ineffective, and that policymakers should support multilateral efforts to address aviation emissions, including CORCIA, research investment in new technologies, and air transport's transition to sustainable aviation fuels (IATA 2019).

Although there are different opinions, according to the ATAG (2020a, b) report, the technology-oriented efficiency of new aircraft and engine models has increased by 80% compared to the first jets produced in 1950. Each new generation aircraft is 15%-20% more fuel efficient than the previous generation. Between 2000 and 2019, the cumulative fuel efficiency increased by 38% and the amount of CO_2 decreased by approximately 54%. Fuel efficiency, which increased by 2% annually between 2009 and 2019, is above the sector target of 1.5% (ATAG 2020b: 34).

Although the increase in air traffic volume generally causes an increase in aviation-related emissions, thanks to the successful steps in technological advancement and improvements in operations and infrastructure, 50% less CO_2 is produced per flight today compared to 1990. Also, aircraft emit 50% less carbon monoxide and 90% less smoke and unburned hydrocarbons, compared to fifty years ago, and emit 40% Nitrogen oxide (NOx) compared to 1981. According to these data, air transport has less impact on local air quality than road transport, thanks to improvements in aircraft technologies (ATAG 2021b).

6.7 Conclusion

The aviation sector is an important component used in a country's economy as an economic, social, and foreign policy tool. Innovation is the leading actor in the emergence, development, and maintenance of this sector. There are various environmental problems caused by this sector, which provides international connectivity. Climate change, which is one of them, is tried to be prevented by policies developed by local and international authorities. There is a consensus among all industry stakeholders that the most effective policy tool in this struggle is sustainable innovation. In this context, research and development on aircraft and engine technology and alternative jet fuels has been considered a priority for sustainable aviation in order to reduce aviation-related GHG emissions.

According to the data obtained from the European Patent Office, the weights of the total number of patents in climate change mitigation technologies related to land-sea-air transportation categories are as follows: 69% aeronautics or air transportation, 19% maritime or waterways transportation and 12% road transportation. According to these values, it is obvious that the aviation industry, which is claimed to cause more environmental pollution than other modes of transport, makes more efforts to compensate for this.

The subcategory weights in climate change mitigation technologies related to aeronautics or air transportation are as follows: efficient propulsion technologies 62%, weight reduction 17%, drag reduction 11%, on board measures aiming to increase energy efficiency 6%, wing lift efficiency 3% and energy efficient operational measures 2%. Accordingly, efficient propulsion technologies are the most innovative areas, followed by weight and drag reduction. The number of patents in the aviation using fuels of non-fossil origin category constitutes only 1% of the total patent. The country with the highest number of patents in all categories is the United States. It is followed by France, Germany, England, and Japan. These countries are already highly developed in airframe, engine, and equipment production. Over the years, significant progress has been made in the field of climate change mitigation technologies on a global scale. The cumulative number of patents between 2010 and 2019 increased by 544% compared to the number of patents in 1990–1999.

The advances mentioned above in sustainable innovation are certainly the result of policy and practice. Thanks to successful steps in technological progress and improvements in operations and infrastructure, 50% less CO_2 is produced per flight today compared to 1990. Aviation fuel use is responsible for about 2–3% of global CO_2 emissions. Under the assumption that it will continue to be a rapidly growing sector (ignoring the impact of the global crisis), aviation will continue to be one of the potential causes of greenhouse gas emissions.

For this reason, "creating new environmental policy regulations, continuing technological research and development at an increasing rate, coordinating these activities in cooperation with sectoral stakeholders on a global scale" is a set of activities that must be continued uninterrupted. By acting within this holistic framework, the aviation industry can continue to strengthen itself and support the sustainability of aviation and therefore sustainable economic development.

References

Abbas A, Schrauf G, Valero E (2011) Aerodynamic technologies for more effective, environmentally friendly air transport system: the KATnet strategy, book chp. In: Knörzer D, Szodruch J (eds) Innovation for sustainable aviation in a global environment, IOS Press BV, pp 82–88

AEF (2020) https://www.aef.org.uk/about/. 20 November 2020

Anser MK, Ali S, Khan MA, Nassani AA, Askar SE, Zaman K, Abro MMQ, Akbey F (2014) A literature review on the relationship between R&D, innovation and development: theoretical brief. Maliye Dergisi 166:1–16

Anser MK, Ali S, Khan MA, Nassani AA, Askar SE, Zaman K, Abro MMQ, Kabbani A (2021) The role of sustainable technological innovations in the relationship between freight pricing and environmental degradation: Evidence from a panel of 39 R & D economies. Atmosfera. https://doi.org/10.20937/ATM.52922

ATAG (2017) Flying in formation air transport and the sustainable development goals, Switzerland

ATAG (2020a) Global fact sheet, ABBB. https://www.atag.org/our-publications/latest-publications.html. 20 April 2021

ATAG (2020b) Aviation benefits beyond borders. www.aviationbenefits.org. 20 April 2021

ATAG (2021a). CORSIA. https://aviationbenefits.org/environmental-efficiency/climate-action/offsetting-emissions-corsia/corsia/corsia-explained/. 20 January 2021a

ATAG (2021b) Environmental efficiency. https://aviationbenefits.org/environmental-efficiency/

Bozoğlan R (2005) Sürdürülebilir Gelişme Düşüncesinin Tarihsel Arka Planı. Sosyal Siyaset Konferansları Dergisi 50:1011–1028

Capoccitti S, Khare A, Mildenberger U (2010) Aviation industry—mitigating climate change impacts through technology and policy. J Technol Manag Innov 5(2):66–75

CATE (2021) Centre for aviation, transport and the environment. https://cate.mmu.ac.uk/. 20 January 2021

Daley B (2008) Is air transport an effective tool for sustainable development? Sustain Dev 17:210–219. https://doi.org/10.1002/sd.383

Desvallées P (2012) CORAC: the concerted approach of french national council for civil aviation. book chp. In: Knörzer D, Szodruch J (eds) Innovation for sustainable aviation in a global environment, IOS Press BV, pp 415–417

Dincer I (2000) Renewable energy and sustainable development: a crucial review. Renew Sustain Energy Rev 4(2):157–175

EASA, EEA, EUROCONTROL (2019). European Aviation Environmental Report 2019. https://ec.europa.eu/transport/sites/transport/files/2019-aviation-environmental-report.pdf. 19 November 2020

EPO (2021) https://worldwide.espacenet.com. 1 June 2021

Eşkinat R (2016) Binyıl Kalkınma Hedeflerinden Sürdürülebilir Kalkınma Hedeflerine. Anadolu Universitesi Hukuk Fakültesi Dergisi 2(3):267–282

EU (2021) EU Emissions Trading System (EU ETS). https://ec.europa.eu/clima/policies/ets_en. 20 January 2021

Fenley CA, Machado WV, Fernandes E (2007) Air transport and sustainability: lessons from Amazonas. Appl Geogr 27:63–77

Forsyth P (2011) Environmental and financial sustainability of air transport: are they incompatible? J Air Transp Manag 17:27–32

GARDN (2021) A quick look at the organization behind green aviation. https://gardn.org/accueil/who-we-are/about/. 20 January 2021

Goetz AR, Graham B (2004) Air transport globalization, liberalization and sustainability: post-2001 policy dynamics in the United States and Europe. J Transp Geogr 12:265–276

Grimme W (2008) Measuring the long-term sustainability of air transport—an assessment of the global airline fleet and its CO_2 emissions up to the year 2050. German Aerospace Center (DLR), Air Transport and Airport Research. https://www.dlr.de/fw/Portaldata/42/Resources/dokumente/pdf_dokumente/MEASURING_THE_LONG_TERM_SUSTAINABILITY_OF_AIR_TRANSPORT.pdf. 20 November 2020

Gümüş Akar P, Seçilmiş N (2019) Enerji ve Çevre İlişkisi Bağlamında Kavramsal Boyut. book chp. In: Manga M, Ballı E (eds) Enerji ve Çevre Ekonomisi. Ekin Pub., ISBN: 978–605–327–927–3, 1–36

Haščič I, Migotto M (2015). Measuring environmental innovation using patent data, OECD Environment Working Papers, No. 89, OECD Publishing, Paris. https://doi.org/10.1787/5js009kf48xw-en

Hodgkinson D, Coram A, Garner R (2007) Strategies for airlines on aircraft emissions and climate change: sustainable, long-term solutions, The Hodgkinson Group, Working Paper (2)

IATA (2019) Taxes & the environment. https://www.iata.org/contentassets/72cc53f0e74e47449f62b0b0de583495/fact-sheet-greentaxation.pdf. 20 January 2021

IATA (2007) Aviation sets a benchmark on environmental performance for other industries to follow. https://www.iata.org/en/pressroom/2007-releases/2007-10-18-01/ 20 November 2020

ICAO (2016a) Aviation and the environment, published in Montréal, Canada. 20 November 2020

ICAO (2019) 2019 environmental report aviation and environment. https://www.icao.int/environmental-protection/Documents/ICAO-ENV-Report2019-F1-WEB%20(1).pdf. 1 December 2020

ICAO (2021a) World aviation and the world economy. https://www.icao.int/sustainability/pages/facts-figures_worldeconomydata.aspx, 20 January 2021a

ICAO (2021b) ICAO'S policies on environmental levies. https://www.icao.int/sustainability/Pages/eap-im-levies.aspx. 20 January 2021b

ICAO (2021c) Committee on aviation environmental protection (CAEP). https://www.icao.int/environmental-protection/pages/caep.aspx. 20 January 2021c

ICAO (2021d) Environmental protection. https://www.icao.int/environmental-protection/Pages/default.aspx. 20 January 2021d

ICSA (2021). About ICSA. https://www.icsa-aviation.org/icsa-aviation-about-us/. 20 January 2021

IPCC (1999) Aviation and the global atmosphere summary for policymakers. https://www.ipcc.ch/site/assets/uploads/2018/03/av-en-1.pdf. 20 October 2020

Janic M (2002) Methodology for assessing sustainability of an air transport system. J Air Transp 7(2):115–152

Janic M (2004) An application of the methodology for assessment of the sustainability of the air transport system. J Air Transp 9(2):40–82

Kaszewski AL, Sheate WR (2004) Enhancing the sustainability of airport developments. Sustain Dev 12:183–199

Kaypak Ş (2011) Küreselleşme Sürecinde Sürdürülebilir Bir Kalkınma İçin Sürdürülebilir Bir Çevre. KMÜ Sosyal Ve Ekonomik Araştırmalar Dergisi 13(20):19–33

Kılıç M, Uyar A, Karaman AS (2019) What impacts sustainability reporting in the global aviation industry? Inst Persp Transp Policy 79:54–65

Kılıç S (2012) Sürdürülebilir Kalkınma Anlayışının Ekonomik Boyutuna Ekolojik Bir Yaklaşım. İ.Ü. Siyasal Bilgiler Fakültesi Dergisi 47:201–226

Lee J, Mo J (2011) Analysis of technological innovation and environmental performance improvement in aviation sector. Int J Environ Res Public Health 8:3777–3795. https://doi.org/10.3390/ijerph8093777

Lee JJ, Lukachko SP, Waitz IA, Schafer A (2001) Historical and future trends in aircraft performance, cost, and emissions. Annu Rev Energy Environ 26:167–200

Longshurst J, Gibbs DC, Raper DW, Conlan DE (1996) Towards sustainable airport development. Environmentalist 16:197–202

MFA (2021) Sustainable development. Republic of Turkey Ministry of Foreign Affairs. https://www.mfa.gov.tr/surdurulebilir-kalkinma.tr.mfa. 21 January 2021

MIT LAE (2020). https://lae.mit.edu/. 21 September 2020

NISA (2021) Aviation electrofuel, the green fuel of the future. https://cleancluster.dk/NISA/. 2 October 2020

NSTC (1995) Goals for a national partnership in aeronautics research and technology. https://clintonwhitehouse2.archives.gov/WH/EOP/OSTP/html/aero/cv-ind.html. 13 September 2020

NSTC (1999) National research and development plan for aviation safety, security, efficiency and environmental compatibility. https://www.hsdl.org. 13 September 2020

OECD (2021) OECD Data. https://data.oecd.org/. 21 January 2021

Palmer WJ (2015) Will sustainability fly? aviation fuel options in a low-carbon world. Published by Ashgate, England

Pastowski, A. (2003). Climate policy for civil aviation: actors, policy instruments and the potential for emissions reductions. book chp. in Towards Sustainable Aviation. Edited by Paul Upham, Janet Maughan, David Raper and Callum Thomas,179–195.

Reynolds-Feighan A (2005) Institutional issues in transatlantic aviation, book chapter in barriers to sustainable transport: institutions, regulation and sustainability, Edited by Rietveld P, Stough RR, by Spon Press, pp 143–157

Rogner HH, Popescu A (2000) An introduction to energy. UNDP, world energy assessment. energy and the challenge of sustainability. New York:UNDP/UNDESA/WEC, 27–37

Seçilmiş N (2020) Ar-Ge, Yenilikçi Yapı ve İktisadi Kalkınma. chanpter in Kalkınma Yazıları Teori ve Uygulama. IKSAD Pub. ISBN: 978–625–7954–82–2, 129–150

Seçilmiş N, ve Konu, A. (2019) An empirical study on R&D incentives and innovation relation in OECD countries. Kahramanmaraş Sütçü İmam Üniversitesi Sosyal Bilimler Dergisi. 16(2):686–702. https://doi.org/10.33437/ksusbd.533175

Sipahi EB (2010) Collective solutions for global environmental problems and governance. Selçuk Üniversitesi Sosyal Bilimler Enstitüsü Dergisi 24:331–344

Štreimikiene D, Mikalauskiene A, Ciegis R (2020) Sustainable development, leadership, and innovations. CRC Press, New York

T&E (1998) Sustainable aviation: the need for a European environmental aviation charge, European Federation for Transport and Environment, T&E 98/1, Brussels

Tıraş HH (2012) Sürdürülebilir Kalkınma ve Çevre: Teorik Bir İnceleme. Kahramanmaraş Sütçü İmam Üniversitesi İİBF Dergisi 2:57–73

UN (1987) Our Common Future, Chapter 2: Towards Sustainable Development. UN Documents. http://www.un-documents.net/ocf-02.htm#III.1. 4 October 2020

UN (2021) Support sustainable development and climate action. https://www.un.org/en/our-work/support-sustainable-development-and-climate-action. 4 October 2020

UNDP (2021) What are the sustainable development goals. https://www.tr.undp.org/content/turkey/tr/home/sustainable-development-goals.html. 1 January 2021

UNFCCC (2017) technological innovation for the Paris agreement, Implementing nationally determined contributions, national adaptation plans and mid-century strategies, Bonn, Germany. https://unfccc.int/ttclear/misc_/StaticFiles/gnwoerk_static/brief10/8c3ce94c20144fd5a8b0c06fefff6633/57440a5fa1244fd8b8cd13eb4413b4f6.pdf. 3 November 2020

Upham P (2003) Introduction: perspectives on sustainability and aviation, book chp. In: Upham P, Maughan J, Raper D, Thomas C (eds) Towards sustainable aviation, pp. 3–18.

Walker S, Cook M (2009) The contested concept of sustainable aviation. Sustainable Developmen 17:378–390. https://doi.org/10.1002/sd.400

Nisa Seçilmiş is an Asst. Prof. in the Department of Aviation Management at the Faculty of Aviation and Space Sciences, Gaziantep University. She received her BSc. and MSc. in Economics at Gaziantep University. She holds a Ph.D. in Economics at Istanbul University, Institute of Social Sciences in 2012. She worked on the relationship between R&D activities and economic growth in her doctoral thesis. Her research interests include economic development, economic growth, environmental sustainability, innovation, R&D and aviation economics.

Part III
Information System and Risk Management in Airlines

Chapter 7
Purchase Intention Toward Green Airlines and Willingness to Pay More: Extending the Theory of Planned Behavior

Şahap Akan⑩, Emircan Özdemir⑩, and Mahmut Bakır⑩

Abstract This research attempts to contribute to understanding the formation of pro-environmental behaviors in the airline industry. To this end, the Theory of Planned Behavior (TPB) was extended by adding environmental consciousness to predict the factors that influence consumers' green airline purchase intention. Another objective of the research was to examine the effect of green purchase intention on willingness to pay more. A web-based questionnaire was developed, and 150 usable questionnaires were collected from the participants. To examine the hypothesized relationships, the partial least squares path modeling (PLS-PM) method was employed by using the "plspm" package in R (a statistical software). The results indicated that the elements of extended TPB (subjective norm, perceived behavioral control, attitude toward green airline, and environmental consciousness) exert a positive effect on green purchase intention. In addition, green purchase intention positively influences willingness to pay more. At the end of the chapter, theoretical contributions and avenues for future research were also discussed.

Keywords Theory of planned behavior (TPB) · Green airline · Environmental consciousness · Willingness to pay more · Partial least squares path modeling (PLS-PM)

Ş. Akan (✉)
Department of Civil Aviation Management, Anadolu University, Eskişehir, Turkey
e-mail: sakan@anadolu.edu.tr

E. Özdemir
Department of Aviation Management, Eskişehir Technical University, Eskişehir, Turkey
e-mail: emircanozdemir@eskisehir.edu.tr

M. Bakır
Department of Aviation Management, Samsun University, Samsun, Turkey
e-mail: mahmut.bakir@samsun.edu.tr

K. Kiracı and K. T. Çalıyurt (eds.), *Corporate Governance, Sustainability, and Information Systems in the Aviation Sector, Volume I*, Accounting, Finance, Sustainability, Governance & Fraud: Theory and Application, https://doi.org/10.1007/978-981-16-9276-5_7

7.1　Introduction

Ecological problems such as air pollution and global warming are among the most important problems in the world (Houghton 2005). These problems, whose effects have been felt more in recent years, are critical for the future of humankind (Tal 2009). Governments and non-governmental organizations have made efforts to increase society's awareness to overcome these destructive issues. Pressures from media and non-governmental organizations have also made ecological issues more visible, leading to the increased environmental awareness of consumers (Çabuk et al. 2019). Thus, sensitive consumers started to express their concerns about environmental problems and their purchasing behavior started to change on the axis of environmental sensitivity. As such, consumers are more environmentally conscious today and prefer to purchase "green" products (Hwang and Lyu 2019). On the other hand, companies are also interested in novel solutions both to contribute to the attenuation of environmental problems and to gain a competitive advantage by meeting the changing customer needs (Yarimoglu and Gunay 2020). Accordingly, various sectors are making their production processes and operations environmentally oriented (Verma and Chandra 2018), and the aviation industry is a clear example.

Currently, the aviation industry accounts for 3.5% of cumulative global warming and 2% of total carbon emissions (Fleming and Lépinay 2019; Lee et al. 2020). Moreover, the International Civil Aviation Organization (ICAO) predicts that aviation emissions will triple by 2050 (Fleming and Lépinay 2019). In this case, the aviation industry attempts to reduce the damage it causes to the environment by applying technological developments and a number of policies. On the technological side, more efficient aircraft designs (Airbus A350 XWB and Boeing 787 Dreamliner) and sustainable aviation fuels initiatives stand out (Anwar 2020). The policy side, on the other hand, implies consumer awareness practices such as carbon emission calculation per passenger (ICAO 2020) and carbon tax regulations for airlines to reduce less efficient aircraft use (Qiu et al. 2020). While all of these efforts are meaningful, it is also worth examining how airline consumers perceive these initiatives and respond to environmental problems. It is increasingly accepted that environmental problems cannot be solved only with technical and economic actions and that consumer attitudes and behaviors should also be taken into account (Chen et al. 2011). In this case, it is very important to understand the perceptions and attitudes of customers toward green airlines.

In the current literature, pro-environmental behaviors and green actions have frequently been discussed (Hagmann et al. 2015; Hwang and Choi 2017; Lu and Shon 2012; Mayer et al. 2012; Niu et al. 2016). Accordingly, research suggests that the most common theory for investigating pro-environmental behaviors at an individual level is the theory of planned behavior (TPB) (Ajzen 1985). The TPB aims to predict behavioral intention for any action based on attitude toward the behavior, subjective norm, and perceived behavioral control variables (Ajzen 1991). The TPB has been used to explain pro-environmental behaviors in many domains including green hotels, eco-friendly restaurants, and green products (Chen and Tung 2014;

Jang et al. 2015; Yadav and Pathak 2016). However, very few studies have investigated airline customers' pro-environmental behaviors (Davison et al. 2014; Han et al. 2018). Moreover, to the authors' knowledge, the formation of green purchase intention in the context of the Turkish airline industry is still unvisited. Therefore, this chapter aims to investigate the purchase intention of consumers toward green airlines by using the TPB. Basic TPB explains fundamental behaviors well, but it may be inadequate depending on the research context. Therefore, in order to customize the TPB and increase the prediction power, the TPB is extended by incorporating environmental consciousness in this chapter. This research investigates the effect of the extended theory of planned behavior elements (subjective norms, perceived behavioral control, attitude toward green airlines, and environmental consciousness) on green purchase intention and ultimately the effect of purchase intention on willingness to pay more. This chapter contributes to the existing literature in a few ways. First, it contributes to the understanding of the pro-environmental purchasing behavior of airline consumers. In doing so, it also highlights its impact on willingness to pay more. Second, it successfully extends the TPB within the airline industry context using environmental consciousness. On the practical level, it helps airlines to develop new environmental strategies and increase their long-term competitiveness.

The remainder of the present chapter will first introduce the extended theory of planned behavior, green airlines, and pro-environmental consumer behavior, followed by the proposed research framework along with proposed hypotheses. Next is a presentation of the measurement and data collection process within the research methodology. The results section includes the sample characteristics and testing of the measurement and structural models. In the last section, research findings and implications are discussed, pointing out avenues for future research.

7.2 Literature Review

7.2.1 Extended Theory of Planned Behavior

TPB is one of the most important social psychology theories used to predict the behavior of individuals under their control (Ajzen 1991). TPB was developed as a derivative of the Theory of Reasoned Action (TRA). TRA explains that engaging in a particular behavior depends on behavioral intention, which is the outcome of pre-existing attitudes and subjective norms (Ajzen and Fishbein 1980). TPB has evolved into a new model by incorporating perceived behavioral control into the TRA to comprehensively predict intention toward particular behaviors. In other words, the behavioral intention or behavior of individuals is not independent of the perceived ease or difficulty of performing the behavior. In this vein, individual motivations for particular behavior can be predicted by three variables; an individuals' attitude toward the behavior, subjective norms, and perceived behavioral control. Note that all three variables affect intention, thus shaping behavior (Ajzen 1985, 1991). Figure 7.1

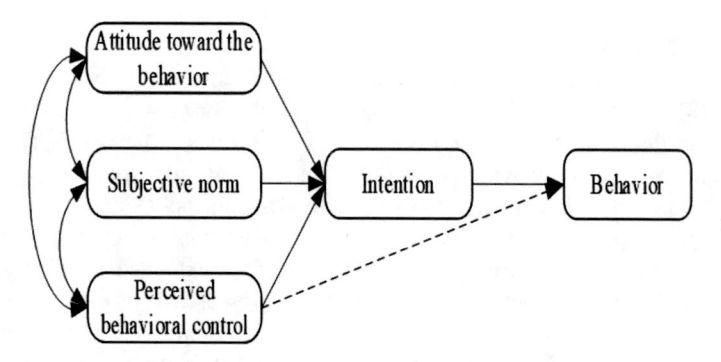

Fig. 7.1 The theory of planned behavior

shows the relationships between these components in the TPB model (Ajzen 1991).

In the model, attitude toward behavior refers to "the degree to which the person has a favorable or unfavorable evaluation of the behavior in question". Subjective norm is a social factor and represents the perceived social pressure to perform or not to perform the behavior. Lastly, perceived behavioral control, as another antecedent of intention, refers to the perceived ease or difficulty of performing the behavior, and it is assumed to reflect anticipated impediments and obstacles with past experiences (Ajzen and Driver 1992). Intention is an indicator of how willing and planning people are to strive to achieve the behavior. As a general rule, the stronger the intention to act, the more likely the behavior is to occur (Ajzen 1991). Behavior is the final outcome of the TPB and implies that individuals take action as a conscious and observable response (Ajzen 1985). While discussing the adequacy of the theory, Ajzen (1991) stated that if it provides meaningful insights, some additional variables can be added and the theory is open to expansion. That is, when examining a specific behavior, potential predictors having additional explanation power can be incorporated into the model (Ajzen 1991). A similar view suggests that more predictors should be added to the TPB to increase explanatory ability (Hagger et al. 2002). In this respect, the TPB has also been extended by adding various variables in numerous studies (Gao et al. 2017; Yang 2012; Yarimoglu and Gunay 2020).

In this research, we incorporated environmental consciousness into the TPB. Environmental consciousness reflects the level of perceptions and concerns toward the environment and expresses attitudes toward reducing environmental problems (Jain et al. 2020). In recent years, more and more consumers are concerned about the environment and prefer to buy alternatives that are less harmful to the environment (Huang et al. 2014). Supporting this argument, Schwartz's (1977) norm activation model (NAM) also outlines that individuals can waive their interests for collective benefits and drive pro-environmental behaviors to protect the environment (Jain et al. 2020). In this respect, we suggest that the NAM provides the underlying rationale for the inclusion of environmental consciousness into the TPB.

7.2.2 Green Airlines and Pro-environmental Consumer Behavior

During the past decades, environmental pollution and global warming have become the primary concern of humanity (Tang and Lam 2017). In parallel with this, academics and practitioners have started to focus on consumers' attitudes toward environmental aspects and consumer environmentalism since the 2000s, and a body of literature has been created in this area (Mayer et al. 2012). The current literature suggests that consumers induced by environmental concerns develop a positive attitude toward environmentally friendly products and that they perceive these products with higher quality and intend to pay more for (Cheah and Phau 2011; Mayer et al. 2012). As a result, most sectors have started to prioritize the issue of sustainability in all stages of production and service delivery.

Today, more environmentally friendly practices have started to come to the fore in the airline industry as in many other industries. In this direction, the concept of green airline was introduced. Green airlines are basically eco-friendly airlines and aim to reduce their carbon footprint (Sarkar 2012). Thus, green airlines focus on supporting sustainable economic development without sacrificing the local and global environment. Although the concept of "green" sounds strange for the aviation industry, the number of studies on green airline practices has increased considerably in recent years (Hagmann et al. 2015). For instance, Abdullah et al. (Abdullah et al. 2016) specified the key criteria of the environmental friendly airlines: operational activities (fuel management program, aircraft weight reduction incentives, flight planning, greening onboard ground operation), corporate environment management practices (Corporate Social Responsibility (CSR), Environment Management System (EMS), and corporate policies/strategic planning (Fleet Renewal, Commute Options Program). Alkhatip and Migdadi (Alkhatib and Migdadi 2020) listed green airline practices under the titles of fuel-saving, energy saving, waste management and recycling, and water management. Despite such managerial efforts, it is observed that the perceptions are not sufficiently understood from the consumer perspective.

Evaluating the existing literature, it is observed that there are limited studies on consumer perceptions about green airlines. In these studies, Benady (2007) revealed that the environmentally friendly image of the airline industry is quite low compared to other sectors. Mayer et al. (2012) stated that there is no perceptual difference between traditional airlines and low-cost carriers (LCCs) in terms of being environmentally friendly. As per Hagmann et al. (2015), airline green image affects passengers' airline choices during booking. In the study investigating willingness to pay for carbon-offsets, Lu and Shon (2012) revealed that passengers' perceptions related to the carbon-offset plan determines willingness to pay more. Hwang and Choi (2017) studied the psychological benefits of green brands and found that the components of warm glow, self-expressive benefits, and nature experiences strengthen the image of eco-friendly airlines. Niu et al. (2016) found that customers tend to choose green airlines and young passengers are more concerned about environmental protection. In summary, airlines have a low environmentally friendly image, but passengers tend to buy from green airlines. In addition, the body of literature on green airlines has generally been centered on developed countries. Therefore, understanding the perceptions of customers in an emerging country such as Turkey is anticipated to offer fresh insights.

7.2.3 Conceptual Model and Hypotheses

The underpinning theory of this research is TPB. Moreover, environmental consciousness has been incorporated into the TPB to customize the research context. The research framework in which willingness to pay more is included as the final outcome is shown in Fig. 7.2.

Subjective norm suggests that people who are important to an individual (e.g., friends, colleagues, family) influence the decision-making process (Verma and Chandra 2018). Coleman et al. (2011) demonstrated that peer pressure has a strong influence on young participants' green consumption. Chen and Tung (2014) stated that individuals' intention to visit green hotels may be higher if subjective norms support them. Likewise, Kalafatis et al. (1999) found that subjective norms are significantly correlated with green product purchase intention. Thus, it can be inferred that other people's opinions shape the purchase intention of consumers (Ha and Janda 2012). Based on the above literature, the following hypothesis was established:

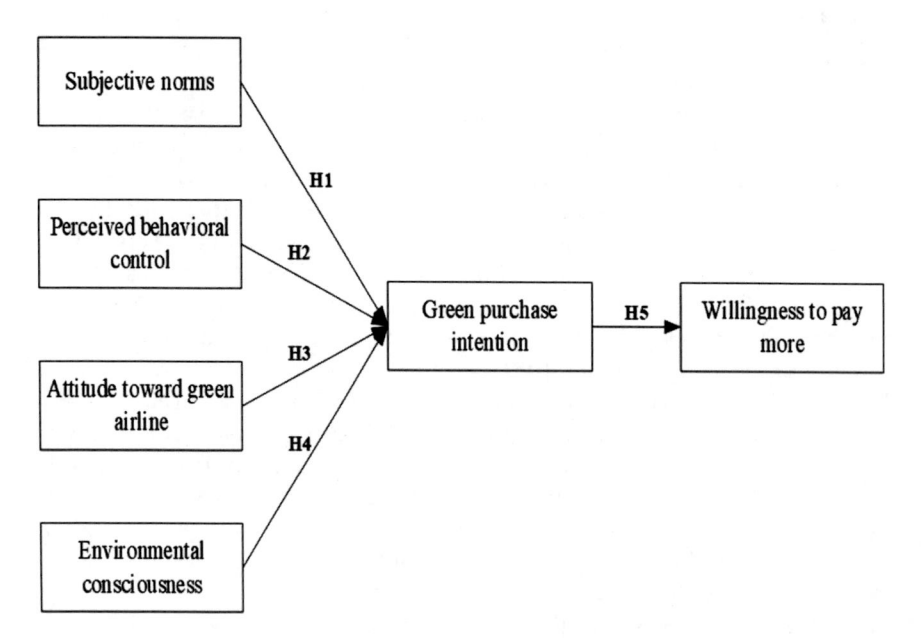

Fig. 7.2 Proposed research framework

H1: Subjective norm positively affects green purchase intention.

Perceived behavioral control represents an individual's belief in the ability to perform a particular behavior (Ajzen 1985). In general, individuals who think they can access resources such as money, time, etc., to perform a behavior are considered to have a high degree of perceived behavioral control (Ajzen 1991; Pan and Truong 2018). In their studies examining green product purchase behavior, Karatu and Mat (2015) revealed that perceived behavioral control is directly related to intention to purchase green products. Another study found that perceived behavioral control significantly predicts purchase intention of green skincare items (Hsu et al. 2017). Bong Ko and Jin (2017) noted that as consumers gain perceived behavioral control, they are more likely to buy green apparel products. Zhou et al. (2013) explored a stronger effect of perceived behavioral control on the intention to buy organic foods compared to other predictors. Thus, we formulated the following hypothesis:

H2: Perceived behavioral control positively affects green purchase intention.

The higher the positive attitude of individuals toward the given behavior, the higher the probability of performing the behavior (Ajzen 1991; Verma and Chandra 2018). In this regard, Aulina and Yuliati (2017) argued that positive attitudes toward the green brand increase green product purchase intention. Han et al. (2009) revealed the positive relationship between attitude toward green hotels and components of behavioral intention. It was also found that attitude toward behavior is the most explanatory variable of the intention to visit within the context of the green hotel selection (Han et al. 2010). In the context of green airlines, the attitude is operationalized as the value and weight that a passenger attributes to green airlines. According to the above arguments, the following hypothesis was formulated:

H3: Attitude toward green airlines positively affects green purchase intention.

There has been a dramatic increase in environmental awareness around the world over the past two decades. Consumers are increasingly concerned about environmental problems such as climate change and ozone depletion. This increase in environmental awareness has also had a profound effect on consumer demands and behavior (Chen et al. 2011). As per Mishal et al. (2017), environmentally conscious customers are more likely to patronize green purchases. Moreover, Iyer et al. (2016) noted that environmental consciousness is positively correlated with purchase intentions toward green products. Leaniz et al. (2018) found that customers with high environmental consciousness are more likely to prefer green hotels. Considering these findings, it can be hypothesized that:

H4: Environmental consciousness positively affects green purchase intention.

Within the scope of TPB, besides the intention of individuals, potential behaviors are also analyzed. One of these potential behaviors is willingness to pay more, which occurs depending on the strong intention (Schniederjans and Starkey 2014). In this chapter, willingness pay more refers to the willingness of a passenger to pay the price difference if they prefer to buy green airlines instead of non-green airlines (Khoiriyah and Toro 2018). Schniederjans and Starkey (2014) examined purchase intention for green transportation products and highlighted the relationship between green purchase intention and willingness to pay more. Focusing on brand crises, Küçük and Toklu (2016) stated that green purchase intention positively affects willingness to pay more. In the study on renewable energies, Irfan et al. (2020) reported a significant association between consumers' intention and willingness to pay more. Based on the above literature, the last hypothesis was devised:

H5: Green purchase intention positively affects willingness to pay more.

7.3 Research Methodology

Methodologically, the present chapter has adopted a cross-sectional research design through a questionnaire technique. The questionnaire items were taken from the scales used and validated in previous studies and were slightly tailored to fit green airlines concept. The items pertaining to subjective norm (e.g., "Many people I care about support me that I travel with a green airline") were borrowed from Jang et al. (2015). Likewise, items measuring perceived behavioral control (e.g., "I have sufficient resources, time and opportunity to travel with a green airline") were taken from scales used in the past literature (Pan and Truong 2018; Verma and Chandra 2018). In order to measure attitude toward green airline, the items (e.g., "I have positive perception toward green airline") in Buaphiban and Truong (2017) were tailored to green airlines. For measuring environmental consciousness, a three-item scale (e.g., "The balance of nature is very delicate and easily can be disturbed") from Prakash et al. (2018) were used. Green purchase intention items were modified from Screen et al. (2018) and Wang et al. (2018) by using three items (e.g., "I am willing to choose green airlines for my travels"). Lastly, to measure willingness to pay more for green airlines, we borrowed three items (e.g., "It is acceptable to me to pay more to travel with a green airline") from Yarimoglu and Gunay (2020). Each item was also measured on a five-point Likert scale ranging from 1 (strongly disagree) to 5 (strongly agree). The first part of the questionnaire form was dedicated to the demographics of the participants, whereas the questions related to the scales were included in the second part. Since the scales are originally in English, the questionnaire was translated into Turkish. In doing so, the back-translation technique was applied to eliminate any semantic discrepancy and provide accuracy of items. The initial version of the questionnaire was finalized after a few corrections as a result of a small-scale pilot study.

The target population of this research was drawn on individuals who are airline users or eligible to use airline transportation. While a convenience sampling technique was used in the data collection process, a web-based questionnaire was developed. In terms of survey design, potential respondents were asked if they were willing to participate in the survey. Following the positive responses, the concept of green airline and green aviation practices were explained at the beginning of the survey, and participants were asked to answer the questionnaire accordingly. The questionnaire was distributed over social media groups and communities to reach the participants. The data collection process took place in October 2020, yielding a final sample size of 150. This sample size was deemed sufficient to perform the analysis since the minimum sample size required to obtain 0.95 power at the level of $f^2 = 0.15$ was found to be 138 through a priori power analysis via G*Power software (Faul et al. 2007).

7.4 Results

This section presents the analysis results regarding the hypothesized relationships. To test the proposed hypotheses, we employed the partial least squares path modeling (PLS-PM) technique. PLS-PM is a variance-based approach that allows analysis of nonlinear structural models as well as linear models and embraces many additional features such as importance–performance analysis (Benitez et al. 2020). The choice of PLS-PM is justifiable since the present research aims to examine the relationships in the proposed framework from a predictive perspective and thus expands the TPB (Hair et al. 2019). As such, "plspm" package in R (a statistical software) was employed (Sanchez et al. 2015). As applied by the previous literature (Ali et al. 2018), a two-stage analytical procedure was followed to examine the hypothesized relationships, namely; measurement model testing and structural model testing.

7.4.1 Sample Profiling

The demographic characteristics of the sample are depicted in Table 7.1. Accordingly, among 150 participants, 72 (48%) were male and 78 (52%) were female. Most of the participants were 26–33 years old (44%), followed by the second largest cohort of 18–25 years old (35.3%). Regarding the education level, most of the participants had university/college diploma (66%), then followed by participants possess a postgraduate degree (29.3%). The income level of the majority was over 6501 TL (29.3%). Moreover, it is worth noting that the sample was highly heterogeneous in terms of income level. Almost half of the respondents flew never or once a year (46%), while the smallest cohort represented those who traveled 5–9 times a year (8.7%).

7.4.2 Common Method Bias

Self-reported studies may cause respondent bias that leads to inflation of observed relationships. Prior to the PLS-PM analysis, Harman's single factor test was conducted to find out any evidence of common method bias in this research. As

Table 7.1 Summary of the sample demographics

Characteristic	Category	Frequency	(%)
Gender	Male	72	48.0
	Female	78	52.0
Age group	18–25	53	35.3
	26–33	66	44.0
	34–41	24	16.0
	Above 42	7	4.7
Education level	Primary school	1	0.7
	High school	6	4.0
	University/College	99	66.0
	Graduate school	44	29.3
Average income	Below 2300 TL	32	21.3
	2301 TL–3500 TL	21	14.0
	3501 TL–5000 TL	26	17.3
	5001 TL–6500 TL	27	18.0
	Above 6501 TL	44	29.3
Travel frequency	Once a year or never	69	46.0
	2–4 times a year	54	36.0
	5–9 times a year	13	8.7
	10 times or more a year	14	9.3

a result of the principal component analysis with the unrotated setting through "FactoMineR" package in R (Lê et al. 2008), common method bias did not seem to be a noticeable concern, as the first factor did not account for more than 50% of the total variance (Explained variance = 41.35%) (Tehseen et al. 2017). We should also note that as a procedural remedy, participants were informed that their confidentiality would be maintained and voluntary participation was asked.

7.4.3 Measurement Model Testing

In the PLS-PM analysis, the measurement model testing is performed first, which calls the confirmatory factor analysis. This step involves reliability analysis, which includes the internal consistency of constructs, and validity analysis called convergent and discriminant analysis. As suggested by Hair et al. (2019), we used the Dillon-Goldstein's rho (ρ) coefficient as a more precise reliability measure. Table 7.2 shows that all Dillon-Goldstein's rho values are above the critical threshold of 0.70, signifying that constructs are strongly reliable (Sanchez 2013; Tenenhaus et al. 2005). As another metric of reliability, we also assessed an eigen-analysis of the correlation matrix of each construct as the unique indicator of the "plspm" package.

Table 7.2 Reliability and validity measures

Construct	Item	Loadings	DG.rho	eig.1st	eig.2nd
Subjective norm (SNO)	SNO1	0.914	0.936	2.49	0.393
	SNO2	0.955			
	SNO3	0.861			
Perceived behavioral control (PBC)	PBC1	0.871	0.882	2.15	0.625
	PBC2	0.903			
	PBC3	0.753			
Attitude toward green airline (AGR)	AGR1	0.833	0.861	2.03	0.657
	AGR2	0.795			
	AGR3	0.807			
Environmental consciousness (ECO)	ECO1	0.642	0.822	1.82	0.665
	ECO2	0.871			
	ECO3	0.792			
Green purchase intention (GPI)	GPI1	0.848	0.918	2.37	0.422
	GPI2	0.906			
	GPI3	0.910			
Willingness to pay more (WPM)	WPM1	0.922	0.948	2.58	0.234
	WPM2	0.932			
	WPM3	0.927			

Based on Sanchez's (2013) guidelines, good reliability has been achieved not only because the value of the first eigenvalue is higher than 1 but also because the second eigenvalue is below 1.

The convergent validity is established when the average variance extracted and item loadings of constructs are greater than 0.50 and 0.60 (Hair et al. 2019; Isa et al. 2020). As seen in Table 7.3, all AVE values are responsible for more than 50% of the variance of the constructs (>0.50). Further, since factor loadings exceed the desired threshold, convergent validity is considered to be achieved.

In recent years, researchers have strongly recommended the use of heterotrait–monotrait ratio of correlations (HTMT), stating that current practices of discriminant validity are underperforming (Ali et al. 2018). The HTMT criterion based on inter-construct indicator correlations should be below the cut-off points of 0.85 or 0.90, thus ensuring that constructs do not have similar characteristics (Hair et al. 2019). As Table 7.3 shows, the highest HTMT value is 0.733, and discriminant validity has been achieved since it does not exceed the more conservative threshold of HTMT$_{.85}$ (Henseler 2020).

Table 7.3 AVE values and HTMT matrix results

Construct	AVE	HTMT results					
		1	2	3	4	5	6
1. SNO	0.829						
2. PBC	0.714	0.515					
3. AGR	0.659	0.540	0.588				
4. ECO	0.600	0.391	0.450	0.649			
5. GPI	0.789	0.730	0.666	0.733	0.629		
6. WPM	0.859	0.474	0.458	0.296	0.526	0.608	

7.4.4 Structural Model Testing

Since the previous section proved the validity and reliability of the constructs, this section further proceeded with testing the structural model. To this end, we first evaluated the in-sample prediction performance of the proposed model using the coefficient of determination(R^2). Sanchez (2013) divided R^2 into three categories, namely; low($R^2 < 0.30$), medium($0.30 < R^2 < 0.60$), and high ($R^2 > 0.60$) effects. Therefore, as seen in Table 7.4, exogenous variables are responsible for 61.3% of the variance of green purchase intention and a high magnitude has been observed($R^2 = 0.613$). Moreover, green purchase intention explains 29.4% of the variance of willingness to pay more, thus indicating medium magnitude($R^2 = 0.294$). Note that the variance explained above 0.20% in consumer research is generally considered significant (Isa et al. 2020). This research also employed the cross-validated redundancy procedure to test the out-of-sample predictive relevance of the model (Ali et al. 2018). Statistically, a strictly defined ideal redundancy index Q^2 does not exist, however, this value must be greater than zero (Hair et al. 2019; Sanchez 2013). From Table 7.4, redundancy values confirmed the predictive relevance of endogenous constructs($Q^2_{GPI} = 0.484 \, and \, Q^2_{WPM} = 0.253$).

Finally, we evaluated the overall model fit using the Goodness-of-fit (GoF) index, which is a measure of pseudo-goodness-of-fit, by Tenenhaus et al. (2005). If the GoF measure derived from the average extracted variance of the constructs and the observed average R^2 values is greater than 0.1, 0.25, and 0.36, the poor, medium,

Table 7.4 Prediction accuracy and GoF index

Construct	Type	R2	Redundancy	GoF
SNO	Exogenous	–	–	0.5802
PBC	Exogenous	–	–	
AGR	Exogenous	–	–	
ECO	Exogenous	–	–	
GPI	Endogenous	0.613	0.484	
WPM	Endogenous	0.294	0.253	

Table 7.5 Path coefficients and hypotheses testing results

Path	Std.Beta	Std.Error	t-value	perc.025	perc.975	Status
H1: SNO -> GPI	0.366	0.0632	5.971**	0.246	0.498	Support
H2: PBC -> GPI	0.197	0.0654	3.198*	0.069	0.322	Support
H3: AGR -> GPI	0.266	0.0747	4.000**	0.116	0.410	Support
H4: ECO -> GPI	0.196	0.0761	3.353*	0.055	0.352	Support
H5: GPI -> WPM	0.542	0.0647	7.856**	0.405	0.664	Support

Note perc.025 versus perc.975 denotes the lower and upper bounds of the 95% bootstrapped confidence interval; * p < 0.01, ** p < 0.001

and good fit is observed, respectively (Al-Ansi et al. 2019). The GoF value of 0.5802 given in Table 7.4 clearly indicates that a satisfactory model fit was observed.

The statistical significance of the hypothesized relationships was tested using the bootstrapping procedure running 5000 resamples. Table 7.5 exhibits the path coefficients and confidence intervals for the hypotheses presented. It should be noted that the non-significant effect occurs when the bootstrapped confidence intervals (perc.025–perc.975) includes the zero (Isa et al. 2020; Sanchez 2013). Accordingly, subjective norm has a positively significant effect on green purchase intention, thus supporting H1 ($\beta = 0.366$, CI: [perc.025 = 0.246, perc.975 = 0.498]). Next, perceived behavioral control also exert a positive effect on green purchase intention ($\beta = 0.197$, CI: [perc.025 = 0.069, perc.975 = 0.322]). Thus, H2 was supported. Third, a significant and positive effects of attitude toward green airline ($\beta = 0.266$, CI: [perc.025 = 0.116, perc.975 = 0.410]) and environmental consciousness ($\beta = 0.196$, CI: [perc.025 = 0.055, perc.975 = 0.352]) on green purchase were found. Thus, H4 and H5 were also supported. Finally, as we hypothesized, green purchase intention was found to positively predict willingness to pay more ($\beta = 0.542$, CI: [perc.025 = 0.405, perc. 975 = 0.664]), thus indicating that H5 found evidence of support.

7.5 Discussion and Implications

In this chapter, airline passengers' green purchase intention and willingness to pay more for green airlines are discussed by extending TPB. In research framework, TPB model previously proposed by Ajzen (1985) is extended by adding one more construct, namely environmental consciousness, in order to explain green purchase intention of passengers toward green airlines more precisely. Moreover, this study assists both scholars and practitioners for understanding premises of pro-environmental behaviors of passengers better by researching passengers' green purchase intention toward green airlines. Ultimate component in the conceptual framework, which is exploratory inherently, is willingness to pay more for green airlines.

In conceptual framework of extended TPB model, potential behavior of passengers is regarded as willingness to pay more. Findings of this study validate the competence of TPB model in context of pro-environmental behavior of airline passengers. Subjective norms, perceived behavioral control, attitude toward green airline and environmental consciousness considerably attribute to the development of green purchase intention as revealed in the results. Subjective norm is the most significant factor that affects green purchase intention of airline passengers positively ($\beta = 0.366$, $p < 0.001$). This finding is consistent with that of previous studies (Chen and Tung 2014; Coleman et al. 2011; Verma and Chandra 2018). The second most significant factor that effects green purchase intention is attitude toward green airline ($\beta = 0.266$, $p < 0.001$). Herein, assumed direction of attitude is in the positive way. As presented in the previous studies, findings supports the relation between attitude and green purchase intention (Han et al. 2010; Verma and Chandra 2018). Nextly, perceived behavioral control has effect on green purchase intention significantly and positively ($\beta = 0.197$, $p < 0.01$). Several previous studies show that perceived behavioral control predicts green behavioral intention and our findings are consistent with them (Bong Ko and Jin 2017; Hsu et al. 2017; Karatu and Mat 2015; Zhou et al. 2013). Ultimately environmental consciousness is the backbone of the proposed extended TPB model, and it significantly and positively enhances green purchase intention ($\beta = 0.196$, $p < 0.01$). This finding is also in line with that of past studies (Iyer et al. 2016; Leaniz et al. 2018). In general, this study has expanded the TPB model by including the environmental consciousness factor. This allow us to understand the relationships between subjective norms, perceived behavioral control, attitude toward green airlines, green purchase intention, and willingness to pay more in the context of pro-environmental behavior.

In this study, the green concept forms around environmental consciousness construct which contains environmental knowledge, environmental concern and environmental values (Corraliza and Berenguer 2000; Sharma and Bansal 2013). On the other hand, higher environmental consciousness intensifies the pro-environmental behavior (Kollmuss and Agyeman 2002; Lee and Holden 1999). As a result of the relationships revealed, it is seen that environmental consciousness of airline passengers develops pro-environmental behavior. In addition, the study has revealed the customer perceptions in the processes ranging from green purchasing intentions to willingness to pay more for green airlines. In this direction, it is envisaged that the chapter will contribute to green marketing research in the airline industry.

Considering the findings of the study in terms of practical application, the development of suitable strategies for the airlines passengers comes first. When airline managers implement environmental strategies by supporting passengers' environmental consciousness, they will also strengthen customers' pro-environmental behavior alongside outputs such as reducing carbon emissions and achieving energy savings. In this case, green marketing strategies can be applied (Polonsky 1994). Green marketing emerges as a viable strategy for airlines. Previous studies on airlines show that green marketing strategies positively affect consumer attitudes and behaviors, and are effective in driving passengers to green consumption. (Hwang and Lyu

2019; Wu et al. 2018). Environmental consciousness rises globally among consumers (Huang et al. 2014).

Considering that the passengers are getting more environmentally conscious day by day, it is obvious that green marketing strategies will increase the operational performance of airlines. The findings also show that airline marketers can target customers who have a positive attitude toward greener airline services. Consistent with TPB, attitude, subjective norms and perceived behavioral control emerged as important positive predictors of purchasing green products or services. According to findings, subjective norm has a stronger effect on green purchase intention than attitude. In light of this, marketers should effectively deliver messages that can strongly influence societal intention toward green airlines. Using persuasive messages like that will help airline marketers to form social norms toward green airlines. On the other hand, airline companies can act in line with a common goal with their passengers within the scope of green marketing strategies in order to strengthen their mutual relations with their passengers. For example, in this process, airlines may carry out activities to increase the environmental awareness of their passengers in the context of corporate social responsibility (F.-Y. Chen et al. 2012). In this way, it contributes to the movement of passengers with this awareness outside the aviation industry.

This will be effective in strengthening the bond between the passenger and the airline. In addition, advertising campaigns and sponsorship programs that airline marketers will carry out to encourage passengers who are concerned about the environment will ensure passengers to prefer green airlines. As long as passengers consider that using green airlines will help protect the environment, their tendency to opt for green airlines will increase. This also would provide an advantage to airlines for eco-friendly customer retention by tapping passengers' volitional, non-volitional, cognitive, and normative decision-making processes (Verma and Chandra 2018).

According to the results from the analysis, as the environmental consciousness of passengers increases, their intention to use green airlines will also increase. Sequentially, the behavior of paying more, which is also named willingness to pay more, for the product or service will develop in relation to the pro-environmental behavior.

However, previous studies presented that there are important discrepancies in terms of willingness to pay more. First, willingness to pay more can only remain in-principle, which is mainly formed by perceived behavioral control and environmental concern. Secondly, amount of price difference affects customers' willingness to pay more for green products or services (Liebe et al. 2011; Royne et al. 2011; Seip and Strand 1992). The proposed TPB model suggests that environmentally conscious passengers are willing to pay more for green airlines. Nevertheless, airline managers should determine the price sensitivity of passengers, which will ensure passengers not just to stay in principle for green airlines.

Finally, improving passengers' positive attitude toward green airlines will increase by their environmental awareness and orientation toward green airlines. At this point, it plays an important role for the government to develop education policies to increase environmental awareness of individuals. In addition, incentive programs to be implemented by the government for green airlines will pave the way for environmental protection practices in the airline industry.

This study is one of the first attempts in the airline industry to associate the purchase intention for green airlines with TPB. However, although the green purchase intentions of passengers are emerging, the number of airlines positioning themselves as green airlines is very few. The outputs of the TBP model reveal the green purchase intention of the airline passengers and their willingness to pay more for green airlines. On the other hand, the study empirically reveals that the expanded TBP model is valid by incorporating the environmental consciousness structure. This additional construct have increased the predictive ability and robustness of the proposed theoretical framework for measuring green purchase intention and willingness to pay more toward green airlines. In conclusion, it is revealed that as the environmental awareness of airline passengers increases, their green purchase intention and willingness to pay more for green airline services increase. Also we can state that passengers began to request airlines to be more environmentally friendly.

Despite the specific contributions of the study to current green airline literature, it has some limitations that will guide future studies. The main limitation of the study is that consumers' preferences for green airlines are examined according to the TPB model and there is no direct assessment of their actual behavior. Although previous studies have shown that behavioral intention models are robust in many behavioral areas, the actual behavior of consumers is always equivalent to their own attitudes (Verma and Chandra 2018). Recollecting actual behavioral data focusing green purchase would be needed for future studies. After that, when the education level of the participants of the study is examined, it is seen that 95.3% of them are graduated from university or higher degree. In research on environmental issues, if participants are well educated, they may tend to give the desired answer (Kaiser et al. 2008). Since the participants are mostly from young educated and adult educated segments, the generalizability of the study can be questioned. Therefore, larger samples should be considered to ensure higher representation power for future research. The predictive ability of the extended theoretical framework is 61.3% (i.e. $R^2 = 0.613$).

For the future studies, it is possible to improve the explanatory power by incorporating additional constructs from relevant literature. All participants of this study are from Turkey, so these findings don't give insights about other countries or cultures. Proposed model is still can be valid for the emerging markets like Turkey, but cultural differences should be discussed further. Lastly, this study was conducted only in the context of the airline industry. Green customer behavior of consumers in different industries may change. In future studies, green consumer behavior in other industries may be researched and compared. Finally, this study hopes that these suggestions will be useful for both academia and practice.

References

Abdullah M-A, Chew B-C, Hamid S-R (2016) Benchmarking key success factors for the future green airline industry. Procedia Soc Behav Sci 224:246–253. https://doi.org/10.1016/j.sbspro.2016.05.456

Ajzen I (1985) From intentions to actions: a theory of planned behavior. In: Kuhl J, Beckmann J (eds) Action control: from cognition to behavior. Springer, pp 11–39

Ajzen I (1991) The theory of planned behavior. Organ Behav Hum Decis Process 50(2):179–211. https://doi.org/10.1016/0749-5978(91)90020-T

Ajzen I, Driver BL (1992) Application of the theory of planned behavior to leisure choice. J Leis Res 24(3):207–224. https://doi.org/10.1080/00222216.1992.11969889

Ajzen I, Fishbein M (1980) Understanding attitudes and predicting social behaviour. Prentice-Hall

Al-Ansi A, Olya HGT, Han H (2019) Effect of general risk on trust, satisfaction, and recommendation intention for halal food. Int J Hosp Manag 83:210–219. https://doi.org/10.1016/j.ijhm.2018.10.017

Ali F, Rasoolimanesh SM, Sarstedt M, Ringle CM, Ryu K (2018) An assessment of the use of partial least squares structural equation modeling (PLS-SEM) in hospitality research. Int J Contemp Hosp Manag 30(1):514–538. https://doi.org/10.1108/IJCHM-10-2016-0568

Alkhatib SF, Migdadi YKAA (2020) A novel technique for evaluating and ranking green airlines: benchmarking-base comparison. Manag Environ Qual Int J. https://doi.org/10.1108/MEQ-04-2020-0065

Anwar N (2020) Eco travel: what are our current options for green aviation. https://www.cnbc.com/2020/02/28/eco-travel-what-are-our-current-options-for-green-aviation.html?

Aulina L, Yuliati E (2017) The effects of green brand positioning, green brand knowledge, and attitude towards green brand on green products purchase intention. Proceedings of the international conference on business and management research (ICBMR 2017), pp 548–557. https://doi.org/10.2991/icbmr-17.2017.50

Benady D (2007) Brands show true colours. Marketing Week. https://www.marketingweek.com/brands-show-true-colours/

Benitez J, Henseler J, Castillo A, Schuberth F (2020) How to perform and report an impactful analysis using partial least squares: Guidelines for confirmatory and explanatory IS research. Inform Manag 57(2). https://doi.org/10.1016/j.im.2019.05.003

Bong Ko S, Jin B (2017) Predictors of purchase intention toward green apparel products: a cross-cultural investigation in the USA and China. J Fash Mark Manag 21(1):70–87. https://doi.org/10.1108/JFMM-07-2014-0057

Buaphiban T, Truong D (2017) Evaluation of passengers' buying behaviors toward low cost carriers in Southeast Asia. J Air Transp Manag 59:124–133. https://doi.org/10.1016/j.jairtraman.2016.12.003

Çabuk S, Güreş N, İnan H, Arslan S (2019). Attitudes of passengers towards green airlines. J Yaşar Univ 14(55):237–250. https://doi.org/10.19168/jyasar.452297

Cheah I, Phau I (2011) Attitudes towards environmentally friendly products: the influence of ecoliteracy, interpersonal influence and value orientation. Mark Intell Plan 29(5):452–472. https://doi.org/10.1108/02634501111153674

Chen F-Y, Chang Y-H, Lin Y-H (2012) Customer perceptions of airline social responsibility and its effect on loyalty. J Air Transp Manag 20:49–51. https://doi.org/10.1016/j.jairtraman.2011.11.007

Chen F-Y, Hsu P-Y, Lin T-W (2011) Air travelers' environmental consciousness: a preliminary investigation in Taiwan. Int J Bus Manag 6(12):78–86. https://doi.org/10.5539/ijbm.v6n12p78

Chen MF, Tung PJ (2014) Developing an extended Theory of planned behavior model to predict consumers' intention to visit green hotels. Int J Hosp Manag 36:221–230. https://doi.org/10.1016/j.ijhm.2013.09.006

Coleman LJ, Bahnan N, Kelkar M, Curry N (2011) Walking the walk: how the theory of reasoned action explains adult and student intentions to go green. J App Bus Res 27(3):107–116. https://doi.org/10.19030/jabr.v27i3.4217

Corraliza JA, Berenguer J (2000) Environmental values, beliefs, and actions. Environ Behav 32(6):832–848. https://doi.org/10.1177/00139160021972829

Davison L, Littleford C, Ryley T (2014) Air travel attitudes and behaviours: the development of environment-based segments. J Air Transp Manag 36:13–22. https://doi.org/10.1016/j.jairtraman.2013.12.007

Faul F, Erdfelder E, Lang AG, Buchner A (2007) G*Power 3: a flexible statistical power analysis program for the social, behavioral, and biomedical sciences. Behav Res Methods 39(2):175–191. https://doi.org/10.3758/BF03193146

Fleming GG, Lépinay I de (2019) Environmental Trends in Aviation to 2050. In 2019 Environmental Report. https://www.icao.int/environmental-protection/pages/envrep2019.aspx

Gao L, Wang S, Li J, Li H (2017) Application of the extended theory of planned behavior to understand individual's energy saving behavior in workplaces. Resour Conserv Recycl 127:107–113. https://doi.org/10.1016/j.resconrec.2017.08.030

Ha HY, Janda S (2012) Predicting consumer intentions to purchase energy-efficient products. J Consum Mark 29(7):461–469. https://doi.org/10.1108/07363761211274974

Hagger MS, Chatzisarantis NLD, Biddle SJH (2002) A meta-analytic review of the theories of reasoned action and planned behavior in physical activity: predictive validity and the contribution of additional variables. J Sport Exerc Psychol 24(1):3–32. https://doi.org/10.1123/jsep.24.1.3

Hagmann C, Semeijn J, Vellenga DB (2015) Exploring the green image of airlines: passenger perceptions and airline choice. J Air Transp Manag 43:37–45. https://doi.org/10.1016/j.jairtraman.2015.01.003

Hair JF, Risher JJ, Sarstedt M, Ringle CM (2019) When to use and how to report the results of PLS-SEM. Eur Bus Rev 31(1):2–24. https://doi.org/10.1108/EBR-11-2018-0203

Han H, Hsu LT (Jane), Lee JS (2009) Empirical investigation of the roles of attitudes toward green behaviors, overall image, gender, and age in hotel customers' eco-friendly decision-making process. Int J Hosp Manag 28(4):519–528. https://doi.org/10.1016/j.ijhm.2009.02.004

Han H, Hsu LT, (Jane), Sheu C (2010) Application of the Theory of Planned Behavior to green hotel choice: testing the effect of environmental friendly activities. Tour Manage 31(3):325–334. https://doi.org/10.1016/j.tourman.2009.03.013

Han H, Yu J, Kim W (2018) Youth travelers and waste reduction behaviors while traveling to tourist destinations. J Travel Tour Mark 35(9):1119–1131. https://doi.org/10.1080/10548408.2018.1435335

Henseler J (2020) HTMT online calculator. http://www.henseler.com/htmt.html

Houghton J (2005) Global warming. Rep Prog Phys 68(6):1343–1403. https://doi.org/10.1088/0034-4885/68/6/R02

Hsu CL, Chang CY, Yansritakul C (2017) Exploring purchase intention of green skincare products using the theory of planned behavior: Testing the moderating effects of country of origin and price sensitivity. J Retail Consum Serv 34:145–152. https://doi.org/10.1016/j.jretconser.2016.10.006

Huang HC, Lin TH, Lai MC, Lin TL (2014) Environmental consciousness and green customer behavior: an examination of motivation crowding effect. Int J Hosp Manag 40:139–149. https://doi.org/10.1016/j.ijhm.2014.04.006

Hwang J, Choi JK (2017) An investigation of passengers' psychological benefits from green brands in an environmentally friendly airline context: the moderating role of gender. Sustainability (switzerland) 10(1):1–17. https://doi.org/10.3390/su10010080

Hwang J, Lyu SO (2019) Relationships among green image, consumer attitudes, desire, and customer citizenship behavior in the airline industry. Int J Sustain Transp, 1–11.https://doi.org/10.1080/15568318.2019.1573280

ICAO (2020) ICAO carbon emissions calculator. ICAO. https://www.icao.int/ENVIRONMENTAL-PROTECTION/CarbonOffset/Pages/default.aspx

Irfan M, Zhao ZY, Li H, Rehman A (2020) The influence of consumers' intention factors on willingness to pay for renewable energy: a structural equation modeling approach. Environ Sci Pollut Res 27(17):21747–21761. https://doi.org/10.1007/s11356-020-08592-9

Isa NF, Annuar SNS, Gisip IA, Lajuni N (2020) Factors influencing online purchase intention of millennials and gen Z consumers. J Appl Struct Equat Modeling 4(2):21–43. https://doi.org/10. 47263/jasem.4(2)03

Iyer P, Davari A, Paswan A (2016) Green products: Altruism, economics, price fairness and purchase intention. Soc Bus 6(1):39–64. https://doi.org/10.1362/204440816x14636485174912

Jain S, Singhal S, Jain NK, Bhaskar K (2020) Construction and demolition waste recycling: Investigating the role of theory of planned behavior, institutional pressures and environmental consciousness. J Cleaner Prod, 263. https://doi.org/10.1016/j.jclepro.2020.121405

Jang SY, Chung JY, Kim YG (2015) Effects of environmentally friendly perceptions on customers' intentions to visit environmentally friendly restaurants: an extended theory of planned behavior. Asia Pacific J Tourism Res 20(6):599–618. https://doi.org/10.1080/10941665.2014.923923

Kaiser FG, Schultz PW, Berenguer J, Corral-Verdugo V, Tankha G (2008) Extending planned environmentalism. Eur Psychol 13(4):288–297. https://doi.org/10.1027/1016-9040.13.4.288

Kalafatis SP, Pollard M, East R, Tsogas MH (1999) Green marketing and Ajzen's theory of planned behaviour: a cross-market examination. J Consum Mark 16(5):441–460. https://doi.org/10.1108/07363769910289550

Karatu VMH, Mat NKN (2015) The Mediating effects of green trust and perceived behavioral control on the direct determinants of intention to purchase green products in Nigeria. Mediterr J Soc Sci 6(4):256–265. https://doi.org/10.5901/mjss.2015.v6n4p256

Khoiriyah S, Toro MJS (2018) Attitude toward green product, willingness to pay and intention to purchase. Int J Bus Soc 19:620–628

Kollmuss A, Agyeman J (2002) Mind the Gap: Why do people act environmentally and what are the barriers to pro-environmental behavior? Environ Educ Res 8(3):239–260. https://doi.org/10. 1080/13504620220145401

Lê S, Josse J, Husson F (2008) FactoMineR: an R package for multivariate analysis. J Statist Softw 25(1):1–18. https://doi.org/10.18637/jss.v025.i01

de Leaniz PMG, Crespo ÁH, López RG (2018) Customer responses to environmentally certified hotels: the moderating effect of environmental consciousness on the formation of behavioral intentions. J Sustain Tour 26(7):1160–1177. https://doi.org/10.1080/09669582.2017.1349775

Lee DS, Fahey DW, Skowron A, Allen MR, Burkhardt U, Chen Q, Doherty SJ, Freeman S, Forster PM, Fuglestvedt J, Gettelman A, De León RR, Lim LL, Lund MT, Millar RJ, Owen B, Penner JE, Pitari G, Prather MJ, … Wilcox LJ (2020). The contribution of global aviation to anthropogenic climate forcing for 2000 to 2018. Atmosph Environ, 117834. https://doi.org/10.1016/j.atmosenv. 2020.117834

Lee JA, Holden SJS (1999) Understanding the determinants of environmentally conscious behavior. Psychol Mark 16(5):373–392. https://doi.org/10.1002/(SICI)1520-6793(199908)16:5%3c373:: AID-MAR1%3e3.0.CO;2-S

Liebe U, Preisendörfer P, Meyerhoff J (2011) To pay or not to pay: competing theories to explain individuals' willingness to pay for public environmental goods. Environ Behav 43(1):106–130. https://doi.org/10.1177/0013916509346229

Lu JL, Shon ZY (2012) Exploring airline passengers' willingness to pay for carbon offsets. Transp Res Part d Transp Environ 17(2):124–128. https://doi.org/10.1016/j.trd.2011.10.002

Mayer R, Ryley T, Gillingwater D (2012) Passenger perceptions of the green image associated with airlines. J Transp Geogr 22:179–186. https://doi.org/10.1016/j.jtrangeo.2012.01.007

Mishal A, Dubey R, Gupta OK, Luo Z (2017) Dynamics of environmental consciousness and green purchase behaviour: an empirical study. Int J Climate Change Strat Manag 9(5):682–706. https://doi.org/10.1108/IJCCSM-11-2016-0168

Niu SY, Liu CL, Chang CC, Ye KD (2016) What are passenger perspectives regarding airlines' environmental protection? An empirical investigation in Taiwan. J Air Transp Manag 55(2016):84–91. https://doi.org/10.1016/j.jairtraman.2016.04.012

Pan JY, Truong D (2018) Passengers' intentions to use low-cost carriers: An extended theory of planned behavior model. J Air Transp Manag 69:38–48. https://doi.org/10.1016/j.jairtraman. 2018.01.006

Polonsky MJ (1994) An introduction to green marketing. Electron Green J 1(2). https://doi.org/10.5070/G31210177

Prakash G, Singh PK, Yadav R (2018) Application of consumer style inventory (CSI) to predict young Indian consumer's intention to purchase organic food products. Food Qual Prefer 68:90–97. https://doi.org/10.1016/j.foodqual.2018.01.015

Qiu R, Xu J, Xie H, Zeng Z, Lv C (2020) Carbon tax incentive policy towards air passenger transport carbon emissions reduction. Transp Res Part D: Trans Environ, 85. https://doi.org/10.1016/j.trd.2020.102441

Royne MB, Levy M, Martinez J (2011) The public health implications of consumers' environmental concern and their willingness to pay for an eco-friendly product. J Consum Aff. https://doi.org/10.1111/j.1745-6606.2011.01205.x

Sanchez G (2013) PLS path modeling with R. Trowchez E. https://www.gastonsanchez.com/PLS_Path_Modeling_with_R.pdf

Sanchez G, Trinchera L, Russolillo G (2015) plspm: tools for partial least squares path modeling (PLS-PM). R Package Version 0.4.7. https://github.com/gastonstat/plspm

Sarkar AN (2012) Evolving green aviation transport system: a Hoilistic approah to sustainable green market development. Am J Clim Chang 01(03):164–180. https://doi.org/10.4236/ajcc.2012.13014

Schniederjans DG, Starkey CM (2014) Intention and willingness to pay for green freight transportation: An empirical examination. Transp Res Part d: Transp Environ 31:116–125. https://doi.org/10.1016/j.trd.2014.05.024

Schwartz SH (1977) Normative influences on altruism. Adv Exp Soc Psycholo 10(C):221–279. https://doi.org/10.1016/S0065-2601(08)60358-5

Seip K, Strand J (1992) Willingness to pay for environmental goods in Norway: a contingent valuation study with real payment. Environ Resource Econ 2(1):91–106. https://doi.org/10.1007/BF00324691

Sharma K, Bansal M (2013) Environmental consciousness, its antecedents and behavioural outcomes. Journal of Indian Bus Res 5(3):198–214. https://doi.org/10.1108/JIBR-10-2012-0080

Sreen N, Purbey S, Sadarangani P (2018) Impact of culture, behavior and gender on green purchase intention. J Retail Consum Serv 41:177–189. https://doi.org/10.1016/j.jretconser.2017.12.002

Tal H (2009) The future is now. Alpha Omegan 102(4):155–156. https://doi.org/10.1016/j.aodf.2009.10.015

Tang CMF, Lam D (2017) The role of extraversion and agreeableness traits on Gen Y's attitudes and willingness to pay for green hotels. Int J Contemp Hosp Manag 29(1):607–623. https://doi.org/10.1108/IJCHM-02-2016-0048

Tehseen S, Ramayah T, Sajilan S (2017). Testing and controlling for common method variance: a review of available methods. J Manag Sci 4(2):142–168. https://doi.org/10.20547/jms.2014.1704202

Tenenhaus M, Vinzi VE, Chatelin YM, Lauro C (2005) PLS path modeling. Comput Stat Data Anal 48(1):159–205. https://doi.org/10.1016/j.csda.2004.03.005

Toklu IT, Küçük HO (2016) The impact of brand crisis on consumers' green purchase intention and willingness to pay more. Int Bus Res. https://doi.org/10.5539/ibr.v10n1p22

Verma VK, Chandra B (2018) An application of theory of planned behavior to predict young Indian consumers' green hotel visit intention. J Clean Prod 172:1152–1162. https://doi.org/10.1016/j.jclepro.2017.10.047

Wang J, Wang S, Wang Y, Li J, Zhao D (2018) Extending the theory of planned behavior to understand consumers' intentions to visit green hotels in the Chinese context. Int J Contemp Hosp Manag 30(8):2810–2825. https://doi.org/10.1108/IJCHM-04-2017-0223

Wu H-C, Cheng C-C, Ai C-H (2018) An empirical analysis of green switching intentions in the airline industry. J Environ Planning Manage 61(8):1438–1468. https://doi.org/10.1080/09640568.2017.1352495

Yadav R, Pathak GS (2016) Young consumers' intention towards buying green products in a developing nation: Extending the theory of planned behavior. J Clean Prod 135:732–739. https://doi.org/10.1016/j.jclepro.2016.06.120

Yang K (2012) Consumer technology traits in determining mobile shopping adoption: an application of the extended theory of planned behavior. J Retail Consum Serv 19(5):484–491. https://doi.org/10.1016/j.jretconser.2012.06.003

Yarimoglu E, Gunay T (2020) The extended theory of planned behavior in Turkish customers' intentions to visit green hotels. Bus Strateg Environ 29(3):1097–1108. https://doi.org/10.1002/bse.2419

Zhou Y, Thøgersen J, Ruan Y, Huang G (2013) The moderating role of human values in planned behavior: the case of Chinese consumers' intention to buy organic food. J Consum Mark 30(4):335–344. https://doi.org/10.1108/JCM-02-2013-0482

Şahap Akan is both a Research Assistant and received his Ph.D. degree in Civil Aviation Management at the Graduate School of Social Sciences, Anadolu University, Turkey. He received his MSc degree in the area of consumer neuroscience in the aviation industry from Anadolu University in 2017. Şahap's master research named "Comparison of airline brand personality perception by face to face interview method and EEG method of neuromarketing: An application" investigated airline brand personality perceptions of passengers by employing a mixed-methods research methodology. His research interests lie in the fields of consumer behavior, consumer neuroscience, service marketing, and airline marketing strategies. In his Ph.D. journey, he aims to undertake research using neuromarketing techniques to provide new insights into the airline consumer market.

Emircan Özdemir is a Research Assistant in the Faculty of Aeronautics and Astronautics, Eskişehir Technical University in Turkey. After graduating with a double major in industrial engineering and aviation management from Anadolu University, he received his MA degree in Civil Aviation Management from Anadolu University in 2016. His master's thesis ("Analysis of check-in counters and the research of the factors affecting service quality perception") was about the factors affecting service quality perceptions of passengers during airport check-in processes. In 2022, he's got a PhD in Civil Aviation Management from Anadolu University. His PhD dissertation ("Analysis of online reviews of airline passengers through text mining") was about determining passenger sentiments toward airline service attributes and examining the relationships between passengers' value for money perception, satisfaction, recommendation behavior and passenger sentiments. His research interests center around service marketing, marketing research, marketing communication, and data mining applications.

Mahmut Bakır received his Ph.D. degree in Civil Aviation Management at the Graduate School of Social Sciences, Anadolu University, Turkey. He has been a Research Assistant in the Department of Aviation Management, Samsun University in Turkey since 2016. Mahmut obtained his MSc degree in Airline Management from Anadolu University in 2017. His Master's degree thesis entitled "An integrated approach to the evaluation of e-service quality in airline companies" investigated the e-service quality performance of scheduled airlines in the Turkish airline industry. His primary research interests include airline business, service marketing, consumer behavior, multi-attribute decision analysis, and multivariate statistics.

Chapter 8
The Factors Affecting the Passengers' Avoidance to Use the Mobile Applications of Airlines

Mutlu Yuksel Avcilar⊙, **Nuriye Günebakan**⊙, **Hilal Inan**⊙, **and Seda Arslan**⊙

Abstract Nowadays, rapid development in information technologies has been experienced in the aviation industry as in other industries. This development has affected both passengers and airlines in different aspects. The use of information technologies enables passengers to carry out flight-related transactions more easily and quickly, without place and time constraints. Information technologies have necessitated the use of different applications in the marketing activities of airlines. So in terms of airlines, it is extremely important to adapt these information technologies to all processes in order to achieve sustainable competitive advantage. Today airlines have also started to use mobile applications (mobile apps) in their marketing activities. They encourage their passengers to use these apps, which require great investment. However, some passengers may avoid using mobile apps due to a variety of reasons such as technology anxiety, privacy concerns, complexity, etc. In this case, high investments made for the development of these apps may not be compensated and the competitiveness of airlines may be weakened. Therefore, the purpose of this study is to determine the factors affecting passengers' avoidance of using airlines' mobile apps. A survey was conducted using a face-to-face interview technique with convenience sampling method to collect data. Questionnaire items were measured

---

This research was presented orally at the 23rd EBES Conference, September 27-29, 2017, Madrid - Spain.

M. Y. Avcilar
Faculty of Economics and Administrative Sciences, Osmaniye Korkut Ata University, Osmaniye, Turkey
e-mail: myukselavcilar@osmaniye.edu.tr

N. Günebakan · S. Arslan (✉)
Faculty of Aeronautics and Astronautics, Iskenderun Technical University, Hatay, Turkey
e-mail: seda.arslan@iste.edu.tr

N. Günebakan
e-mail: nuriye.gunebakan@iste.edu.tr

H. Inan
Faculty of Economics and Administrative Sciences, Cukurova University, Adana, Turkey
e-mail: ihilal@cu.edu.tr

K. Kiracı and K. T. Çalıyurt (eds.), *Corporate Governance, Sustainability, and Information Systems in the Aviation Sector, Volume I*, Accounting, Finance, Sustainability, Governance & Fraud: Theory and Application, https://doi.org/10.1007/978-981-16-9276-5_8

by a five-point Likert scale ranging from 1 (strongly disagree) to 5 (strongly agree). The survey was conducted in the two busiest airports in Turkey (Atatürk and Sabiha Gökçen Airports in Istanbul). A total of 450 respondents have been reached but 400 valid questionnaires were analyzed by using Partial Least Squares method. A pilot study was conducted on 20 volunteer participants before the field research was initiated. In this context, the questionnaire was finalized by taking into consideration the suggestions expressed in order to maintain the coherence. The survey questionnaire consisted of two sections. The first section comprised of perceived irritation, perceived lack of utility, perceived lack of incentive, technology anxiety, privacy concern, and mobile app usage avoidance. In the second section, there were questions about passenger's demographic characteristics such as; age, gender, education level, average income level, and flight frequency. Respondents were also asked about their general usage related to smartphones and mobile apps. According to the results; perceived irritation, perceived lack of utility, perceived lack of incentive, technology anxiety, and privacy concerns have a positive and significant effect on mobile app usage avoidance. Among these variables, technology anxiety is the most affected variable on passengers' avoidance of using mobile apps. Both in national and international literature, there are numerous studies on mobile internet, apps adoption, and actual usage of mobile apps in different industries. Previous studies are especially related to the factors accepting the mobile apps. In addition to this, there are limited studies including avoidance of using airline mobile apps. In aviation industry also, there are a few studies both national and international literature on passengers' avoidance of using airline mobile apps. Therefore, this study will be one of the few studies both shedding light on literature and airlines to increase the use of mobile apps.

Keywords Passengers · Mobile applications · Avoidance · Airlines

8.1 Introduction

Intense developments in the field of information and communication technologies have led to the emergence of smart technologies (Caragliu et al. 2011; Wang et al. 2016). The most widely used devices in smart technologies are smartphones. Smartphones are platforms where other functions such as phones, personal data assistants, and music players work together in an integrated way (Amadeus 2011). According to the Statista (2020), there are 3.5 billion smartphone users worldwide. In addition 5.2 billion people own any mobile device (Bankmycell 2020). As the use of smartphones increases, the use of mobile apps that give functionality to these devices has also become widespread (Liu et al. 2017). Thanks to these apps, mobile phones can provide users with both entertainment and productivity (Chen et al. 2017). According to Schmitz et al. (2016: 231), mobile apps are *"end-user software apps that are designed for a cell phone operating system and which extend the phone's capabilities by enabling users to perform particular tasks."* These apps, parallel to technological developments, have become the primary tools used daily in both personal and professional life. Mobile apps play critical roles in facilitating processes such as communication, education, work, entertainment, health, finance, travel, and public services (Siuhi and Mwakalonge 2016). So people use mobile apps an average of 30 h per month (Chen et al. 2019a).

Mobile apps, which are designed for personal or commercial use, have been offered to users through the app stores (paid or free) (Lim et al. 2015). There are numerous mobile apps developed by many countries and companies in the app stores. So there is a tough competition in this market (Monno and Xiao 2014). In 2019, 100 billion mobile apps were downloaded to people's devices worldwide. It is estimated that more than 590 million apps will be downloaded daily by the end of 2020 and the number of downloaded applications will exceed 252 billion in 2022 (WebsiteBuilder 2020). Developing a mobile app is a rather expensive and long-lasting process. A mobile app is costing between \$ 50,000 and \$ 1,000,000 and is being developed for over six months to 1 year (Blair 2019). As the number of mobile apps in the app market increases, these applications' usage by the customers in the long run may become more difficult (Kumar et al. 2018). In addition it is not possible to say that all mobile apps are always successful and some users still avoid using mobile apps due to different reasons. In some cases, the mobile apps are either not downloaded at all or is deleted immediately after it has been downloaded and does not meet the expectations. Sometimes, the mobile apps may no longer be needed. For example, a passenger can download the mobile apps before traveling. Afterward, although the apps meets the expectations of the passenger, the need may disappear when the journey is completed, and the passenger may choose to delete the apps. The deletion of apps due to various reasons means failure for companies (Lim et al. 2015). The existence of such problems has revealed the need to answer the question of why users avoid the use of mobile apps belonging to companies.

There is an increasing demand for air transport in Turkey. This increase has been predicted to continue in the future. (Vakıf Investment Company 2018). In line with increasing demand, air transport has been growing more than in Europe and other countries in the world. DGCA (2018). So, airlines originated in Turkey compete to provide better services to passengers as a result of high competitiveness and increasing demand. Therefore, airline companies to gain more passengers, use mobile apps technology broadly. However, it is very costly to design and maintain these apps. If passengers do not use these apps, negative consequences may occur for airlines. This is also the case for airlines originated in Turkey, which have been trying to maintain their operations with the high cost and low-profit margins (Gerede and Orhan 2015). This situation may damage the overall brand image of the airlines as well (Amadeus 2011). Therefore, to investigate the causes of passengers' avoiding the use of mobile apps designed by the airline companies operating in Turkey have become more important in the sense of the Turkish aviation industry where competition is intense and profit margins are low.

In international literature, there are numerous studies on mobile Internet, app adoption and usage in different industries (Egger 2013; Richard and Meuli 2013; Şanlıöz et al. 2013; Im and Hancer 2014; Monno and Xiao 2014; Dickinson et al. 2015; Hsu and Lin 2015; Khalid et al. 2015; Rivera et al. 2015; Chang et al. 2016; Madan and Yadav 2016; Pentina et al. 2016; Schmitz et al. 2016; Siuhi and Mwakalonge 2016; Veríssimo 2016; Fang et al. 2017; Liu et al. 2017; Unal et al. 2017), including aviation (Budd and Vorley 2013; Koch and Tritscher 2016; Martin-Domingo and Martín 2016; Suki and Suki 2017). Besides, the avoidance behaviors of customers, especially unwanted advertisements on different media platforms (i.e., TV, radio, internet, SMS, mobile), are frequently examined in the literature (Greyser 1973; Aaker and Bruzzone 1985; Speck and Elliott 1997; De Pelsmacker and Van den Bergh 1998; Fennis and Bakker 2001; Edwards et al. 2002; Li et al. 2002; Morimoto and Chang 2006; Bellman et al. 2010; Dix and Phau 2010; Kelly et al. 2010; Suher and İspir 2011; Rau et al. 2013). When the studies were examined, it was seen that the national and international literature generally focused on the adoption of information technologies and mobile apps. There are limited studies on avoiding the use of technologies (Hsu et al. 2019), resistance (Chen et al. 2019b; Yang and Park 2019), and discontinuous behavior (Chen et al. 2019a). However, there are relatively limited studies in the national and international literature on passengers' avoidance of using airline mobile apps in Turkey. Therefore, with this study, it may be learned why passengers refrain from using the mobile apps. Thus, it may be possible to guide the airlines aiming to provide better service quality to their passengers. In addition, it is considered that the suggestions presented within the scope of the study will be beneficial in order to obtain high value for the apps developed by bearing large costs. In addition, research on this subject may also contribute to the literature and close the gap in the field.

8.2   Theoretical Framework and Hypotheses

In this section, first of all, the importance of using mobile app in airline transportation has been mentioned. Afterwards, sector applications and studies in the literature of this field have been given. Then, avoidance behavior has been defined and avoidance in mobile app usage has been mentioned. In addition, the variables determined as antecedents of airline mobile app usage avoidance have been defined and hypotheses containing the relevant variables have been included.

8.2.1   Mobile Applications in Aviation

Technology is an inseparable part of the aviation industry (Bogicevic et al. 2017). Mobile apps are heavily used by companies operating in many different industries (Siuhi and Mwakalonge 2016). Particularly the airline industry attaches great importance to information technology. Therefore, the airline industry has adapted to these apps much faster than other industries as tourism (Buhalis 2004). Airline companies often use self-service technologies, biometrics, wearable technologies, and mobile apps to support their services (Bogicevic et al. 2017). Thanks to these apps, airline companies have been able to obtain new opportunities in service operations and distribution activities (Dickinson et al. 2015; Fang et al. 2017). The presence of low-cost airline companies in the sector has made the competition more challenging. Thus, airline companies will also be able to gain competitive advantage as a result of successful mobile app designs (Yang and Park 2019). Mobile apps can provide many opportunities to airline companies (Clarke 2001; Koch and Tritscher 2016). Airlines can enrich their business operations with mobile technologies (Suki and Suki 2017) and interact more with their passengers during their travels (Budd and Vorley 2013; Fang et al. 2017). In addition, mobile apps can also provide benefits to airlines in accessing more information about their passengers, creating customer satisfaction, and reducing operating costs (Yang and Park 2019).

As this is the case, the use of a mobile app for travel and aviation companies and issues for the adoption of mobile apps by the passengers were examined in the literature (Budd and Vorley 2013; Koch and Tritscher 2016; Suki and Suki 2017; Fang et al. 2017, etc.). For example, Budd and Vorley (2013), by using content analysis, identified and classified the extent of the information and functionality of the official iOS apps of the world's 25 largest passenger airlines, and evaluated user experiences of these apps. Accordingly, the authors identified seven functionalities: flight search, flight booking, manage booking, mobile check-in, mobile boarding pass, flight status, and frequent flyer programming. They found that almost all airline mobile apps can be downloaded free of charge and they are available in English. Moreover, almost all of the seven features identified as functionalities were found to be present in most airline mobile apps. Koch and Tritscher (2016) did a scientific study for the adoption of the app called "social seating" based on the Technology Acceptance Model (TAM). According to the results, it has been found a consistent influence of perceived ease of use on perceived usefulness. Suki and Suki (2017) explored Malaysian travelers' intention to use mobile apps for booking air tickets. According to the results, it was concluded that the perceived benefit was the variable that had the most significant impact on the adoption of airline mobile apps by passengers.

Fang et al. (2017) explored the impact of design and performance features of mobile apps on mobile app adoption. According to the results, it was concluded that two app design features (i.e., user interface attractiveness and privacy/security) and three app performance attributes (i.e., compatibility, ease of use, and relative advantages) had significant impacts. Im and Hancer (2014) examined the issue of the direct and indirect relationship of utilitarian motivation, hedonic motivation, and self-identity to travelers' attitudes toward travel mobile app usage. According to the results, utilitarian motivation was the most critical factor in shaping the attitude in using travel mobile apps. However, hedonic motivation played a role as an essential catalyst for utilitarian motivation.

Apart from the studies examining the adoption behavior toward airline mobile apps mentioned above, there are also studies examining the factors affecting the resistance behavior toward using airline mobile apps (Chen et al. 2019b; Yang and Park 2019). Chen et al. (2019b), based his study investigating why users show resistance toward mobile apps designed by airlines on adoption barriers (usage barrier, image barrier, value barrier) and knowledge of alternatives quality variables. According to the research results; usage barrier (difficulty in use), image barrier (technology anxiety) and value barrier (lack of utility and relative advantages) positively affect the behavior of refusing to use the airline mobile app. In other words, the presence of these barriers causes passengers to refuse to use mobile apps. Yang and Park (2019), on the other hand, while investigating the reasons for the resistance of passengers to use the airline mobile apps, selected the user characteristics (attitude toward change, mobile literacy, and mental model) as the baseline. The authors, by including perceived ease of use and perceived usefulness variables in the research model, argued that these two elements of the Technology Acceptance Model will reduce the resistance to use the airline mobile app. According to the results of this study conducted by Yang and Park (2019), users' perception of the airline mobile app as easy to use and useful reduces the behavior of rejecting mobile app usage.

8.2.2   Avoidance to Use Mobile Apps

If the attitude of individuals toward technological innovations is negative and their attitude toward traditional methods is positive, they may show resistance toward these innovations (Yang and Park 2019). Innovation resistance is the negative situation that individuals feel when they encounter a method / innovation different from traditional methods (Yang and Park 2019). Individuals who resist by not adopting technological innovations may be classified into three different groups as postponers, opponents, and rejectors (Laukkanen et al. 2008). Avoidance of using technological innovations behaviors may be displayed in all three groups.

Avoidance refers to "*an express when the user consciously and intentionally seeks to avoid a stimulus*" (Koshksaray et al. 2015: 39). Avoidance can be seen in different ways (Koshksaray et al. 2015). For example, physical avoidance (Abernethy 1991), cognitive, affective, and behavioral avoidance (Cho and Cheon 2004; Speck and Elliot 1997).

The rapid progress in mobile phones and apps used in these phones have led many companies to use these apps. Due to the presence of many mobile app providers in the market, the competition is quite high, and users have many options (Ding and Chai 2015). However, most mobile apps developed by companies may not be successful (Chen et al. 2019b). For example, the mobile app, which allows users to share live video feeds and photos with friends and cost $ 41 million, was abandoned due to insufficient numbers of users and high levels of customer loss (Lim et al. 2015). According to a study by Flurry (2011), individuals stop using the mobile apps to avoid negative emotions as a result of using a low-quality mobile app. Only about one-third of users continue to use the mobile app one month after the first use, and this rate drops to 4 percent after one year. According to another study (Lim et al. 2015), the most common reasons for mobile app users to leave an app are: the app is no longer needed, better alternatives are available, and users are bored with the apps. According to the same study, users are more likely to be bored with an app and therefore quit it as their age decreases. Performance, reliability, and usability are crucial for mobile app users (Lim et al. 2015). More than 30% of users give up the apps due to the crashes, insufficient features, slow operation and not being user-friendly. Seventeen percent of the participants in the study stopped using the mobile app due to privacy concerns (Lim et al. 2015). Similar results were found by Khalid et al. (2015). According to this, the complaints of 6390 people who gave low scores to 20 free mobile apps were examined and the most common complaints were functional errors, feature requests, and app crashes. Complaints about privacy, ethical issues, and hidden app costs have the most negative impacts on the rating of an app. Finally, 11% of respondents stated that the last update failed.

Therefore, users will have negative feelings against mobile apps that cannot provide a successful service and will avoid using them (Hsu et al. 2019). In some cases, although users download mobile apps to their devices, they do not display continuous usage behavior due to various reasons (perceived information overload, perceived intrusiveness, perceived reward, etc.) (Chen et al. 2019a). Sometimes, people may prefer traditional methods instead of using mobile apps (Lian 2018). Therefore, for whatever reason it is highly likely that users avoid mobile apps that do not meet their expectations and preferences.

8.2.3 Antecedents of Airline Mobile App Usage Avoidance

8.2.3.1 Perceived Irritation

If people feel threatened by any event, human or innovation they encounter, they may want to protect themselves. This theory, called the Protection Motivation Theory (PMT), developed by Rogers (1975), has become frequently used in studies related to the adoption of information technologies in later years (Guo et al. 2015; Tsai et al. 2016; Karahoca et al. 2018; Miraja et al. 2019; Mousavi et al. 2020; Wu 2020). In other words, if individuals do not feel positive feelings toward technology and perceive the use of technology as an irritating situation, they may go into self-protection motivation and avoid using technology (Chen et al. 2019a).

Perceived irritation has frequently been searched in avoidance studies in the literature and has been found as an essential factor on avoidance (Bauer and Raymond 1968; Greyser 1973; Aaker and Bruzzone 1985; De Pelsmacker and Van den Berg 1998; Fennis and Bakker 2001; Edwards et al. 2002; Morimoto and Chang 2006; Morimoto and Chang 2009). According to the study of Liu et al. (2012), irritation has been defined as the extent to which the web or mobile platform is messy and irritating to consumers. Irritation is defined as the negative, impatient, and displeasing feeling consumer experience (Aaker and Bruzzone 1985) as a result of mobile app usage. Irritation may involve the mobile app's ability to annoy, offend, insult, or overly manipulate individuals (Richard and Meuli 2013). If individuals are uncomfortable with the mobile app they will use or they have been already using, they may display the behavior of not using or avoiding the app (Dennison et al. 2013; Chen et al. 2019a). Similar situation may be valid for airline mobile apps. In light of the information above, hypothesis H1 is as follows:

H1: Perceived irritation affects passengers' avoidance of using mobile apps of airlines.

8.2.3.2 Perceived Lack of Utility

In general, the perceived utility is one of the main determinants of the technology acceptance model (Davis 1989; Koch and Tritscher 2016). The utility that users perceive for new technologies can increase their value in the eyes of the user (Davis 1989). As a result of this situation, the user can be able to think that this technology is necessary for him/her to increase his / her performance (Çakar 2017). According to Davis (1989: 320), perceived utility refers to "*the degree to which individuals believe that using a particular system will enhance their performance.*" Mobile apps that can be beneficial to their users can be downloaded more and the rate of continuous use will increase (Ding and Chai 2015). This result is supported with the studies in the literature (Kim et al. 2009; Suki and Suki 2011, 2017; Yang 2012).

However, the more negative the perception of the individuals toward the use of mobile apps, the more the opinion that the app will not be useful may increase (Yang and Park 2019) and this may cause individuals to avoid using the mobile app. Perceived lack of utility is considered to be one of the most important reasons of avoidance behavior toward innovation. (Chen et al. 2019b). Because perceived lack of utility may cause customer dissatisfaction (Shin and Lin 2016). When the studies in the literature were examined, the attitude toward mobile apps that were perceived as useful was positively affected and using behavior was displayed (Lu et al. 2015; Shen 2015; Byun et al. 2018; McLean 2018). So it would not be surprising to avoid a mobile app that is not perceived as useful. In this context, passengers will be able to avoid using the mobile app if they believe that the airline mobile app will not save money / time and provide them with useful information about airline travel. In this context, hypothesis H2 is as follows:

H2: Perceived lack of utility affects passengers' avoidance of using mobile apps of airlines.

8.2.3.3 Perceived Lack of Incentive

Promotion is one of the elements of marketing mix that is under the control of the companies in order to meet and satisfy the wants and needs of the target market (Eser et al. 2011). Promotion can be defined as a set of activities carried out by companies to adopt an idea / innovation, to persuade a product or service to be purchased, and to create a channel for information flow (Eser et al. 2011). Sales promotion activity, which is one of the promotion activities, offers an incentive to the target market. Digital coupons are the fastest growing incentives in this field (Gülso 2018).

This situation may also be evaluated in terms of mobile apps. In order to accept mobile apps, today's customers may demand some benefits (Carter and Yeo 2016). Therefore, financial incentives such as online coupons, discounts should be offered to customers to use the mobile app designed by the companies (Hsu and Tang 2020). Thus, avoiding the use of mobile apps could be prevented (Lee et al. 2010; Chen et al. 2019a; Wang 2019).

In the literature, there are studies that the incentives offered have a positive impact on the adoption and continuous use of mobile apps such as mobile wallet (Madan and Yadav 2016), QR code in mobile apps (Watson et al. 2013), near field communication (NFC) mobile payment (Zhao et al. 2019), mobile GPS system (Xu and Yuan 2009). Therefore, if no incentives are provided for the use of airline mobile apps, users may prefer traditional methods or the website, in other words, they may display avoidance of using mobile apps. In this context, hypothesis H3 is as follows:

H3: Perceived lack of incentive affects passengers' avoidance of using mobile apps of airlines.

8.2.3.4 Technology Anxiety

Along with the rapid development in the field of information technologies, it is also quite essential to know the customers' ability and willingness to use these technologies (Meuter et al. 2003).The number of people using new technologies, whether for personal or professional purposes, is quite high. However, there are a considerable amount of people who experience technology anxiety. Accordingly, it is estimated that technology anxiety affects about 30% of the general population, depending on factors such as gender, age, education, and personality (Nimrod 2018). Anxiety, being one of the most important determinants of behavioral intention (Yang and Forney 2013) and main reasons for avoiding the use of information technologies (Compeau et al. 1999; Keikhosrokiani et al. 2019), is generally defined as *"the degree to which the usage or idea of using the technology in question arouses unfavorable feelings and fear"* (Sintonen 2010: 837).

Technology anxiety variable may also be an important obstacle in the adoption of mobile apps (Lian 2018). Due to the small screen of the mobile devices on which the apps are located and not equipped with simple functions, users may experience anxiety due to reasons such as not being able to complete the transaction through the app (for example, leaving the app before completing the transaction while purchasing a flight ticket) or fear that their important information (personal information, credit card information) will be lost. (Yang and Forney 2013). Individuals experiencing such anxiety may perceive it as a threat. Moreover, individuals who think they cannot cope with this situation may avoid using the mobile app (Celik 2016). There are also studies in the literature proving that technology anxiety negatively affects the attitude toward mobile apps, causing the app usage to be avoided (Gupta and Arora 2017; Patil et al. 2020). So the fact that passengers experience technology anxiety may lead them to avoid using the mobile app. In this context, hypothesis H4 is as follows:

H4: Technology anxiety affects passengers' avoidance of using mobile apps of airlines.

8.2.3.5 Privacy Concerns

Today, companies tend to offer customized products/services to their customers as a consequence of the change in marketing activities and increasing competition (Marangoz and Aydın 2018). Therefore, more personal data is collected and used (Bozacı 2015). Many users are uncomfortable with their personal information being monitored by commercial companies, public institutions, or malicious people and perceive this as a threat. In this case, users may have privacy concerns and this variable is often used in the literature (Bozacı 2015; Pentina et al. 2016; Gu et al. 2017; Marangoz and Aydın 2018; Feng and Xie 2019; Hassandoust et al. 2020). Privacy concern can be defined as the concern individuals experience about the ability of technology to protect their personal information (Kim et al. 2008). In this context, privacy concerns can occur in two different ways: a) general privacy concerns and b) privacy concerns specific to information systems (Morosan 2016).

Privacy issues in the online setting have been shown to influence attitudes toward and intentions to use different information systems, such as websites (Gao et al. 2010) or biometric e-gates in airports (Morosan 2016). In the mobile age, protecting users' personal information is one of the most critical points (Gu et al. 2017). Mobile app privacy/security refers to the perceived ability of individuals to control when, how, and to what extent their data in the mobile apps are accessed, modified, or disclosed (Fang et al. 2017). The expansion in the use of mobile devices and related services has led to increased privacy risks, and the internal conflict of consumers may increase in the mobile environment (Lee and Rha 2016). In other words, while trying to use the app, they may have negative feelings toward the apps due to the privacy concern experienced (Lin et al. 2017) and display avoidance of mobile app usage (Feng and Xie 2019). When the studies in the literature were examined, it was observed that privacy concerns regarding mobile apps have a negative effect on the download (Gu et al. 2017; Hassandoust et al. 2020), usage (Pentina et al. 2016; Bailey et al. 2017) and adoption of these apps (Wei et al. 2009; Fox and Connolly 2018; Zhang et al. 2018). From this viewpoint, hypothesis H5 is as follows:

H5: Privacy concern affects passengers' avoidance of using mobile apps of airlines.

In this context, the proposed conceptual model and hypotheses are shown in Fig. 8.1.

8.3 Research Methodology

Survey research was conducted to collect empirical data. During the research process, we used convenience sampling method and a face-to-face interview technique to collect data. The analysis was performed by using Partial Least Squares. The Smart

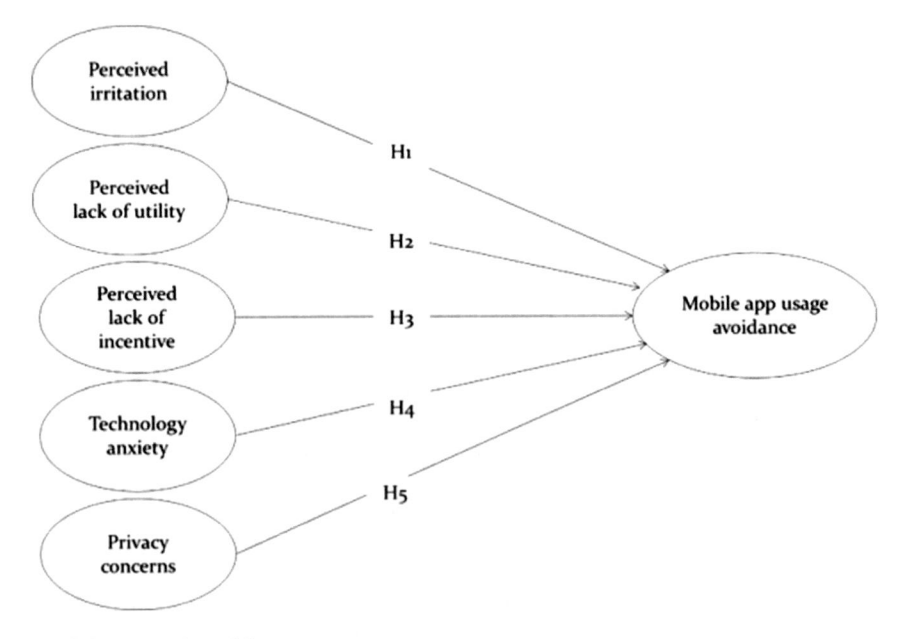

Fig. 8.1 Research model

PLS 3.0 software (Ringle et al. 2015) was used to assess the measurement and structural model. The following sections describe the sampling and data collection process, questionnaire design and measurement instrument, and analysis method.

8.3.1 Sampling and Data Collection

The survey was conducted in the two busiest airports in Turkey (Atatürk and Sabiha Gökçen airports in Istanbul). A questionnaire was applied to Turkish passengers who avoid to use airlines' mobile apps. This research includes a face-to-face survey method to collect passengers' perceptions of data between July and August 2017. There are ten airlines originated in Turkey licensed by the Directorate General of Civil Aviation-DGCA, operating scheduled and non-scheduled passenger transportation business (DGCA 2017). Seven of these ten airlines carry passengers. Only five of the seven airlines (Turkish Airlines, Pegasus, Anadolujet, Atlasglobal, Onur airlines) have mobile apps for passengers. Therefore, these five airlines were included in the study.

In order to use probability sampling methods, there should be a sampling frame (Kinnear and Taylor 1991). So, the sampling frame could not be reached in this study. For this reason, convenience sampling, one of the non-probability sampling methods, has been used. This is one of the limitations of the study. Questionnaire items were measured by a five-point Likert scale ranging from 1 (strongly disagree) to 5 (strongly agree). Four hundred fifty passengers were searched during the data collection. However, for different reasons (i.e., multiple responses to statements and incomplete responses), the validity of 400 questionnaires was accepted and included in the analysis.

8.3.2 Questionnaire Design and Measures

Mobile app usage avoidance (11-items) (Chen et al. 2002; Verkasalo et al. 2010; Baek and Morimoto 2012; Hsu and Lin 2015; Nyheim et al. 2015; Stoyanov et al. 2015 and Google 2016), perceived irritation (9-items) (Edwards et al. 2002; Baek and Morimoto 2012; Nyheim et al. 2015 and Mani and Chouk 2017), perceived lack of utility (5-items) (Cho and Cheon 2004; Shin and Lin 2016), perceived lack of incentive (4-items) (Cho and Cheon 2004; Atchariyachanvanich et al. 2006), technology anxiety (9-items) (Meuter et al. 2003) and privacy concern (4-items) (Bauer et al. 2005; Baek and Morimoto 2012; Nyheim et al. 2015) variables have been adapted to the aviation industry utilizing studies of different authors.

As the original language of the scales included in the study was English, the scales were translated into Turkish. The experts examined the statements of the scales translated into Turkish by taking into account the meaning errors due to possible cultural differences. A pilot study was conducted on 20 volunteer participants before the field research was initiated. In this context, the questionnaire was finalized by taking into consideration the suggestions expressed in order to maintain the coherence. Details of the scales have been provided in Appendix A.

The second section contains questions about passenger's demographic characteristics such as; age, gender, education level, average income level, and flight frequency. Respondents were also asked about their general usage related to smartphones and mobile apps (which airline company's mobile app he/she uses, the frequency of Internet use through smartphones and how long he/she uses the airliner's mobile app).

8.3.3 Data Analysis

In order to analyze the data, Partial Least Squares (PLS-SEM), structural equation modeling technique was used. PLS-SEM estimates the parameters of a set of equations in a structural equation model by combining principal-component analysis with regression-based path analysis (Mateos-Aparicio 2011; Sarstedt et al. 2017). PLS-SEM is a variance-based method to estimate structural equation models. The aim of this analysis is to maximize the explained variance of the endogenous latent variables (Hair et al. 2014). PLS-SEM technique defines a prediction-oriented approach that is suitable for research contexts that have not been well investigated yet (Gefen et al.

2011). Hence, in order to learn the avoidance of passengers to use the mobile apps of airlines, PLS-SEM technique is utilized in this study.

8.4   Results

8.4.1   Demographic Characteristics of the Respondents

According to the demographic characteristics of the respondents, 51% were male; 49% were female. The age of the vast majority of the respondents was between 26 and 55 years (60 percent) and they had a university degree or higher educational level (54 percent). Respondents' average monthly income mostly ranged between $ 350 and 1100 (56 percent). Twenty-five percent of the participants travel by air every six months and 22% every three months. The majority of respondents (89%) enter the Internet every day with their smartphones. Twenty-two percent of the participants use the mobile app of Turkish Airlines, 21% of Pegasus Airlines, 19% of Anadolujet, 18% of Atlasglobal, and 20% of them use Onurair's mobile app. Finally, the participants were asked this question: "How long have you been using the mobile app of the airline company?" The majority of the respondents (43 percent) answered this question as "0–1 years" (Table 8.1).

8.4.2   Measurement Model Assessment

PLS-PM consists of the outer model and the inner model (Vinzi et al. 2010). The outer model states the relationship between latent variables which are observed variables, however the inner model designates the relationship between unobserved variables (Henseler et al. 2009). In the PLS-PM analysis, firstly the measurement model is evaluated. In this study, the procedure offered by Hair et al. (2016) was used to measure the psychometric properties of the reflective measurement model. In order to evaluate the measurement model, primarily, reliability was assessed according to the criterion as Cronbach's α should be higher than 0.70. Then item loadings (λ) (it should be higher than 0.708 as statistical significance), composite construct reliability (it should be higher than 0.80), and average variance extracted (AVE) (it should be higher than 0.50) (Fornell and Lacker 1981; Chin 1998; Hair et al. 2016) were evaluated for convergent validity. Lastly, discriminant validity was evaluated according to the criterion, as the square root of AVE for each latent construct should be larger than its correlations with all other latent constructs (Fornell and Lacker 1981).

As indicated in Table 8.2, standardized item loadings are statistically significant

Table 8.1 Demographic characteristics of the sample (n = 400)

Gender	N	%	Flight frequency	N	%
Female	196	49	A few times in a month	62	16
Male	204	51	Once a month	78	19
Age	N	%	Once a quarter	87	22
18–25	87	22	Once every six months	101	25
26–35	85	21	Once a year	72	18
36–45	78	20	Frequency of internet usage via smartphone	N	%
46–55	76	19	Everyday	354	89
56 and above	74	18	Almost everyday	40	10
Education Level	N	%	Sometimes	3	0,75
Secondary education	35	9	Seldom	3	0,75
High school	47	12	Airline companies	N	%
Associate degree / Vocational school	99	25	Turkish airlines	84	22
Undergraduate	124	31	Pegasus	83	21
Postgraduate	61	15	Anadolujet	80	19
Doctorate	34	8	Atlas global	78	18
Monthly average income ($)	N	%	Onur airlines	75	20
Less than $ 350	58	14	Airlines' mobile app usage length of time	N	%
$350–$600	67	17	0–1 year	170	43
$601–$850	79	20	2 years	118	29
$851–$1100	76	19	3 years	59	15
$1101–$1350	67	17	4 years	37	9
$1351 and above	53	13	5 years and above	16	4

($p < 0.001$) on their constructs and range from 0.731 to 0.946. Cronbach's α coefficient range from 0.904 to 0.928 and all are above the recommended level of 0.70. Analysis result indicates that all reflective constructs have a satisfactory level of internal consistency reliability. Composite construct reliabilities range from 0.924 to 0.948 and all above the recommended level of 0.80. Analysis result indicates that all the reflective constructs have a high level of internal consistency. The average variance extracted (AVE) values range from 0.645 to 0.843 and all above the minimum recommended level of 0.50. This indicates the convergent validity of all constructs (Table 8.2).

In Table 8.3, diagonal elements (values in parentheses) are the square root of the AVE; the off-diagonal values are the correlations between the latent constructs. Analysis result reveals that the square root of the AVE for each construct is higher than its correlations with all other constructs. The analysis result supports reflective measurement constructs' discriminant validity. Thus, analysis results show an

Table 8.2 Constructs' convergent validity and reliability

Latent variables	Item	Outer loadings (Min–Max)*	AVE	Cronbach's Alpha	Composite reliability
Perceived irritation (PI)	9	0.770–0.916	0.758	0.928	0.942
Perceived lack of utility (PLU)	5	0.731–0.857	0.645	0.909	0.927
Perceived lack of incentive (PLI)	4	0.750–0.942	0.776	0.904	0.932
Technology anxiety (TA)	9	0.830–0.946	0.843	0.917	0.948
Privacy concern (PC)	4	0.852–0.929	0.801	0.920	0.941
Mobile app usage avoidance (MAUA)	11	0.774–0.925	0.771	0.907	0.924

Note *Factor loadings statistically significant ($P < 0.001$); Values in parentheses (diagonal elements) are the square root of the AVE values. The off-diagonal values are the correlations between the latent constructs

Table 8.3 Constructs' discriminant validity

Constructs	Mean	STDEV	PI	PLU	PLI	TA	PC	MAUA
Perceived irritation (PI)	2.183	0.903	*0.871*					
Perceived lack of utility (PLU)	2.734	0.930	0.516	*0.803*				
Perceived lack of incentive (PLI)	2.124	1.069	0.468	0.575	*0.881*			
Technology anxiety (TA)	2.232	1.089	0.665	0.509	0.404	*0.918*		
Privacy concern (PC)	2.006	0.929	0.476	0.420	0.412	0.444	*0.895*	
Mobile app usage avoidance (MAUA)	2.765	1.129	0.693	0.742	0.459	0.780	0.475	*0.878*

acceptable level of reliability, convergent, and discriminant validity. Therefore, analysis results imply that our data and measurement model is suitable for the structural model assessment and hypothesis testing process.

8.4.3 Structural Model Assessment

After the measurement model has been confirmed, the next step is to assess the structural model results, which involve the relationships between latent variables.

Table 8.4 PLS-PM hypothesis test results

	Path coefficient	STDEV	T—statistics*	P—values	Hypothesis	Test result
PI → MAUA	0.280	0.045	6.222	0.000	H1	Supported
PLU → MAUA	0.309	0.052	5.942	0.000	H2	Supported
PLI → MAUA	0.161	0.047	3.425	0.001	H3	Supported
TA → MAUA	0.424	0.056	7.571	0.000	H4	Supported
PC → MAUA	0.187	0.058	3.224	0.001	H5	Supported

Note t-values for two-tailed test; * 1.96 (sig. level 5%), ** 2.58 (sig. level 1%). MAUA: Mobile app usage avoidance, PI: Perceived irritation, PLU: Perceived lack of utility, PLI: Perceived lack of incentive, Technology anxiety: TA, Privacy concern: PC

According to Hair et al. (2014), the critical criteria for evaluating the structural model in PLS-PM are the significance of path coefficient, the level of latent variables R^2 values, the predictive relevance (Q^2) values (Table 8.4).

Perceived irritation, perceived lack of utility, perceived lack of incentive, technology anxiety, and privacy concern have a positive and significant effect on mobile app usage avoidance. The acceptable fit of this research model supports the stated hypotheses, and mobile app usage avoidance is composed of five latent variables.

PLS-PM analysis result reveals that the perceived irritation construct affects passengers' avoidance of using mobile apps of airlines ($\beta = 0.280$, $p < 0.01$). This result empirically supports Hypothesis 1. In other words, if the passengers perceive the mobile app as irritating, they may avoid using the app in question. Besides, the perceived lack of utility construct affects passengers' avoidance of using mobile apps of airlines ($\beta = 0.309$, $p < 0.01$). This result empirically supports Hypothesis 2. From this point of view, if the airline mobile app does not provide the desired benefit related to the air travel, passengers may display an avoidance behavior. Moreover, the perceived lack of incentive construct affects passengers' avoidance of using mobile apps of airlines ($\beta = 0.161$, $p < 0.01$). This result empirically supports Hypothesis 3. This result may indicate that another factor causing passengers to avoid using the airline mobile app is the perceived lack of incentives. According to PLS-PM results, technology anxiety construct affects passengers' avoidance to use mobile apps of airlines ($\beta = 0.424$, $p < 0.01$). This result empirically supports Hypothesis 4. Therefore, if passengers are scared of using the technology, they may avoid to use the airline mobile app. Finally, the privacy concern construct affects passengers' avoidance of using mobile apps of airlines ($\beta = 0.187$, $p < 0.01$). This result empirically supports Hypothesis 5. If users have concerns that their personal information will be shared with third parties without their permission, this may cause avoidance of using mobile app. The current study found that perceived irritation, perceived lack of utility, perceived lack of incentive, technology anxiety and privacy concern are

Table 8.5 PLS results for endogenous latent construct predictive accuracy and relevance

Dependent endogenous latent construct	Predictive accuracy (R^2)	Predictive relevance effect size (Q^2)	Effect size
Mobile app usage avoidance	0.795	0.573	Large

Note * Assessing predictive relevance (Q2) value of the effect size: $0.02 = $ Small, $0.15 = $ Medium, $0.35 = $ Large

significant determinants of passengers' avoidance of using mobile apps of airlines. In terms of the relative importance of the predictive variables on the response variable, the technology anxiety exhibited the most potential predicting power of mobile app usage avoidance, followed by perceived lack of utility, perceived irritation, privacy concern and perceived lack of incentive.

Finally, to examine the predictive accuracy and predictive relevance of the research model: R^2 (coefficient of determination) and Stone-Geisser's Q^2 values were also obtained by using the PLS algorithm procedure. As can be seen in Table 8.5, the R^2 values for the endogenous latent construct are 0.795, which indicates the structural model's predictive accuracy. The R^2 value represents the amount of explained variance of the endogenous constructs in the structural model. The Q^2 value is an indicator of the model's predictive relevance and Q^2 value bigger than zero for a specific reflective endogenous construct indicates the path model's predictive relevance for a particular construct (Hair et al. 2014). Also, from the table, the Q^2 value is above zero (0.573), which supports the model's predictive relevance for the endogenous construct.

8.5 Conclusions and Implications

In recent years, most of the airlines have utilized mobile app and tried to increase the number of passengers using an app due to its advantages to both airlines and passengers. However, research shows that mobile apps do not meet the demands and needs of many users (Lim et al. 2015). As in the literature, it was determined that airlines, like other companies, do not know exactly what passengers expect from a mobile app (Lee and Kim 2019: 437) and many apps are not as successful as expected (Chen et al. 2019b). In such a case, passengers may show avoidance of using mobile apps. Thus, the purpose of this study is to determine the factors that may have an impact on passengers' avoidance of using mobile apps of airlines.

Technology anxiety was found as the most critical dimension of passengers' avoidance of using mobile apps of airlines. In other words, if passengers individually experience technology anxiety, they will refrain from using the airline's mobile app. This conclusion is also supported by other studies in the literature (Meuter et al. 2003; Lee et al. 2010; Gupta and Arora 2017; Lian 2018; Patil et al. 2020). Passengers' anxiety about the use of technology can negatively affect their intention and attitude to use mobile apps and cause them to show resistance. Therefore, special attention should be given to those who prefer traditional methods, and both airlines and app designers should take responsibility to reduce the level of anxiety. Therefore, airline companies should design the interface of a mobile app as simple, user-friendly, and the terms used in the app in an understandable manner. The transition between menus in the mobile app should be quick and easy. At this point, airline companies should frequently receive feedback from their passengers. They should continuously update their mobile app's design, based on the ideas of passengers who experience technology anxiety. In this context, frequently asked questions or videos should be added in order to ensure that the users who experience anxiety do not avoid the use of mobile apps. This information should also be announced to existing and potential users on all platforms (social media, websites, etc.). It should be designed to allow the users, who are afraid of making a mistake in the purchase phase, to return to the previous menu easily. The possibility of encountering an error or loss (especially a material error or loss) during the use of service via mobile apps may increase technology anxiety (Yang and Forney 2013). Due to the possibility of such losses, passengers who experience anxiety should be informed that they will not encounter any problems, and if they do, the problem can be solved with 24/7 live support systems. Thus, avoidance behavior from using the airline mobile app due to technology anxiety may be reduced.

Perceived lack of utility is the second important factor that affects passengers' avoidance of using mobile apps of airlines. This conclusion is also supported by studies in the literature (Lu et al. 2015; McLean 2018; Chen et al. 2019b). That is, if individuals perceive a mobile app as useful, there would be less tendency to avoid using the app or display continuous use behavior (Chen et al. 2019a). In addition, current and potential users' attitude toward the mobile app with high utilitarian features may be more positive (Shen 2015), and the intention to display the app use behavior may be higher (Lu et al. 2015; Byun et al. 2018). However, perceived lack of utility for airline mobile app may cause passenger dissatisfaction (Chen et al. 2019b). This may cause the app not to be used continuously or even to be uninstalled even if the app is downloaded (Hsiao et al. 2016). Providing personalized service through the mobile app is very important in adopting the app and creating loyalty. In addition, personalized services can highlight the utilitarian features of the app (McLean 2018). So, airline companies should first provide personalized information / services via mobile apps in order to increase perceived utility. For example, in this context, personalized information such as flight time, check-in time, departure gate should be provided to users via a mobile app. In addition, the app should be designed to help ensure complete airline service and improve service performance. In this context, by developing the content management system (Lu et al. 2015), sufficient and correct information may be obtained. Also, the feedback system should be run actively in order for the contents to be useful. Within this scope, a mobile app should provide an entertaining experience for its users. Because, if users find the mobile app enjoyable, their satisfaction level will increase (Lee and Kim 2019). The perception of the mobile apps as fun also positively affects the continuous use intention of the apps (Hsiao et al. 2016). It may be helpful to use vivid colors and animated icons in order to provide an entertaining mobile app experience. Besides, passengers should be able to perform their operations faster, more efficiently and effortlessly with the airline mobile app. Thus, the number of people who avoid using the mobile apps may be reduced.

According to the results of the study; perceived irritation is another factor affecting passengers' avoidance of using mobile apps of airlines. This result is similar to Liu et al. (2012); Dennison et al. (2013); Nyheim et al. (2015); Chen et al. (2019a). From this point of view, if individuals perceive an irritation toward the mobile apps themselves or their content, they may avoid to use the app with the motivation to protect themselves. Therefore, airline companies should ensure that their mobile apps are attractive and not distracting passengers from using their mobile apps. Besides, the mobile app's appearance should be pleasing and attractive to reduce the mobile app's irritability. Especially the presence of advertisements is often seen as one of the main causes of avoidance behavior (Morimoto and Chang 2009; Baek and Morimoto 2012; Liu et al. 2012; Guardia 2014; Nyheim et al. 2015). Therefore, if advertisements such as ticket, hotel and car rental are used in airline mobile apps, these advertisements should be used in a frequency / way that will not irritate the users. In addition, banner ads that will reduce interaction with the app should be avoided while purchasing airline services, collecting information, or check-in / boarding with the mobile app. In fact, if the app user does not want it, information and advertisements should not be presented to that passenger group. In other words, control must be with the passengers themselves. Otherwise, passengers may have negative feelings toward the mobile app and avoid using it. Therefore, airline companies should correctly determine all the features of the mobile app that can irritate the user. In this context, airlines may periodically conduct researches on their target market and redesign their apps according to the issues that are perceived as irritating. Thus, passengers will not find the app irritating and avoid using the airline's mobile app.

Besides, the perceived lack of incentive has an impact on passengers' avoidance of using mobile apps of airlines. This result is also in line with previous studies on mobile apps (Madan and Yadav 2016; Chen et al. 2019a; Zhao et al. 2019). Today, customers generally expect financial incentives to use new technologies (Khajeh et al. 2015; Kim and Han 2014) such as mobile applications. In this context, people's attitude toward mobile app may be shaped by perceived incentives (Carter and Yeo 2016) and decided according to these incentives (Chen et al. 2019a). In particular, the frequency of app usage is significantly affected by the incentives offered to users (Hsu and Tang 2020). Therefore, the incentives offered to use the mobile app may increase usage motivation of mobile app. At the same time, it may positively affect app use by increasing the level of customer satisfaction. Moreover, positive word-of-mouth communication may be achieved, thus, it will contribute to the use of the mobile app by more people (Wang et al. 2019). However, it is important that these incentives are presented in the right place at the right time and in the right way (Hsu and Tang 2020). In this context, airlines should offer some personalized financial incentives such as free or discount flight tickets, air miles and coupons, etc. to their passengers. In addition, value added services should be presented through the application. Also, in order to increase the continuous use of the mobile app, competitions can be organized through the apps and users can be given discounts or free air tickets based on the points obtained from the competition. In order to eliminate the perceived lack of incentives and therefore avoidance usage, the airline companies may announce what incentives will be obtained as a result of the use of the app in different media environments with the phenomena that are important for the passengers. Thus, the motivation to use airline mobile apps will be further increased (Escobar-Rodríguez and Carvajal-Trujillo 2013, 2014).

Lastly, the study concluded that privacy concern has a significant effect on passengers' avoidance to use mobile apps. This result is similar to Bailey et al. (2017), Gu et al. (2017), Zhang et al. (2018), Feng and Xie (2019), and Hassandoust et al (2020). Individuals' high level of sensitivity regarding information that is important to them also increases the privacy concern (Hassandoust et al. 2020). In addition, the less the potential mobile app user has control over privacy, the higher the anxiety level may be. In such a case, the attitude toward app usage may be adversely influenced (Feng and Xie 2019). Therefore, it may be beneficial for airlines to set a clear, transparent, and understandable privacy policy and establish appropriate control mechanisms. In addition, the determined privacy policies should be announced to the current and potential airline mobile app users through different media tools and these people should be given the image that the app usage is quite safe. In addition, airline companies should not share the personal information of mobile app users with third parties without their consent. Apps must have the required certificates in order to increase the user's security perception. It should also be ensured that it is visible that the app has a certificate. If airline companies increase the popularity of their mobile apps, users' privacy concerns may be reduced (Gu et al. 2017). If airline companies adapt all these features to their mobile app and update it continuously, passengers may not avoid using airlines' mobile apps and more passengers will be able to utilize more it in the future. Thus, passengers may prefer these airlines in their future travels. Therefore, the website, forum, or mobile apps that individuals prefer to use should ensure user privacy (Pentina et al. 2016).

The results of the present study highlight several implications for research and practice in the context of passengers' avoidance of using mobile app of airlines. From the management perspective, the airline companies operating in a competitive market must follow the technology closely in order to be successful. Airline companies should design the technology they use to meet the passengers' demands and needs and adapt them to all marketing processes. It would be helpful to equip the app with the necessary features in order to ensure that passengers do not avoid using the airline mobile app and to ensure their continuous usage. Airline mobile apps designed in line with the expectations and needs of the passengers will be able to provide passenger satisfaction. When passenger satisfaction is achieved, avoidance behavior from using the airline mobile app may decrease, positive word-of-mouth communication will occur, and thus the app will be used by more passengers. Airlines will also be able to more effectively and successfully implement marketing activities such as customer relationship management, customer segmentation and targeting, and loyalty programs through the mobile app (Kumar et al. 2018).

From the academic perspective, in this study, perceived irritation, perceived lack of utility, perceived lack of incentive, technology anxiety, privacy concerns, and mobile app usage avoidance have been integrated and searched with a model. It has also adopted the PLS-SEM approach for the analysis of data. So, this research may fill the gap in the literature since it has been studied less and the theoretical model developed may be used as a baseline model in future researches. Besides, the study has some limitations. The convenience sampling method was used as one of the non-probability sampling methods. The use of probability sampling methods in future studies may be useful for increasing the ability of the study to represent the population. Also, the comparisons between generations (X, Y, and Z generations) on this topic may be searched to see the difference among generations about avoidance of mobile apps. Another limitation is that the study was applied only to Turkish passengers. In this context, applying the research to the passengers of different countries may contribute to the literature in terms of identifying intercultural differences.

Appendix

Measurement of Instruments

Perceived irritation (PI)	I think the mobile app of this airline is… – disturbing – ridiculous – negative – uninviting – distracting – retarding – interfering – unsettling – unnecessary
Perceived lack of utility (PLU)	The mobile app of this airline does not … – provide personalized flight service information – allow me to buy tickets at the best price – save money – save time – provide an entertaining experience
Perceived lack of incentive (PLI)	I was not offered any … – incentives to use the mobile app of this airline – financial benefits from using the mobile app of this airline – incentive/reward such as gifts, coupons, points to use the mobile app of this airline – discount or offer to use the mobile app of this airline

(continued)

(continued)

Perceived irritation (PI)	I think the mobile app of this airline is... − disturbing − ridiculous − negative − uninviting − distracting − retarding − interfering − unsettling − unnecessary
Technology anxiety (TA)	− I think I cannot learn to use mobile app technology − I have difficulty understanding mobile app technology − I am worried about using mobile app technology − I am afraid to use mobile app technology because I think I might harm − I am not sure I can interpret the results of mobile app technology − The terms used by mobile app technology may confuse me − Mobile app technology is unfamiliar to me, so I avoid using it − I think I cannot keep up with mobile app technology − I hesitate using mobile app technology since I am afraid of making mistakes
Privacy concerns (PC)	− I would be uncomfortable with the unauthorized sharing of my personal information when I use the mobile app of this airline − I would be uncomfortable with the exploitation of my personal information when I use the mobile app of this airline − I would be uncomfortable with the sharing of my personal information with others when I use the mobile app of this airline − Since I am using the mobile app of this airline, I believe that the company shares my personal information without permission

(continued)

(continued)

Perceived irritation (PI)	I think the mobile app of this airline is… – disturbing – ridiculous – negative – uninviting – distracting – retarding – interfering – unsettling – unnecessary
Mobile app usage avoidance (MAUA)	– I avoid using the mobile app of this airline business on purpose – I do not like the mobile app of this airline – It would be nice if there were no mobile apps for this airline – I avoid using this airline's mobile app because frequent updates are required – I avoid using this airline's mobile app because of its low performance – I avoid using this airline's mobile app because it is not fast – I avoid using this airline's mobile app because of the difficulty of reading the text on the screen – I avoid using this airline's mobile app because it is not well designed – I avoid using this airline's mobile app because it is difficult to use – I avoid using this airline's mobile app because it takes a long time to switch between menus – I avoid using this airline's mobile app because its users give it low points

References

Aaker D, Bruzzone D (1985) Causes of irritation in advertising. J Mark 49(2):47–57. https://doi.org/10.1177/002224298504900204

Abernethy AM (1991) Physical and mechanical avoidance of television commercials: an exploratory study of zipping, zapping and leaving. In Holman, R. (Eds.), Proceedings of the American academy of advertising, pp 223–231. NY: The American Academy of Advertising

Amadeus (2011) The always-connected traveler: How mobile will transform the future of air travel. http://www.amadeus.com/web/binaries/1333089230782/blobheader=application/pdf&blobheadername1=Content-Disposition&blobheadervalue1=inline%3B+filename%3DAlways+Connected+Travler_REAL+US+VERSION.pdf

Atchariyachanvanich K, Okada H, Sonehara N (2006) What keeps online customers repurchasing through the internet? ACM SIGecom Exchanges 6(2):47–57. https://doi.org/10.1145/1228621.1228626

Baek TH, Morimoto M (2012) Stay away from me: examining the determinants of consumer avoidance of personalized advertising. J Advert 41(1):59–76. https://doi.org/10.2753/JOA0091-3367410105

Bailey AA, Pentina I, Mishra AS, Mimoun MSB (2017) Mobile payments adoption by US consumers: an extended TAM. Int J Retail Distr Manag 45(6):626–640. https://doi.org/10.1108/IJRDM-08-2016-0144

Bankmycell (2020) How many smartphones are in the world?. https://www.bankmycell.com/blog/how-many-phones-are-in-the-world

Bauer R, Raymond A (1968) Advertising in America. Harvard University, Graduate School of Business Administration, Division of Research, The consumer view

Bauer HH, Reichardt T, Barnes SJ, Neumann MM (2005) Driving consumer acceptance of mobile marketing: a theoretical framework and empirical study. J Electron Commer Res 6(3):181–192

Bellman S, Schweda A, Varan D (2010) The residual impact of avoided television advertising. J Advert 39(1):67–82. https://doi.org/10.2753/JOA0091-3367390105

Blair I (2019) 10 Biggest hidden costs of developing an app & how to handle them. https://buildfire.com/hidden-app-development-costs/

Bogicevic V, Bujisic M, Bilgihan A, Yang W, Cobanoglu C (2017) The impact of traveler-focused airport technology on traveler satisfaction. Technol Forecast Soc Chang 123:351–361. https://doi.org/10.1016/j.techfore.2017.03.038

Bozacı İ (2015) Determining the factors related with customer privacy behaviors. J Human Soc Sci Res 4(3):612–633. https://doi.org/10.15869/itobiad.11614

Budd L, Vorley T (2013) Airlines, apps, and business travel: a critical examination. Res Transp Bus Manag 9:41–49. https://doi.org/10.1016/j.rtbm.2013.08.004

Buhalis D (2004) eAirlines: strategic and tactical use of ICTS in the airline industry. Inform Manag 41(7):805–825. https://doi.org/10.1016/j.im.2003.08.015

Byun H, Chiu W, Bae JS (2018) Exploring the adoption of sports brand apps: an application of the modified technology acceptance model. Int J Asian Bus Inform Manag (IJABIM) 9(1):52–65. https://doi.org/10.4018/IJABIM.2018010105

Caragliu A, Bo CD, Nijkamp P (2011) Smart cities in Europe. J Urban Technol 18(2):65–82. https://doi.org/10.1080/10630732.2011.601117

Carter S, Yeo ACM (2016) Mobile apps usage by Malaysian business undergraduates and postgraduates. Internet Res 26(3):733–757. https://doi.org/10.1108/IntR-10-2014-0273

Celik H (2016) Customer online shopping anxiety within the unified theory of acceptance and use technology (UTAUT) framework. Asia Pac J Mark Logist 28(2):278–307. https://doi.org/10.1108/APJML-05-2015-0077

Chang IC, Chou PC, Yeh R, Tseng H-T (2016) Factors influencing Chinese tourists' intentions to use the Taiwan medical travel app. Telematics Inform 33(2):401–409. https://doi.org/10.1016/j.tele.2015.09.007

Chen Y, Lou H, Luo W (2002) Distance learning technology adoption: a motivation perspective. J Comput Inf Syst 42(2):38–43. https://doi.org/10.1080/08874417.2002.11647485

Chen Q, Zhang M, Zhao X (2017) Analysing customer behaviour in mobile app usage. Ind Manag Data Syst 117(2):425–438. https://doi.org/10.1108/IMDS-04-2016-0141

Chen JV, Tran A, Nguyen T (2019a) Understanding the discontinuance behavior of mobile shoppers as a consequence of technostress: An application of the stress-coping theory. Comput Hum Behav 95:83–93. https://doi.org/10.1016/j.chb.2019.01.022

Chen Q, Lu Y, Tang Q (2019b) Why do users resist service organization's brand mobile apps? The force of barriers versus cross-channel synergy. Int J Inf Manage 47:274–282. https://doi.org/10.1016/j.ijinfomgt.2018.07.012

Chin WW (1998) The partial least squares approach to structural equation modeling. In: Marcoulides GA (ed) Modern methods for business research. Lawrence Erlbaum Associates Publisher, London, pp 295–336

Cho C-H, Cheon JH (2004) Why do people avoid advertising on the internet? J Advert 33(4):89–97. https://doi.org/10.1080/00913367.2004.10639175

Clarke IIII (2001) Emerging value propositions for m-commerce. Journal of Business Strategies 18(2):133–148

Compeau D, Higgins CA, Huff S (1999) Social cognitive theory and individual reactions to computing technology: a longitudinal study. MIS Q 23(2):145–158. https://doi.org/10.2307/249749

Çakar MM (2017) Reasons for using information technologies by entrepreneurs, analysis technology acceptance model form: Manisa city exemplary [Unpublished master's thesis]. İzmir Kâtip Çelebi University

Davis FD (1989) Perceived usefulness, perceived ease of use, and user acceptance of information technology. MIS Q 13(3):319–340. https://doi.org/10.2307/249008

De Pelsmacker P, Van den Bergh J (1998) Advertising content and irritation: a Study of 226 TV commercials. J Int Consum Mark 10(4):5–27. https://doi.org/10.1300/J046v10n04_02

Dennison L, Morrison L, Conway G, Yardley L (2013) Opportunities and challenges for smartphone applications in supporting health behavior change: Qualitative study. J Medi Internet Res 15(4), Article e86. https://doi.org/10.2196/jmir.2583

Dickinson JE, Cherrett T, Hibbert JF, Winstanley C, Shingleton D, Davies N, Norgate S, Speed C (2015) Fundamental challenges in designing a collaborative travel app. Transp Policy 44:28–36. https://doi.org/10.1016/j.tranpol.2015.06.013

Ding Y, Chai KH (2015) Emotions and continued usage of mobile applications. Ind Manag Data Syst 115(5):833–852. https://doi.org/10.1108/IMDS-11-2014-0338

Directorate General of Civil Aviation—DGCA (2017) Aviation companies in Turkey. http://web.shgm.gov.tr/tr/havacilik-isletmeleri/2063-hava-tasima-isletmeleri

Directorate General of Civil Aviation—DGCA (2018) We are the fastest developing country in Europe. http://web.shgm.gov.tr/documents/sivilhavacilik/files/GM/Sn_Bahri_Kesici_Yanki_Dergisi.pdf.

Dix S, Phau I (2010) Television advertising avoidance: advancing research methodology. J Promot Manag 16(1–2):114–133. https://doi.org/10.1080/10496490903574013

Edwards SM, Li H, Lee J-H (2002) Forced exposure and psychological reactance: antecedents and consequences of the perceived intrusiveness of pop-up ads. J Advert 31(3):83–95. https://doi.org/10.1080/00913367.2002.10673678

Egger R (2013) The impact of near field communication on tourism. J Hosp Tour Technol 4(2):119–133. https://doi.org/10.1108/JHTT-04-2012-0014

Escobar-Rodríguez T, Carvajal-Trujillo E (2013) Online drivers of consumer purchase of website airline tickets. J Air Transp Manag 32:58–64. https://doi.org/10.1016/j.jairtraman.2013.06.018

Escobar-Rodríguez T, Carvajal-Trujillo E (2014) Online purchasing tickets for low cost carriers: An application of the unified theory of acceptance and use of technology (UTAUT) model. Tour Manage 43:70–88. https://doi.org/10.1016/j.tourman.2014.01.017

Eser Z, Korkmaz S, Öztürk SA (2011) Pazarlama: Kavramlar, ilkeler, kararlar. 2. Baskı. Siyasal Kitapevi

Fang J, Zhao Z, Wen C, Wang R (2017) Design and performance attributes driving mobile travel application engagement. Int J Inf Manage 37:269–283. https://doi.org/10.1016/j.ijinfomgt.2017.03.003

Feng Y, Xie Q (2019) Privacy concerns, perceived intrusiveness, and privacy controls: an analysis of virtual try-on apps. J Interact Advert 19(1):43–57. https://doi.org/10.1080/15252019.2018.1521317

Fennis BM, Bakker AB (2001) "Stay tuned-We will be back right after these messages": Need to evaluate moderates the transfer of irritation in advertising. J Advert 30(3):15–25. https://doi.org/10.1080/00913367.2001.10673642

Flurry (2011) iOS & Android apps challenged by traffic acquisition not discovery. http://blog.flurry.com/bid/76874/iOS-Android-Apps-Challenged-by-Traffic-AcquisitionNot-Discovery

Fornell C, Larcker DF (1981) Evaluating structural equation models with unobservable variables and measurement error. J Mark Res 18(1):39–50. https://doi.org/10.1177/002224378101800104

Fox G, Connolly R (2018) Mobile health technology adoption across generations: narrowing the digital divide. Inf Syst J 28(6):995–1019. https://doi.org/10.1111/isj.12179

Gao T, Sultan F, Rohm AJ (2010) Factors influencing Chinese youth consumers' acceptance of mobile marketing. J Consum Mark 27(7):574–583. https://doi.org/10.1108/07363761011086326

Gefen D, Rigdon EE, Straub DW (2011) An update and extension to SEM guidelines for administrative and social science research. MIS Q 35(2):iii–xiv. https://doi.org/10.2307/23044042

Gerede E, Orhan G (2015) Türk havayolu taşımacılığındaki ekonomik düzenlemelerin gelişim süreci. In E. Gerede (Eds.), Havayolu taşımacılığı ve ekonomik düzenlemeler: Teori ve Türkiye uygulaması (pp 163–208). Ankara: Sivil Havacılık Genel Müdürlüğü Yayınları

Google. (2016). Next generation travel habbits. https://storage.googleapis.com/think-v2.../27821_Yeni%20Nesil%20Seyahat-1.pdf

Greyser SA (1973) Irritation in advertising. J Advert Res 13(1):3–10

Gu J, Xu Y, Xu H, Zhang C, Ling H (2017) Privacy concerns for mobile app download: an elaboration likelihood model perspective. Decis Support Syst 94:19–28. https://doi.org/10.1016/j.dss.2016.10.002

Guardia FR (2014) Generalization of advertising avoidance on social networks. http://dee.uib.es/digitalAssets/312/312676_rejon.pdf

Guo X, Han X, Zhang X, Dang Y, Chen C (2015) Investigating m-health acceptance from a protection motivation theory perspective: gender and age differences. Telemed e-Health 21(8):661–669. https://doi.org/10.1089/tmj.2014.0166

Gupta A, Arora N (2017) Understanding determinants and barriers of mobile shopping adoption using behavioral reasoning theory. J Retail Consum Serv 36:1–7. https://doi.org/10.1016/j.jretconser.2016.12.012

Gülsoy TY (2018) Kitle iletişiminin yönetimi: Reklam, satış tutundurma, etkinlikler ve deneyimler, halkla ilişkiler (Philip Kotler—Kevin Lane Keller). In İ. Kırcova (Eds.), Pazarlama Yönetimi: Kotler—Keller (pp 584–613). İstanbul: Beta Yayınevi

Hair JF, Hult GTM, Ringle C, Sarstedt M (2014) A primer on partial least squares structural equation modeling (PLS-SEM). Sage Publications

Hair Jr JF, Hult GTM, Ringle C, Sarstedt M (2016). A primer on partial least squares structural equation modeling (PLS-SEM)(2nd ed.). Sage Publications

Hassandoust F, Akhlaghpour S, Johnston AC (2020) Individuals' privacy concerns and adoption of contact tracing mobile applications in a pandemic: A situational privacy calculus perspective. J Amer Med Inform Assoc, Article ocaa240. https://doi.org/10.1093/jamia/ocaa240

Henseler J, Ringle CM, Sinkovics RR (2009) The use of partial least squares path modeling in international marketing. In: Sinkovics RR, Ghauri PN (eds) New challenges to international marketing advances in international marketing. Emerald Group Publishing Limited, UK, pp 227–319

Hsiao CH, Chang JJ, Tang KY (2016) Exploring the influential factors in continuance usage of mobile social Apps: satisfaction, habit, and customer value perspectives. Telematics Inform 33(2):342–355. https://doi.org/10.1016/j.tele.2015.08.014

Hsu C, Lin J (2015) What drives purchase intention for paid mobile apps? An expectation confirmation model with perceived value. Electron Commer Res Appl 14(1):46–57. https://doi.org/10.1016/j.elerap.2014.11.003

Hsu LC, Wang KY, Chih WH, Lin WC (2019) Modeling revenge and avoidance in the mobile service industry: Moderation role of technology anxiety. Serv Indus J, 1–24. https://doi.org/10.1080/02642069.2019.1585428

Hsu TH, Tang JW (2020) Development of hierarchical structure and analytical model of key factors for mobile app stickiness. J Innov Knowl 5(1):68–79. https://doi.org/10.1016/j.jik.2019.01.006

Im JY, Hancer M (2014) Shaping travelers' attitude toward travel mobile applications. J Hosp Tour Technol 5(2):177–193. https://doi.org/10.1108/JHTT-11-2013-0036

Karahoca A, Karahoca D, Aksöz M (2018) Examining intention to adopt to internet of things in healthcare technology products. Kybernetes 47(4):742–770. https://doi.org/10.1108/K-02-2017-0045

Keikhosrokiani P, Mustaffa N, Zakaria N, Abdullah R (2019) Assessment of a medical information system: the mediating role of use and user satisfaction on the success of human interaction with the mobile healthcare system (iHeart). Cogn Technol Work 22:281–305. https://doi.org/10.1007/s10111-019-00565-4

Kelly L, Kerr L, Drennan J (2010) Avoidance of advertising in social networking sites: the teenage perspective. J Interact Advert 10(2):16–27. https://doi.org/10.1080/15252019.2010.10722167

Khajeh F, Nayebzadeh S, Sadeghian A (2015) Determinants of consumer perceptions toward mobile advertising by confirmatory factor analysis and structural equation modeling techniques (SEM). Cumhuriyet Univ Faculty Sci, Sci J (CSJ) 36(4):926–936.

Khalid H, Shihab E, Nagappan M, Hassan AE (2015) What do mobile app users complain about? IEEE Softw 32(3):70–77. https://doi.org/10.1109/MS.2014.50

Kim J, Brewer P, Bernhard B (2008) Hotel customer perceptions of biometric door locks: Convenience and security factors. J Hosp Leis Mark 17(1–2):162–183. https://doi.org/10.1080/10507050801978323

Kim HB, Kim T, Shin SW (2009) Modeling roles of subjective norms and e-trust in customers' acceptance of airline B2C ecommerce websites. Tour Manage 30(2):266–277. https://doi.org/10.1016/j.tourman.2008.07.001

Kim YJ, Han JY (2014) Why smartphone advertising attracts customers: a model of web advertising, flow, and personalization. Comput Hum Behav 33:256–269. https://doi.org/10.1016/j.chb.2014.01.015

Kinnear TC, Taylor JR (1991) Instructor's manual and test bank to accompany: marketing research an applied approach (4th ed.). McGraw Hill

Koch S, Tritscher F (2016) Social media in the airline industry: acceptance of social seating. J Hosp Tour Technol 8(2):256–279. https://doi.org/10.1108/JHTT-11-2016-0078

Koshksaray AA, Franklin D, Hanzaee KH (2015) The relationship between e-lifestyle and internet advertising avoidance. Australas Mark J 23(1):38–48. https://doi.org/10.1016/j.ausmj.2015.01.002

Kumar DS, Purani K, Viswanathan SA (2018) Influences of 'appscape' on mobile app adoption and m-loyalty. J Retail Consum Serv 45:132–141. https://doi.org/10.1016/j.jretconser.2018.08.012

Laukkanen P, Sinkkonen S, Laukkanen T (2008) Consumer resistance to internet banking: postponers, opponents and rejectors. Int J Bank Mark 26(6):440–455. https://doi.org/10.1108/02652320810902451

Lee HJ, Cho HJ, Xu W, Fairhurst A (2010) The influence of consumer traits and demographics on intention to use retail self-service checkouts. Mark Intell Plan 28(1):46–58. https://doi.org/10.1108/02634501011014606

Lee JM, Rha JY (2016) Personalization-privacy paradox and consumer conflict with the use of location-based mobile commerce. Comput Hum Behav 63:453–462. https://doi.org/10.1016/j.chb.2016.05.056

Lee Y, Kim HY (2019) Consumer need for mobile app atmospherics and its relationships to shopper responses. J Retail Consum Serv 51:437–442. https://doi.org/10.1016/j.jretconser.2017.10.016

Li H, Edwards SM, Lee J-H (2002) Measuring the intrusiveness of advertisements: scale development and validation. J Advert 31(2):37–47. https://doi.org/10.1080/00913367.2002.10673665

Lian JW (2018) Why is self-service technology (SST) unpopular? Extending the IS success model. Library Hi Tech. https://doi.org/10.1108/LHT-01-2018-0015

Lim SL, Bentley PJ, Kanakam N, Ishikawa F, Honiden S (2015) Investigating country differences in mobile app user behavior and challenges for software engineering. IEEE Trans Software Eng 41(1):40–64. https://doi.org/10.1109/TSE.2014.2360674

Lin WR, Wang YH, Shih KH (2017) Understanding consumer adoption of mobile commerce and payment behaviour: an empirical analysis. Int J Mobile Commun 15(6):628–654. https://doi.org/10.1504/IJMC.2017.086880

Liu CLE, Sinkovics RR, Pezderka N, Haghirian P (2012) Determinants of consumer perceptions toward mobile advertising—a comparison between Japan and Austria. J Interact Mark 26:21–32. https://doi.org/10.1016/j.intmar.2011.07.002

Liu F, Zhao S, Lia Y (2017) How many, how often, and how new? A multivariate profiling of mobile app users. J Retail Consum Serv 38:71–80. https://doi.org/10.1016/j.jretconser.2017.05.008

Lu J, Mao Z, Wang M, Hu L (2015) Goodbye maps, hello apps? Exploring the influential determinants of travel app adoption. Curr Issue Tour 18(11):1059–1079. https://doi.org/10.1080/136 83500.2015.1043248

Madan K, Yadav R (2016) Behavioural intention to adopt mobile wallet: a developing country perspective. J Indian Bus Res 8(3):227–244. https://doi.org/10.1108/JIBR-10-2015-0112

Mani Z, Chouk I (2017) Drivers of consumers' resistance to smart products. J Mark Manag 33(1–2):76–97. https://doi.org/10.1080/0267257X.2016.1245212

Marangoz M, Aydın AE (2018, April 20–24) A discussion on consumer privacy concern and privacy protection [Conference Presentation]. 5th International conference on social sciences and education research, Antalya, Turkey

Martin-Domingo L, Martín JC (2016) Airport mobile internet and innovation. J Air Transp Manag 55:102–112. https://doi.org/10.1016/j.jairtraman.2016.05.002

Mateos-Aparicio G (2011) Partial least squares (PLS) methods: origins, evolution, and application to social sciences. Commun Stat Theory Methods 40(13):2305–2317. https://doi.org/10.1080/03610921003778225

McLean G (2018) Examining the determinants and outcomes of mobile app engagement-a longitudinal perspective. Comput Hum Behav 84:392–403. https://doi.org/10.1016/j.chb.2018.03.015

Meuter ML, Ostrom AL, Bitner MJ, Roundtree R (2003) The influence of technology anxiety on consumer use and experiences with self-service technologies. J Bus Res 56(11):899–906. https://doi.org/10.1016/S0148-2963(01)00276-4

Miraja B, Persada S, Prasetyo Y, Belgiawan P, Redi AAN (2019) Applying protection motivation theory to understand generation z students intention to comply with educational software anti piracy law. Int J Emerg Technol Learn (iJET) 14(18):39–52

Monno P, Xiao D (2014) Mobile commerce app adoption: consumer behavior differences between Europe and Asia (Lund University). http://lup.lub.lu.se/student-papers/record/4468608

Morimoto M, Chang S (2006) Consumers' attitudes toward unsolicited commercial e-mail and postal direct mail marketing methods: intrusiveness, perceived loss of control and irritation. J Interact Advert 7(l):1–11. https://doi.org/10.1080/15252019.2006.10722121

Morimoto M, Chang S (2009) Psychological factors affecting perceptions of unsolicited commercial e-mail. J Current Issues Res Advert 31(1):63–73. https://doi.org/10.1080/10641734.2009.105 05257

Morosan C (2016) An empirical examination of US travelers' intentions to use biometric e-gates in airports. J Air Transp Manag 55:120–128. https://doi.org/10.1016/j.jairtraman.2016.05.005

Mousavi R, Chen R, Kim DJ, Chen K (2020). Effectiveness of privacy assurance mechanisms in users' privacy protection on social networking sites from the perspective of protection motivation theory. Decision Support Systems, Article 113323.https://doi.org/10.1016/j.dss.2020.113323

Nimrod G (2018) Technophobia among older internet users. Educ Gerontol 44(2–3):148–162. https://doi.org/10.1080/03601277.2018.1428145

Nyheim P, Xu S, Zhang L, Mattila AS (2015) Predictors of avoidance towards personalization of restaurant smartphone advertising: a study from the millennials' perspective. J Hosp Tour Technol 6(2):145–159. https://doi.org/10.1108/JHTT-07-2014-0026

Patil P, Tamilmani K, RanaNP, Raghavan V (2020) Understanding consumer adoption of mobile payment in India: Extending Meta-UTAUT model with personal innovativeness, anxiety, trust, and grievance redressal. Int J Inform Manag 54, Article 102144. https://doi.org/10.1016/j.ijinfo mgt.2020.102144

Pentina I, Zhang L, Bata B, Chen Y (2016) Exploring privacy paradox in information-sensitive mobile app adoption: a cross-cultural comparison. Comput Hum Behav 65:409–419. https://doi.org/10.1016/j.chb.2016.09.005

Rau PLP, Liao Q, Chen C (2013) Factors influencing mobile advertising avoidance. Int J Mobile Commun 11(2):123–139. https://doi.org/10.1504/IJMC.2013.052637

Richard JE, Meuli PG (2013) Exploring and modelling digital natives' intention to use permission-based location-aware mobile advertising. J Mark Manag 29(5–6):698–719. https://doi.org/10.1080/0267257X.2013.770051

Ringle CM, Wende S, Becker JM (2015) SmartPLS 3. SmartPLS GmbH

Rivera M, Gregory A, Cobos L (2015) Mobile application for the timeshare industry: the influence of technology experience, usefulness, and attitude on behavioral intentions. J Hosp Tour Technol 6(3):242–257

Rogers RW (1975) A protection motivation theory of fear appeals and attitude change1. J Psychol 91(1):93–114. https://doi.org/10.1080/00223980.1975.9915803

Sarstedt M, Ringle CM, Hair JF (2017) Partial least squares structural equation modeling. In: Homburg C, Klarmann M, Vomberg A (eds) Handbook of Market Research. Springer, Heidelberg, pp 1–40

Schmitz C, Bartsch S, Meyer A (2016) Mobile app usage and its implications for service management—empirical findings from German public transport. Procedia Soc Behav Sci 224:230–237. https://doi.org/10.1016/j.sbspro.2016.05.492

Shen GCC (2015) Users' adoption of mobile applications: Product type and message framing's moderating effect. J Bus Res 68(11):2317–2321. https://doi.org/10.1016/j.jbusres.2015.06.018

Shin W, Lin TT-C (2016) Who avoids location-based advertising and why? Investigating the relationship between user perceptions and advertising avoidance. Comput Hum Behav 63:444–452. https://doi.org/10.1016/j.chb.2016.05.036

Sintonen S (2010) Socio-economic effects on mobile phone adoption behavior among older consumers (Encyclopedia of E-Business Development and Management in the Global Economy). https://www.igi-global.com/chapter/socio-economic-effects-mobile-phone/41245

Siuhi S, Mwakalonge J (2016) Opportunities and challenges of smart mobile applications in transportation. J Traffic Transp Eng 3(6):582–592. https://doi.org/10.1016/j.jtte.2016.11.001

Speck PS, Elliott MT (1997) Predictors of advertising avoidance in print and broadcast media. J Advert 26(3):61–76. https://doi.org/10.1080/00913367.1997.10673529

Statista (2020) Number of smartphone users from 2016 to 2021. https://www.statista.com/statistics/330695/number-of-smartphone-users-worldwide/

Stoyanov SR, Hides L, Kavanagh DJ, Zelenko O, Tjondronegoro D, Mani M (2015) Mobile app rating scale: a new tool for assessing the quality of health mobile apps. JMIR Mhealth Uhealth 3(1), Article e27. https://doi.org/10.2196/mhealth.3422

Suher HK, İspir NB (2011) Permission based mobile marketing and sms ad avoidance. J Yasar Univ 21(6):3633–3647

Suki NM, Suki NM (2011) Exploring the relationship between perceived usefulness, perceived ease of use, perceived enjoyment, attitude and subscribers' intention towards using 3G mobile services. J Inform Technol Manag 22(1):1–7

Suki NM, Suki NM (2017) Flight ticket booking app on mobile devices: examining the determinants of individual intention to use. J Air Transp Manag 62:146–154. https://doi.org/10.1016/j.jairtraman.2017.04.003

Şanlıöz HK, Dilek SE, Koçak N (2013) Changing world, transformed marketing: a pioneering mobile marketing example from Turkish tourism sector. Anatolia J Turizm Araştırmaları Dergisi 24(2):250–260

Tsai HYS, Jiang M, Alhabash S, LaRose R, Rifon NJ, Cotten SR (2016) Understanding online safety behaviors: a protection motivation theory perspective. Comput Secur 59:138–150. https://doi.org/10.1016/j.cose.2016.02.009

Unal P, Taskaya Temizel T, Eren PE (2017) What installed mobile applications tell about their owners and how they affect users' download behavior. Telematics Inform 34(7):1153–1165. https://doi.org/10.1016/j.tele.2017.05.005

Vakıf Investment Company (2018) Turkish aviation industry report. http://www.vkyanaliz.com/Files/docs/sektor-raporu-havacilik-1537766932.pdf

Veríssimo JMC (2016) Enablers and restrictors of mobile banking app use: a fuzzy set qualitative comparative analysis (fsQCA). J Bus Res 69(11):5456–5460. https://doi.org/10.1016/j.jbusres.2016.04.155

Verkasalo H, López-Nicolás C, Molina-Castillo FJ, Bouwman H (2010) Analysis of users and non-users of smartphone applications. Telematics Inform 27(3):242–255. https://doi.org/10.1016/j.tele.2009.11.001

Vinzi EV, Trinchera L, Amato S (2010) PLS path modeling: From foundations to recent developments and open issues for model assessment and improvement. In: Vinzi VE, Chin WW, Henseler J, Wang H (eds) Handbook of partial least squares: Concepts, methods and applications. Springer handbooks of computational statistics series, Heidelberg, pp 47–82

Wang X, Li XR, Zhen F, Zhang JH (2016) How smart is your tourist attraction? Measuring tourist preferences of smart tourism attractions via a FCEM-AHP and IPA approach. Tour Manage 54(June):309–320. https://doi.org/10.1016/j.tourman.2015.12.003

Wang YS, Tseng TH, Wang WT, Shih YW, Chan PY (2019) Developing and validating a mobile catering app success model. Int J Hosp Manag 77:19–30. https://doi.org/10.1016/j.ijhm.2018.06.002

Watson C, McCarthy J, Rowley J (2013) Consumer attitudes towards mobile marketing in the smart phone era. Int J Inf Manage 33(5):840–849. https://doi.org/10.1016/j.ijinfomgt.2013.06.004

WebsiteBuilder (2020). 30 Truly fascinating app usage statistics to know in 2020. https://websitebuilder.org/app-usage-statistics/#:~:text=Consumers%20downloaded%20204%20billion%20apps,exceed%20250%20billion%20in%202022

Wei TT, Marthandan G, Chong AYL, Ooi KB, Arumugam S (2009) What drives Malaysian m-commerce adoption? An empirical analysis. Ind Manag Data Syst 109(3):370–388. https://doi.org/10.1108/02635570910939399

Wu D (2020) Empirical study of knowledge withholding in cyberspace: Integrating protection motivation theory and theory of reasoned behavior. Comput Human Behav 105, Article 106229. https://doi.org/10.1016/j.chb.2019.106229

Xu Z, Yuan Y (2009) The impact of context and incentives on mobile service adoption. Int J Mobile Commun 7(3):363–381. https://doi.org/10.1504/IJMC.2009.023677

Yang K (2012) Consumer technology traits in determining mobile shopping adoption: an application of the extended theory of planned behavior. J Retail Consum Serv 19(5):484–491. https://doi.org/10.1016/j.jretconser.2012.06.003

Yang K, Forney JC (2013) The moderating role of consumer technology anxiety in mobile shopping adoption: Differential effects of facilitating conditions and social influences. J Electron Commer Res 14(4):334–347

Yang HS, Park JW (2019) A study of the acceptance and resistance of airline mobile application services: with an emphasis on user characteristics. Int J Mobile Commun 17(1):24–43. https://doi.org/10.1504/IJMC.2019.096514

Zhang T, Lu C, Kizildag M (2018) Banking "on-the-go": examining consumers' adoption of mobile banking services. Int J Qual Serv Sci 10(3):279–295. https://doi.org/10.1108/IJQSS-07-2017-0067

Zhao H, Anong ST, Zhang L (2019) Understanding the impact of financial incentives on NFC mobile payment adoption. Int J Bank Market 37(5):1296–1312. https://doi.org/10.1108/IJBM-08-2018-0229

Dr. Mutlu Yuksel Avcilar is an Associated Professor in Management Information Systems Department at Osmaniye Korkut Ata University., Osmaniye, south of Turkey where he has been a faculty member since 2010. He completed his Ph.D. degree in Business department of the Institute of Social Sciences in Nigde University. He completed his Master degree in Business department of the Institute of Social Sciences in Cukurova University. Also he graduated from Business department of the Faculty of Economics and Business Sciences in Cukurova University. He has been teaching mainly in the area of Marketing. Principles of Marketing, Marketing Management, Marketing Research, International Marketing and Research Methods in Social Sciences Marketing are some of the lectures he has been teaching. In recent years, his research interests focus on collaborative consumption, sharing economy, services marketing, innovation-technology, augmented reality and airline marketing.

Dr. Nuriye Günebakan is a Professor in the Department of Aviation Management at Iskenderun Technical University, Hatay, south of Turkey where she has been a faculty member since 2007. She completed her Ph.D. and Master degree in Business department of the Institute of Social Sciences in Cukurova University. Also she graduated from Business department of the Faculty of Economics and Business Sciences in Cukurova University. She has been teaching mainly in the area of Marketing. Principles of Marketing, Introduction to Business, Services Marketing, Airline Marketing, Customer Relationship Management, Introduction to Communication and Statistical Data Analysis are some of the lectures she has been teaching. In recent years, her research interests focus on airline marketing related to passengers' behaviour, technology adoption of passengers and innovation in aviation industry.

Dr. Hilal Inan is a Professor in Business Department at Cukurova University, Adana, south of Turkey where she has been a faculty member since 1996. She completed her Ph.D. and Master degree in Business department of the Institute of Social Sciences in Cukurova University. Also she graduated from Business department of the Faculty of Economics and Business Sciences in Cukurova University. She has been teaching mainly in the area of Marketing. Principles of Marketing, Services Marketing, Customer Relationship Management, Marketing Management, Sports Marketing are some of the lectures she has been teaching. In recent years, her research interests focus on green marketing, services marketing and airline marketing related to passengers' behaviour.

Seda Arslan is a Research Assistant in the Department of Aviation Management at Iskenderun Technical University, Hatay, south of Turkey where she has been a faculty member since 2009. She continues to her Ph.D. program in Business department of the Institute of Social Sciences in Osmaniye Korkut Ata University. She completed her Master degree in Aviation Management department of the Institute of Social Sciences in Anadolu University. Also she graduated from Aviation Management department of the School of Civil Aviation in Anadolu University. Her research interests mainly focus on airline marketing related to passengers' behaviour, technology adoption and use of intention and innovation in aviation industry.

Chapter 9
Supplier Performance Evaluation Using Cluster Analysis and Artificial Neural Networks in a MRO Business in Aviation Sector

Muhammet Mikdat Akbaba and Onur Çetin

Abstract Supplier performance evaluation includes the assessment of suppliers according to several criteria. Supplier performance evaluation, which is a part of purchasing activities, has gained more importance in recent years. Performance evaluation practices exist in both manufacturing and service sectors. The maintenance and repair overhaul (MRO) sector in the service sector has different characteristics compared to other sectors in terms of purchasing and supply management. MRO enterprises in the aviation sector have fewer suppliers and the pressure on supply management of MRO companies in the aviation sector is higher when compared with other sectors. The literature on supplier evaluation in both the MRO sector and the aviation industry is limited. In this study, it is aimed to propose a new supplier evaluation and classification system using cluster analysis (CA) and artificial neural networks (ANN) methods in an MRO company operating in the aviation industry. Initially, the supplier evaluation and classification system of the company was analyzed. Then a new clustering has been proposed for the company. It is shown that the suppliers can be classified using the proposed methodology for an alternative supplier evaluation.

Keywords MRO · Supplier performance evaluation · Service systems · Aviation logistics · Cluster analysis

M. M. Akbaba (✉)
Turkish Airlines Technic, Bursa, Turkey
e-mail: mmikdatakbaba@gmail.com

O. Çetin (✉)
Faculty of Economics and Administrative Sciences, Department of Business Administration, Trakya University, Edirne, Turkey
e-mail: onurcetin@trakya.edu.tr

9.1 Introduction

Businesses are looking for ways to carry out their operations at minimum cost in order to gain competitive advantage (Toker and Görener 2013, p. 17). Considering that approximately 60–80% of the expenses of enterprises in the medium term consist of purchasing activities (Murphy and Knemeyer 2016, p. 94), it can be clearly understood that companies have to focus on purchasing. The concept of purchasing which can be viewed as a business function includes activities such as supplier identification and selection, purchasing operations, negotiation, contracting, supplier measurement, evaluation, and development (Monczka et al. 2009, pp. 8–9). Improving supplier relations is of great importance in improving procurement-related activities (Yurdakul 2015, p. 9). In supplier relations, supplier selection is the determination of the suppliers to be worked with, while supplier performance measurement and evaluation is the monitoring and evaluation of the performance of existing suppliers (Gemici 2009, p. 6). Determining the right performance indicators for suppliers and evaluating suppliers using an effective performance evaluation system are important not only for reducing costs but also for improving other performance dimensions (Wisner et al. 2012, p. 82).

Since the supplier selection studies also include an evaluation among potential suppliers, the research regarding supplier performance evaluation and supplier selection has been carried out with similar methods. Methods such as AHP (analytical hierarchy process), analytical network process (ANP), TOPSIS, analysis of variance, decision trees, data envelopment analysis (DEA), fuzzy logic, and cluster analysis (CA) were used in supplier evaluation.

Although supplier performance evaluation studies have been conducted in many sectors, studies in the service sector are limited compared to the manufacturing sector. MRO (maintenance, repair, overhaul) sector has an important place in the service sector and in this sector purchases are carried out to produce repair and maintenance services (Pooler et al. 2004, p. 335). For MRO companies in the service sector, the literature on supplier performance evaluation is much more limited. This study aims to create a model for supplier performance evaluation in an MRO company operating in the aviation industry in order to contribute to this gap in the literature. CA and ANN were used in the study. The study is expected to contribute to the literature, as studies on supplier performance assessment are limited in both in the aviation sector and in the MRO sector (Önen 2018, pp. 84–109).

9.2 MRO Sector

Purchases are carried out to produce repair and maintenance services in the MRO sector (Pooler et al. 2004, p. 335). MRO activities are important for the transport sector as well as for many other sectors. From the perspective of the aviation industry, the MRO operations of a factory or a machine and the MRO operations of an aircraft are very different. Outsourcing and purchasing in the aviation industry increased from 52 to 75% in 10 years between 2002 and 2012. The main customers of MRO businesses are large aviation companies. The suppliers of MRO companies are OEM (Original Equipment Manufacturers) companies, component and part suppliers, and repair companies. The MRO supply chain has some differences compared to other supply chains. First of all, critical industries such as airlines should perform their MRO operations in a timely and complete manner (Rezaei Somarin et al. 2016). Therefore, time pressure on MRO companies is more than many other industries. According to this fact, time-based performance of suppliers became critical. Besides this, the suppliers of the MRO companies must have the necessary certificates. Therefore, MRO businesses operating in the aviation sector have a narrower supplier pool compared to other sectors. This situation reduces the flexibility of the supply chains of MRO companies (Sahay 2012). Therefore, it may be thought that such enterprises should focus more on performance monitoring, measurement, and development activities of existing suppliers, rather than supplier selection activities.

9.3 Supplier Performance Evaluation

Determining and evaluating the performance of suppliers has an important place in supply management (Monczka et al. 2009, pp. 8–9). Nearly two decades ago, Spekman et al. (1999) proposed that activities for the continuous evaluation of the supplier list are among the ten basic principles for an effective procurement and procurement management. In the following years, businesses placed more emphasis on management of supplier relationships, as they experienced that successful management of their relations with suppliers can result in better quality, more flexible delivery, better material, and information flows (Wisner et al. 2012, p. 75).

Measuring performance will provide information that will both strengthen the communication between the supplier and support the business decisions. General criteria used to measure the performance of procurement plans are price, quality, on-time delivery, technology and innovation, environment and safety, revenue management, integrity of the supply chain, efficiency, social and social factors, and internal customer satisfaction. However, business world faced some problems in measuring performance. These problems can be summarized as having too much data, using wrong criteria, focusing on the short-term criteria (Monczka et al. 2009, pp. 705–715). Considering that performance measurement is a problematic process, it should be managed correctly. According to Gordon (2008), supplier performance management is the process of monitoring, assigning, and evaluating supplier performance with procurement business processes and practices in order to ensure continuous improvement and to keep costs and risks at a minimum. There is a considerable literature on the management and evaluation of supplier performance (Erdal 2014, p. 295).

9.4 Literature Research

First studies on supplier performance evaluation focus on determining the criteria. The most comprehensive study on setting the criteria was done by Dickson (1966) with 273 procurement managers selected from the US and Canadian National Association of Purchasing Managers. At the end of the study, 23 basic criteria regarding the selection and evaluation of the supply were determined (Gemici 2009, p. 14).

Woodside and Vyas (1987) state that management can agree to pay 4–6% more than the lowest cost supplier to a supplier that offers significantly higher performance than other suppliers. This shows that price is not the most important criteria. In another study which supported this, Weber et al. (1991) stated that the most important criterion for supplier evaluation is quality, followed by shipping performance and price (Talluri and Narasimhan 2003, p. 544). The literature contains many researches with several methods including ranking and selection methods and methods analyzing causal relationships.

Some research analyzed causal effects. Kannan and Tan (2002) measured the relationship between supplier selection and evaluation and business performance. Prahinski and Benton (2004) examined the relationship between supplier evaluation and supplier performance using the Structural Equation Model (SEM). Kaynak and Aytekin (2005) examined the effect of supplier performance on purchasing behavior in their study in the manufacturing sector. Modi and Mabert (2007) examined the relationship between knowledge transfer and supplier performance improvement. Cousins et al. (2008) examined the role of social mechanisms in strategic supplier–buyer relationships. Uzun and Karataş (2012) aimed to measure the effect of supplier–customer relationships on performance. Kaynak and İyigün (2016) examined the relationship between corporate social responsibility and long-term relationships with suppliers in the logistics sector. Paparoidamis et al. (2017) analyzed the effect of supplier performance on customer loyalty by using SEM in different countries. Some research used DEA. Liu et al. (2000) measured the performance of suppliers using DEA. Sezen (2004) evaluated the performance of the units that make up the supply chain with DEA in the service sector.

Some research used multi-criteria decision-making methods. Akman and Alkan (2006) evaluated the performance of suppliers in the production sector using the fuzzy AHP method. The AHP method and supplier evaluation criteria are listed in the study conducted by Akdeniz and Turgutlu (2007) in the service sector. In addition, a ranking regarding the suppliers was also made. Küçük and Ecer (2007) aimed to evaluate and rank the suppliers in the service sector with the fuzzy TOPSIS method. Öztürk et al. (2011) aimed to make supplier evaluation and selection using AHP. In the Turkey public sector, Arıkan and Küçükçe (2012) conducted supplier evaluation, Özçakır and Yurdakul (2014) provided models for the supplier selection. Govindan et al. (2013) examined the sustainability performance of the supplier using the fuzzy TOPSIS method. Dey et al. (2015) conducted a strategic supplier performance assessment with Quality Function Deployment and AHP in a manufacturing enterprise. Altınok and Görener (2016) used AHP and TOPSIS methods to measure supplier performance.

Maestrini et al. (2018) examined supplier performance evaluation systems on case studies from the perspective of suppliers and buyers. Er Kara and Oktay Fırat (2018) used the best–worst method, factor analysis, and k-means methods for supplier risk assessment. Gündoğan and Güner (2018) measured the performance of supplier agility with AHP method. Mohammed et al. (2018) measured supplier performance with AHP Fuzzy-TOPSIS method based on green and flexibility criteria.

There are also many studies on supplier clustering. Che and Wang (2010) performed supplier clustering using k-means method, simulated annealing method, particle swarm optimization, and Taguchi method. Keskin et al. (2010) proposed a fuzzy logic-based method for supplier evaluation and classification. Şahin and Supçiller (2015) proposed a decision support system using AHP, TOPSIS, and CA methods together. Kuo et al. (2018) conducted supplier clustering using fuzzy k-mode clustering method and metaheuristic methods. Sabbagh and Ameri (2018) carried out supplier clustering in the manufacturing sector using the k-means method and topical modeling methods. Galo et al. (2018) made supplier categorization using the unstable fuzzy method and ElECTRE TRI methods.

9.5 Supplier Assessment Using Clustering Analysis and Artificial Neural Networks

The study aims to analyze the supplier evaluation and classification system in an enterprise operating in the aviation maintenance and repair sector and to propose a new supplier evaluation and classification system using CA and ANN methods.

9.5.1 Data Collection

In line with the purpose of the research, data on supplier performance were collected. The data was obtained from a company, which is one of the leading aircraft maintenance and repair companies in Turkey. The company operates in MRO sector. Service, control, maintenance, repair, modification, and overhaul activities are performed in the company (www.shgm.gov.tr).

The activities of aircraft maintenance include line maintenance, heavy maintenance, in-cabin and structural-composite maintenance as well as VIP and business jet maintenance. In addition to maintenance activities, the company carries out aircraft painting operations. The company provides 24/7 service to the customers. Purchasing transactions and operations in the business are carried out within the Department of Purchasing and Logistics. Most of the materials are purchased from abroad, however materials should be purchased from approved suppliers.

9.5.2 Current Supplier Performance Evaluation System

In the purchasing process of the company, performance evaluation is made for each supplier on a periodic basis. The person performing the performance review is the employee who is responsible for purchases made from the relevant supplier. Performance evaluation is carried out in a standard manner for suppliers with an annual order number of 50 or more. According to this evaluation, suppliers are evaluated with ten criteria prepared by the company. The criteria are scored between 1 (very bad) and 10 (very good). Each supplier with an annual order number of 50 or more is evaluated by the employee responsible for that supplier according to ten criteria. The criteria in the Supplier Evaluation Form are as follows:

- Criterion1: lead time (k1).
- Criterion 2: compliance with given delivery times (k2).
- Criterion 3: material pricing policy, product price (k3).
- Criterion 4: performance for discount (k4).
- Criterion 5: communication performance (k5).
- Criterion 6: overall support and job tracking performance (k6).
- Criterion 7: ability to provide required referral documents (k7).
- Criterion 8: flexibility against problems (k8).
- Criterion 9: elasticity of payments (k9).
- Criterion 10: performance of providing requested documents (k10).

The company has 141 suppliers with more than 50 orders annually. The company evaluates and classifies these suppliers according to the ten criteria mentioned above. The total score they get from ten criteria can be at most 100 and at least 10. The average of their scores from ten criteria is the main performance indicator for classification. The supplier has determined their average scores as very good (9–10), good (7–8), medium (5–6), bad (3–4), and very bad (1–2), and accordingly the suppliers are divided into five classes as very good, good, medium, bad, and very bad.

Data on 141 suppliers were obtained for analysis. According to the information obtained from the company managers, it was determined that some of the suppliers could not be evaluated properly according to the ten criteria due to their special circumstances. As a result of this, 16 suppliers were excluded among all 141 suppliers. Accordingly, data of 125 suppliers were used, which constitutes 88% of the total suppliers. The number of annual order operations of 125 suppliers with healthy data is 53053 and this number constitutes 80% of the total number of order operations. In this respect, it can be said that 125 suppliers, which will be subject to analysis, reflect all suppliers sufficiently. A total of 125 supplier evaluation forms have been filled in by personnel who have been responsible for these suppliers for at least 1 year.

9.5.3 Method

In the study, descriptive statistics of the data were analyzed first. Subsequently, suppliers were clustered using CA. Then, it was examined whether the obtained clusters could be classified with ANN.

Cluster Analysis

CA, one of the multivariate statistical analysis techniques, aims to collect units in groups according to various similarity criteria. The aim is to create groups based on similarity and define these groups. The method started to be used in the field of marketing especially after the 1960s (Kurtuluş 2004, p. 409). CA is used for classification purposes rather than for generalizing (Hair et al. 2014, pp. 417–418).

CA methods are divided into two as hierarchical and non-hierarchical methods. In the hierarchical cluster analysis, a tree type structure based on a hierarchy is formed. In non-hierarchical methods, the number of clusters must be known in advance. For hierarchical clustering, it is not necessary to predict the number of clusters or to know them. The non-hierarchical approach may be appropriate in situations where the number of clusters is certain (Malhotra and Birks 2000, pp. 604–610).

In non-hierarchical methods, the data is divided into a predetermined number of clusters. The nodes of the cluster centers are calculated. Each unit is assigned to a cluster according to an algorithm. These steps continue until each unit is assigned to a cluster. Since the number of clusters is determined, units can be assigned to a different cluster from the cluster they were first assigned until the clustering algorithm is terminated (Doğan 2008, p. 103).

Sharma (1996) states the stages in non-hierarchical cluster analysis as follows (Ergün 2008, p. 107):

- The average of the clusters for K-clusters is determined by the researcher.
- Observation units are assigned to a cluster according to the value of the unit and cluster means. The unit becomes an element of the cluster.
- Average values of the clusters are recalculated.
- If there is no change in the elements of clusters, the process is completed. If there is a change, return to the second step.

The most known non-hierarchical clustering analysis method is the k-means algorithm (Şimşek 2006, p. 34).

Artificial neural networks

ANN is inspired by biological neural networks and consists of connecting artificial nerve cells with each other in various ways (Silahtaroğlu 2008, p. 66). These are distributed information processing structures that are connected to each other through weighted links and consist of processing units, each with its own memory. In other words, they are computer programs that imitate biological neural networks. Although ANN has many practical uses from industrial applications to fingerprint recognition, the areas where it is used can be grouped as classification, estimation, and modeling (Elmas 2010, pp. 23–25).

The most basic task of ANNs is to determine an output set that can correspond to the input sets shown to it. They can learn by using examples so they have the ability to organize and learn by themselves. They have a certain memory, they store information. Even if some of the information is missing, ANNs can work with connections that are not missing. ANNs can be designed as single layer or multi-layer. Single-layer ANN has only input and output layers. Each network can have one or more inputs and outputs. Output units are connected to all input units. Each link has a weight. In multi-layer networks, it is located in the intermediate layer next to the input and output layer. Intermediate layers can be more than one. Multi-layer networks have a learning structure. Outputs can be obtained with examples to be given to the network. The network generates solution space from the given examples. Later, for similar examples to be shown to the relevant network, the relevant solution space and solutions can be produced (Öztemel 2006, pp. 75–77).

9.5.4 Analysis of Data and Findings

The data were analyzed with SPSS 23 and Weka 3.8 package programs. First of all, descriptive statistics were created.

Descriptive Statistics

The business currently evaluates its suppliers with 10 criteria. Descriptive statistics about this are shown in Table 9.1.

It has been investigated that the criterion with the lowest score is k4 (price elasticity to the demand for discount). The company stated that the suppliers do not have legal

Table 9.1 Descriptive statistics

	N	Min	Max	Mean	Standard deviation
k1	125	1	10	7.18	1.98
k2	125	1	10	7.38	2.13
k3	125	2	10	7.00	1.85
k4	125	1	10	5.76	2.42
k5	125	1	10	7.20	2.21
k6	125	1	10	7.26	2.00
k7	125	2	10	8.18	1.80
k8	125	1	10	7.15	2.04
k9	125	1	10	7.47	1.88
k10	125	2	10	7.75	1.71
Total	125				

Table 9.2 Supplier clusters according to the current evaluation system

Supplier class	Number of suppliers	Average point	Standard deviation
Very good	46	8.67	0.567
Good	51	7.15	0.529
Medium	23	5.35	0.537
Bad	5	3.46	0.610
Total	125		

obligations such as discounts, since they have agreed with the suppliers on exact prices in advance and they are registered with legal contracts. For this reason, the price elasticity they show to the discount request is low. The highest scoring k7 (mandatory shipping documents) is due to the requirement that the materials provided by the suppliers in the aviation sector have certain certificates and documents. It is the necessity to provide these documents in order to use the products of the suppliers and to enter them in the business records.

According to the ten criteria used, the company determined the supplier average scores as very good (9–10), good (7–8), medium (5–6), bad (3–4), and very bad (1–2). Table 9.2 shows the number of clustered suppliers in which cluster according to this evaluation.

As seen in Table 9.2, there are no suppliers in the "very bad" class, and only five in the bad class. 97 of the suppliers (77.6%) are in the good or very good class. The company classifies the suppliers in this way and takes this classification into consideration in the agreements to be made in the next period.

Creating Clusters with Cluster Analysis

Since the number of classes is certain in the current system of the company, the k-means method, one of the non-hierarchical clustering methods, which is the clustering method that takes the number of classes as input, has been used in order not to leave this classification (Kalaycı 2010: 360). Although there are five classes in the current system, since the total number of classes is four, the number of classes in the k-means method has been tested separately as four and five. The k-means method was applied with the SPSS 23 package program. In the clustering of the suppliers, clustering was carried out with 10 criteria, just as the business does.

As a result of all criteria and four-cluster CA, the number of suppliers in each class and its comparison with the current situation are shown in Table 9.3. As a result of the CA, it was observed that there were 57 suppliers in the first cluster with the highest average, 48 in the second cluster, 3 in the third cluster, and 17 in the last cluster with the lowest average. In the four-cluster analysis, it was observed that the suppliers were distributed proportionally to clusters, except for the third cluster. The first cluster can be matched with the cluster names of the original evaluation as "very good", the second cluster as "good", the third cluster as "medium", and the fourth cluster as "bad". When the suppliers are examined from which cluster they switch to which cluster, it is seen that the clusters to which 92 are currently assigned are

Table 9.3 Comparison of current clustering and proposed K-means cluster analysis

Current classification of the company			K-means cluster analysis		
Supplier class	Number of suppliers	Mean point	Supplier class	Number of suppliers	Mean point
Very good	46	8.67	Very good	57	8.47
Good	51	7.15	Good	48	6.81
Medium	23	5.35	Medium	3	5.27
Bad	5	3.46	Bad	17	4.6
Total	125		Total	125	

the same with the clusters assigned by the k-means technique. Therefore, 74% of the suppliers remained in the same cluster.

When Table 9.3 is analyzed, it is seen that the number of suppliers in medium cluster decrease and the number of suppliers in good, very good, and bad clusters increase. Accordingly, while the middle class suppliers decreased, the number of suppliers on the two extremes, bad and very good, increased.

Table 9.4 demonstrates the analysis of the 33 suppliers whose cluster is changed. According to Table 9.4, there are transitions from the middle class to the good and the bad class, and from the good class to the very good class. It can be seen that, nine suppliers existing in the medium cluster due to current evaluation system passed into the good cluster according to the results of CA analysis.

From a managerial point of view, although the number of suppliers in the classes and the number of suppliers obtained as a result of CA are not the same in the current situation, the classes obtained using CA may be an alternative to the existing supplier evaluation system of the company, since the classes obtained match the existing evaluation system of the enterprise and can separate the middle class.

The number of clusters is set as five and CA is repeated. When the number of clusters is selected as five, only one supplier existed in the "*very good*" cluster. In addition, 76 suppliers existed in the "*good*" cluster and this number is more than half of the total number of clustered suppliers. Suppliers with an average score above 5, which is the midpoint, are included in the "*bad*" cluster. At the end of these

Table 9.4 Number of suppliers whose cluster has changed

Current classification of the company	K-means cluster analysis	Number of suppliers with changing set
Medium	Good	9
Medium	Bad	12
Good	Very good	11
Good	Medium	1
Total		33

evaluations, a cluster with four clusters was deemed more suitable for the system of the enterprise and the analysis was continued with the four-cluster model.

Classification with Artificial Neural Networks

The performance of the clusters obtained by ANN can be taken as an indicator of consistency of the clustering performed. The purpose of using ANN in this study is not a better classification, but to compare the clustering approaches by comparing the estimation success of the classes created with different inputs and clustering approaches using the same methods and parameters with the ANN.

In the cluster structure obtained by CA, the scores of each supplier from ten criteria were used as ten input variables, and the cluster to which each supplier was assigned was used as the output variable, and the ANN model was created.

In order to create the ANN model, respectively, "choose", "function", and "multi-layer perceptron" were selected from the classify tab of the Weka program. Thus, a multi-layer ANN has been created. Various parameters are used while creating ANN. One of them is learning coefficients and momentum (support) coefficients. Although these coefficients are usually taken between 0.2 and 0.4 according to experience, they can be changed specifically for the research (Öztemel 2006, s. 99). In the ANN model created, the learning rate was determined as 0.3, the support coefficient as 0.2, the number of repetitions as 500, and the number of hidden layers was left as default. In this case, the ANN model was made with the Weka Multi-layer Perceptron algorithm.

There are four different features supported in the Weka program. The first of these features, the "use training set" feature, is about using all of the selected data for educational purposes. After that, a second set in which learning will be tested can be selected (supplied test set) and control can be made. The third option, "Cross-validation", divides the data set into a number of subsets specified as "fold" and uses one of these subsets as a learning set and the other as a test set. The last method is the "percentage split" method in which the data set is divided into two separate parts (Şeker 2013, s. 191). The first method used is Cross-examination, the "Cross-validation" method. Here, the subset "fold" number has been selected as five. Other parameters remain the same. The results are shown in Fig. 9.1.

According to the results obtained, clusters can be predicted 92.8% correctly. Mean absolute error value was obtained as 0.043 and RMSE value was obtained as 0.16. FPR was obtained as 4.3% (Fig. 9.1).

The second method used is the "percentage split" method in which the data set is divided into two separate parts. In this method, the separation ratio was chosen as 0.6. 60% of 125 data, that is, 75 of them were used for learning and 50 of them were used for testing. According to the results obtained, the output could be estimated at 92%. The MAE value was 0.061 and the RMSE value was 0.179. TPR was obtained as 90.9%. FPR was obtained as 3.6%. Results are shown in Fig. 9.2.

Based on the results obtained with both methods, the newly created model can be successfully predicted by ANN. Thus, the probability of obtaining the four sets obtained by the k-means method with ANN was investigated. Considering all the

```
=== Stratified cross-validation ===
=== Summary ===
Correctly Classified Instances        116          92.8  %
Kappa statistic                       0.8853
Mean absolute error                    0.0432
Root mean squared error                0.164
Relative absolute error               13.7072 %
Root relative squared error           41.4332 %
Total Number of Instances             125
=== Detailed Accuracy By Class ===
          TP Rate FP Rate Precision Recall F-Measure MCC   ROC Area PRC
Area Class
   1      0,947  0,044  0,947   0,947 0,947   0,903 0,992   0,990  1
   2      1,000  0,000  1,000   1,000 1,000   1,000 1,000   1,000  2
   3      0,896  0,052  0,915   0,896 0,905   0,847 0,980   0,971  3
   4      0,941  0,019  0,889   0,941 0,914   0,901 0,998   0,990  4
Weighted 0,928  0,043  0,928   0,928 0,928   0,884 0,988   0,983
Avg.
=== Confusion Matrix ===
 a b c d  <-- classified as
54 0 3 0| a = 1
 0 3 0 0| b = 2
 3 0 43 2| c = 3
 0 0 1 16| d = 4
```

Fig. 9.1 Results using cross-validation

criteria, it has been determined that ANN can obtain the same clusters with a probability of 92–92.8% according to different methods.

9.6 Conclusion

The research aims to analyze the supplier assessment and classification system in an enterprise operating in the aviation MRO sector. The research is important in that the supplier evaluation studies related to the MRO sector are generally limited in the literature and very limited in the domestic literature. The methodology used was created by CA and ANN methods. A supplier evaluation and classification system in which these two methods are used sequentially is proposed.

In the cluster analysis performed for all criteria, results were obtained to match the current system of the enterprise, but it was observed that the middle class was

```
Correctly Classified Instances      46        92    %
Kappa statistic                     0.8703
Mean absolute error                 0.0617
Root mean squared error             0.1793
Relative absolute error             19.5275 %
Root relative squared error         45.3323 %
Total Number of Instances           50
=== Detailed Accuracy By Class ===


        TP Rate  FP Rate  Precision  Recall  F-Measure  MCC    ROC Area  PRC Area
Class
1       0,909    0,036    0,952      0,909   0,930      0,878  0,989     0,986
2       0,000    0,000    0,000      0,000   0,000      0,000  ?         ?
3       0,950    0,100    0,864      0,950   0,905      0,839  0,978     0,972
4       0,875    0,000    1,000      0,875   0,933      0,924  0,997     0,986
Weighted
Avg.    0,920    0,056    0,925      0,920   0,921      0,870  0,986     0,980



=== Confusion Matrix ===
 a b c d <-- classified as
 20 0 2 0| a = 1
  0 0 0 0| b = 2
  1 0 19 0| c = 3
  0 0 1 7| d = 4
```

Fig. 9.2 Results with proportional division method

better analyzed because the suppliers in the middle class moved to the upper and lower classes. Therefore, from a managerial point of view, classifying suppliers with CA instead of average score may be an alternative to the company's existing supplier evaluation system.

It has been tried to estimate the clusters created with CA by using the same parameters and methods with ANN. In addition to the proposed managerial results, CA and ANN can be used sequentially in future studies and can be used in supplier performance evaluation. In addition, criteria that have been weighed or grouped can be compared with the solutions obtained with this methodology.

Among the limitations of the study are the fact that data were obtained from a single business, and the data were subjective, although they were obtained from experts who worked with suppliers for at least 1 year and 50 operations. In addition, since the purpose of ANN is not a better classification, but the success of the classifications estimation, the fact that the ANN has been tested with a single algorithm and that the weights and parameters have not been adjusted can be considered as a separate constraint.

References

Akdeniz HA, Turgutlu T (2007) Türkiye'de perakende sektöründe analitik hiyerarşik süreç yaklaşımıyla tedarikçi performans değerlendirmesi. Dokuz Eylül Üniversitesi Sosyal Bilimler Enstitüsü Dergisi Cilt 9, Sayı: 1 (2007)

Akman G, Alkan A (2006) Tedarik Zinciri Yönetiminde Bulanık AHP yöntemi kullanılarak tedarikçilerin performansının ölçülmesi: Otomotiv Yan Sanayiinde bir uygulama. İstanbul Ticaret Üniversitesi Fen Bilimleri Dergisi 5(9):23–46

Altınok E, Görener A (2016) Tedarikçi Performansı Değerlendirmesi İçin Bütünleşik Bir Model Önerisi.2. Üretim EKonomisi Kongresi, 11–12 Nisan (2016)

Arıkan F, Küçükçe YS (2012) Satın Alma Faaliyeti İçin Bir Tedarikçi Seçimi Değerlendirme Problemi Ve Çözümü. Gazi Üniversitesi Mühendislik- Mimarlık Fakültesi Dergisi 27(2)

Che ZH, Wang HS (2010) A hybrid approach for supplier cluster analysis. Comput Math Appl 59(2):745–763

Cousins PD, Lawson B, Squire B (2008) Performance measurement in strategic buyer-supplier relationships: the mediating role of socialization mechanisms. Int J Oper Prod Manag 28(3):238–258

Dey PK, Bhattacharya A, Ho W, Clegg B (2015) Strategic supplier performance evaluation: a case-based action research of a UK manufacturing organisation. Int J Prod Econ 166:192–214

Doğan B (2008) Bankaların Gözetiminde Bir Araç Olarak KA: Türk Bankacılık Sektörü İçin Bir Uygulama (Yayınlanmamış Doktora Tezi), Kadir Has Üniversitesi Sosyal Bilimler Enstitüsü Finans Ve Bankacılık Bölüm Dalı, İstanbul

Dickson G. W. (1966) An analysis of vendor selection: systems and decisions. J Purch 5–17

Elmas Ç. (2010) Yapay Zeka Uygulamaları, 2. Baskı, Seçkin Yayıncılık, Ankara

Er Kara M, Oktay Fırat S (2018) Supplier risk assessment based on best-worst method and k-means clustering: a case study. Sustainability 10(4):1066

Erdal M (2014) Satınalma ve Tedarik Zinciri Yönetimi, Beta Yayıncılık, İstanbul

Ergün E (2008) Ürün Kategorileri Arasındaki Satış İlişkisinin Birliktelik Kuralları ve KA ile Belirlenmesi Ve Perakende Sektöründe Bir Uygulama (Yayınlanmamış Doktora Tezi), Afyon Kocatepe Üniversitesi Sosyal Bilimler Enstitüsü, Afyon

Galo NR, Calache LDDR, Carpinetti LCR (2018) A group decision approach for supplier categorization based on hesitant fuzzy and ELECTRE TRI. Int J Prod Econ 202:182–196

Gemici MF (2009) Tedarik Zincirinde Veri Zarflama Analitik Hiyerarşi Prosesi Yöntemiyle Perakende Sektöründe Tedarikçi Performans Değerlendirmesi (Yayınlanmamış Yüksek Lisans Tezi), İstanbul Teknik Üniversitesi Fen Bilimleri Enstitüsü, İstanbul

Gordon SR (2008) Supplier evaluation and performance excellence: a guide to meaningful metrics and successful results. J. Ross publishing

Govindan K, Khodaverdi R, Jafarian A (2013) A fuzzy multi criteria approach for measuring sustainability performance of a supplier based on triple bottom line approach. J Clean Prod 47:345–354

Gündoğan T, Güner S (2018) Tedarikçi Çevikliğinin Ölçülmesine Yönelik Bir Yaklaşim Önerisi: Otomotiv Sektörü Uygulamasi. İşletme Bilimi Dergisi 6(1):1–26

Hair JF, Black WC, Babin BJ, Anderson RE (2014) Multivariate data analysis: Pearson new international edition. Pearson Higher Ed.

Kalaycı Ş (2010) SPSS Uygulamalı Çok Değişkenli İstatistik Teknikleri. Asil Yayın Dağıtım LTD. Şti, Ankara, 5. Baskı, S:322

Kannan VR, Tan KC (2002) Supplier selection and assessment: their impact on business performance. J Supply Chain Manag 38(3):11–21

Kaynak R, Aytekin M (2005) Makine Halı Sektöründe Satın Alma Davranışına Etki Eden Tedarikçi Performans Kriterleri. V. Ulusal Üretim Arastırmaları Sempozyumu, _stanbul Ticaret Üniversitesi, 25–27 Kasım 2005

Kaynak R, İyigün İ (2016) The effecets of corporate governance: implications for third party logistics providers' marketing capability and long-term relationships. Atatürk Üniversitesi Sosyal Bilimler Enstitüsü Dergisi, Nisan 20:665–676

Keskin GA, İlhan S, Özkan C (2010) The Fuzzy ART algorithm: a categorization method for supplier evaluation and selection. Expert Syst Appl 37(2):1235–1240

Küçük O, Ecer F (2007) Bulanık TOPSIS kullanılarak tedarikçilerin değerlendirilmesi ve Erzurum'da bir uygulama. Ekonomik ve Sosyal Araştırmalar Dergisi. Cilt 43 ıl 3 sayı 1, 45–65

Kuo RJ, Potti Y, Zulvia FE (2018) Application of metaheuristic based fuzzy K-modes algorithm to supplier clustering. Comput Ind Eng 120:298–307

Kurtuluş K (2004) Pazarlama Araştırmaları, Literatür Yayıncılık, 7. Baskı, İstanbul

Liu J, Ding FY, Lall V (2000) Using data envelopment analysis to compare suppliers for supplier selection and performance improvement. Supply Chain Manag: Int J 5(3):143–150

Maestrini V, Maccarrone P, Caniato F, Luzzini D (2018) Supplier performance measurement systems: communication and reaction modes. Ind Mark Manage 74:298–308

Malhotra KN, Birks DF(2000) Marketing research an applied approach, European edn. Financial Times, Prentice Hall

Modi SB, Mabert VA (2007) Supplier development: improving supplier performance through knowledge transfer. J Oper Manag 25(1):42–64

Mohammed A, Harris I, Soroka A, Naim MM, Ramjaun T (2018) Evaluating green and resilient supplier performance: ahp-fuzzy topsis decision-making approach. In: ICORES, pp 209–216

Monczka RM, Handfield RB, Giunipero LC, Patterson JL (2009) Purchasing and supply chain management, 4th edn. South-western Cengage Learning

Murphy PR, Knemeyer AM (2016) Contemporary logistics. Nobel Yayın Dağıtım, İstanbul

Önen V (2018) Uçak Bakım Kuruluşlarında Tedarikçi Değerlendirmeleri: Analitik Hiyerarşi Prosesiyle Tedarikçi Seçimine Yönelik Bir Bakım Kuruluşu Uygulaması. Uluslararası Sosyal Bilimler Dergisi 2(13):84–109

Özçakar, Necdet ve Yurdakul, Halim (2014) Türk Kamu İhale Kanununda Fiyat ile Birlikte Fiyat Dışı Unsurların da Dikkate Alındığı İhale ve Kazanan Teklif. i.Ü. İşletme Fakültesiİşletme İktisadi Enstitüsü Yönetim Dergisi, Sayı 76, Haziran 2014, İstanbul

Öztemel E (2006) Yapay Sinir Ağları. Papatya Yayıncılık, 2. Baskı, İstanbul

Öztürk A, Erdoğmuş Ş, Arikan VS (2011) Analitik Hiyerarsi Süreci (Ahs) Kullanilarak Tedarikçilerin Degerlendirilmesi: Bir Tekstil Firmasinda Uygulama. Dokuz Eylül Üniversitesi İktisadi İdari Bilimler Fakültesi Dergisi 26(1):93–112

Paparoidamis NG, Katsikeas CS, Chumpitaz R (2017) The role of supplier performance in building customer trust and loyalty: a cross-country examination. Ind Mark Manag 78:183–197

Pooler VH, Pooler DJ, ve Farney SD (2004) Kluwer Academic Publishers, New York

Prahinski C, Benton WC (2004) Supplier evaluations: communication strategies to improve supplier performance. J Oper Manag 22(1):39–62

Rezaei Somarin A, Asian S, Jolai F, Chen S (2016) Flexibility in service parts supply chain: a study on emergency resupply in aviation MRO. Int J Prod Res 56(10):3547–3562

Sabbagh R, Ameri F (2018) Supplier clustering based on unstructured manufacturing capability data. In: ASME 2018 international design engineering technical conferences and computers and information in engineering conference, August 2018. American Society of Mechanical Engineers, pp V01BT02A036-V01BT02A036

Sahay A (2012) Leveraging information technology for optimal aircraft maintenance, repair and overhaul (MRO). Elsevier

Şahin Y, Supçiller A (2015) Tedarikçi seçimi için bir karar destek sistemi. Mühendislik Bilimleri Ve Tasarım Dergisi 3(2):91–104

Şeker SE (2013) İş Zekası ve Veri Madenciliği, Cinius Yayınları, İstanbul

Sezen B (2004) Veri Zarflama Analizi İle Tedarik Zinciri Ortaklarının Performans Değerlendirmesi. YA/EM, 15–18

Sharma S (1996) Applied multivariate techniques. John Wiley & Sons, Inc, ABD

Silahtaroğlu G (2008) Kavram ve Algoritmalarıyla Temel veri Madenciliği, Papatya Yayınevi, İstanbul

Şimşek D (2006) Kümeleme Analizi, Çok Boyutlu Ölçekleme, Doğrulayıcı ve Açıklayıcı Faktör Analizi İle Elde Edilen Yapı Geçerliği Kanıtlarının Karşılaştırılması (Yayınlanmamış Yüksek Lisans Tezi), Hacettepe Üniversitesi Sosyal Bilimler Enstitüsü, Ankara

Spekman RE, Kamauff J, Spear J (1999) Towards more effective sourcing and supplier management. Eur J Purchas Supply Manag 5(2):103–116

Talluri S, Narasimhan R (2003) Vendor evaluation with performance variability: a max–min approach. Eur J Oper Res 146(3):543–552

Toker K, ve Görener A (2013) Lojistik Yönetimi Kapsamında Ulaştırma Modunun Seçimi: Tekstil Sektöründe Bir Uygulama. İ.Ü. İşletme Fakültesi İşletme İktisadı Enstitüsü Yönetim Dergisi 74:16–20

Uzun A, Karataş E (2012) Tedarikçi-Müşteri İlişkilerinin Stoklar Açısından Tedarik Zinciri Performansına Etkisi. Atatürk Üniversitesi İktisadi Ve İdari Bilimler Dergisi 26(2):257–271

Weber CA, Current JR, Benton WC (1991) Vendor selection criteria and methods. Eur J Oper Res 50:2–18

Wisner JD, Tan K-C, Keong Leong G (2012) Principles of supply chain management a balanced approach, 3e, South-western Cengage Learning (2012)

Woodside AG, Vyas N (1987) Industrial purchasing strategies: recommendations for purchasing and marketing managers. Lexington books

Yurdakul H (2015) Satınalma Yönetimi Süreçler ve Uygulamalar. Nobel Yayınevi, Ankara

Muhammet Mikdat Akbaba has bachelors degree of industrial engineering in 2012, masters degree of Business Administration in 2017. His master thesis was about Supply Chain Management. His master thesis was about Examination of supplier performance in Avviation Industry using cluster analysis and artificial neural networks. He is Industrial Engineer in aviation MRO (Maintenance, Repair and Overhaul) industry since 2013. His working areas are purchasing management in aviation industry, supply chain management and quality management.

Onur Çetin Ph.D., has bachelors degree of industrial engineering in 2002, masters degree of Business Administration in 2006. His master thesis was about agile manugacturing. He was visiting scholar in University of Wales for one year during 2006–2007. He had Ph.D. in production management in 2013. His PhD thesis was about Vehicle Routing Problem in Fuel Distribution. He has a one year job experince as industral engineer at Akkanat Holding, four years experience as research assistant at Trakya University and, six years experience as research assistant at Istanbul University. He is Asst. Prof. İn Trakya University since 2015. His working areas are operations management in service sector, supply chain management and logistics and quality management.

Part IV
Corporate Governance and Accountability in Airlines

Chapter 10
Strategic Outsourcing in Airline Business

Dilek Erdoğan

Abstract Outsourcing has become a vital source of competitive advantage because it provides several benefits for organizations, such as reducing the cost of ownership of products/services, resolving technical problems without increasing the number of staff, and enabling the company to focus on its core business. Airlines typically outsource a broad range of business processes such as aircraft maintenance, ground handling, IT systems, crew training, maintenance training, catering services, and aircraft leasing. Some of the airline business outsourcing decisions are tactical and some are strategic. In this section, the decision of airlines to outsource aircraft maintenance, ground handling, and information technologies is examined within the framework of transaction costs and the research dependence theory. Moreover, it was discussed whether the outsourcing decision for these functions is strategic or not.

Keywords Transaction cost · Research dependency · Ground handling · Aircraft maintenance · Information technologies

10.1 Introduction

Liberalization, which started in the United States in 1978, led to some radical changes and developments in the airline industry. The increase in new entries to the market, the entry of low-cost carriers, the start of airlines to operate in international markets, the expansion of flight networks, and the mergers between airline organizations are examples of these developments (Ghobrial 2005). Since the late 1970s, the main strategic driving forces in the airline industry have been the liberalization of the sector and the increasing competition accompanying the reduction of economic and trade barriers around the World. After the liberalization, legacy carriers found their markets increasingly open to both international competitors and new entrants such as low cost carriers (Morrell 2005). Against globalization and competition, many legacy airlines have started to outsource some parts of their business to suppliers to

D. Erdoğan (✉)
Department of Aviation Management, Tarsus University, Mersin, Turkey
e-mail: dilekerdogan@tarsus.edu.tr

© The Author(s), under exclusive license to Springer Nature Singapore Pte Ltd. 2022   195
K. Kiracı and K. T. Çalıyurt (eds.), *Corporate Governance, Sustainability, and Information Systems in the Aviation Sector, Volume I*, Accounting, Finance, Sustainability, Governance & Fraud: Theory and Application, https://doi.org/10.1007/978-981-16-9276-5_10

achieve greater efficiency and competitiveness. Outsourcing provides many benefits to airline companies such as developing core competencies, lowering labor, and capital costs due to economies of scale, increasing operational efficiency, increasing the success of process renewal activities, sharing risks on non-specialized issues, benefiting from the supplier's experience, and accessing up-to-date and modern technology. Despite all these benefits, organizations cannot always achieve the results they want with outsourcing because outsourcing also includes many risks such as increased dependency on the supplier, loss of business knowledge, hidden ownership costs, and opportunistic behavior of the supplier (Barthelemy 2003). For example, according to the research by Willcocks and Lacity (1999, p. 48), 38% of DKK agreements result in success, 35% do not achieve the desired goals, and 27% carry both positive and negative results. Here, the first critical question is the decision about which activities will be kept in-house and which activities will be outsourced.

In addition to the above-mentioned benefits, some unique features of the aviation sector also lead airlines to outsource. Air transport is inherently a capital and labor-intensive service industry. Airlines require enormous investment costs to be able to provide all the resources they need at all airports where they operate, to have their personnel, and to carry out all activities within their structure. The second specific feature of the airline industry is seasonality. Due to the seasonal nature of the airline industry, passenger demands are constantly changing. Airline organizations' revenues fluctuate throughout the year due to the variability in passenger demands. Another reason for outsourcing is the feature of airline transportation to provide the place and time benefits. The timely transfer of passengers, cargo, or mail from one place to another requires many operations. Problems in these operations can cause delays that increase the costs of airline organizations. In this respect, it is extremely important for airline organizations to carry out all operations effectively and efficiently (Doganis 2006). The competence of the supplier ensures that the operations are carried out effectively and efficiently.

According to Rutner and Brown (1999), the six factors listed below are effective in the airline's decision to use outsourcing:

- Institutionalization degree of airline.
- The current state of the economy.
- The attitude of owners or partners of the airline company.
- The existence of financial support of the airline company.
- Age of the airline company.
- The degree of complexity of the outsourced activity.

Considering all these, airline organizations tend to use different degrees of outsourcing in many areas from information technologies to check-in operations, from cargo services to flight operations. Table 10.1. classifies the functions based on the likelihood an airline chooses to outsource.

Table 10.1 Functions likely to be outsourced by airlines	Very likely	• Ticket sales and distribution • Aircraft leasing • Airport gates • Complimentary limousine pick-up • Food services • Ticketing • Baggage handlers • Aircraft interior cleaning
	Likely	• Engine overhaul or rework • Maintenance training • Information systems and technology • Pilot training • Advertising
	Moderate	• Counter personnel • Airframe maintenance • Spare parts inventory • Feeder operations • Gate personnel • C and D level maintenance checks
	Unlikely	• Cargo handling and operations • Marketing • Human resources management and recruitment
	Very unlikely	• Pilots • Strategic management • Flight attendants • Accounting • Routine hanger maintenance

Resource Rutner & Brown

This section will discuss aircraft maintenance outsourcing, ground handling outsourcing, and information technology outsourcing, which covers most of the functions used at a very likely and moderate level by airlines mentioned in the table above. This discussion will be made within the framework of the transaction cost theory, which is often accepted as the framework to explain the issue of outsourcing.

10.2   Tactical and Strategic Outsourcing

Outsourcing decisions can be tactical or strategic. A strategy is a managerial tool that controls innovation, progress, and changes that occur by ensuring continuous adaptation to the environment or mutual harmony with the environment. The tactic is the strategy's plan of distribution of available forces or resources to achieve its goals. Tactics are about mobilizing, in other words, applying these placed forces. In other words, the concept of "strategic" can be defined as the ways that the business will gain structurally competitive advantages over its competitors in the long run.

Tactics, on the other hand, mostly refer to daily and operational goals, plans, and processes. Tactical outsourcing is a type of outsourcing that involves more cost-oriented decisions and ignores other benefits and risks that may occur. Another important feature of tactical outsourcing is that it covers shorter term agreements compared to strategic outsourcing and its most important advantage can be easy and fast applicable. Services that are subject to tactical outsourcing are generally related to activities that are far from the main activity of the enterprise and that are very difficult to manage. A tactical outsourcing relationship is a dicrete relational exchange. Discrete exchange relations are more based on economic evaluations. The focus of this outsourcing is contracting based on the market mechanism. Usually, the contract simply consists of the fees for the services and the social relations between the buyer and supplier are limited (Fink et al. 2006).

Activities subject to outsourcing have increased over time and outsourcing has ceased to be tactical and becomes a strategic tool. Managers realized that outsourcing gave them greater control over their areas of responsibility and allowed them to focus on more strategic issues related to their business. Strategic outsourcing arises with the decision to outsource other activities that are very close to the core business in order to focus on key resources and distinctive skills.

In activities that are the subject of strategic outsourcing, businesses prefer to work with a smaller number of service providers who are the best in their field, instead of working with too many suppliers. In strategic outsourcing, these relationships have evolved from vendor–supplier arrangements to long-term partnerships based on mutual benefits (Brown and Wilson 2005). Strategic outsourcing is a hybrid governance form, unlike market exchange. Here, the relationship between the buyer and the supplier is established for a longer term, based on mutual interests and in which relational norms such as trust and solidarity are intense. In some cases, even strategic partnerships between buyer and seller can be established.

Outsourcing aircraft maintenance services, information technologies, and ground services in airline companies that are the subject of this chapter are strategic decisions. Aircraft maintenance activities are not the core business of the airline business but are critical to airline businesses as they are one of the biggest cost items and also affect safety. Therefore, the decision to outsource these services is strategic. Similarly, ground handling services are not the core business of airlines, but when passengers experience a supplier-related dissatisfaction with these services, they directly blame the airline company. Satisfaction with these services affects airline satisfaction. Therefore, it is a strategic decision about whether the airline company will use internal resources or outsource ground handling services. Airline companies outsource many information technology functions such as system integration, data center management, cloud services, packaged software, customized software, and cloud services. Some of these services, for example, accounting software may be a tactical outsourcing decision, while computer reservation system outsourcing decision is a strategic decision. Because the computer reservation system is an IT function that is directly related to the core business of the airline business.

10.3 Theoretical Underpinnings of Outsourcing

The issue of outsourcing is a multidisciplinary subject, so the subject has been approached from different perspectives such as economics, purchasing, operations research, information technologies accounting, and strategic management in the literature. These approaches basically fall into two categories. While one of them evaluates outsourcing entirely in terms of cost, the other stream takes into account other factors in addition to cost by evaluating outsourcing from a strategic perspective. There is no single, generally accepted theory of outsourcing, but the most widely used theories to explain outsourcing are transaction cost theory, resource-based view, and contingency theory (Barrar and Gervais 2006). In this book chapter, strategic outsourcing decisions made by airline companies for aircraft maintenance, ticketing and reservation systems, and ground handling services functions will be examined in terms of benefits and risks in the light of transaction cost theory and resource dependency theory.

10.3.1 Transaction Cost Theory

According to the transaction cost theory, which was founded by Coase and developed by Williamson, the unit of microanalytical analysis in transaction cost economics is transaction (Williamson 1981, p. 552). When a good or service is transferred over a technologically separable interface, a transaction takes place. According to the theory, there are three different governance structures: market, hybrid, and hierarchy, which minimize the transaction cost according to the characteristics of each transaction (Williamson 1996). In other words, "transactions that differ in their qualities are harmonized with governance structures that differ in cost and ability, as they affect the result of minimizing transaction costs" (Williamson 1991, p. 79). When organizations choose between "market" and "organization" as specified alternative governance mechanisms, they make this choice largely based on which one minimizes transaction costs. Market exchange is generally based on price, competition, and contracts that keep the parties aware of their responsibilities. In market exchange, the parties are completely independent from each other. The hierarchy mechanism (vertical integration) at the other end of governance differs from market exchange. In this form of exchange, transaction parties are dependent and transactions take place within the organization. There is a form of governance called hybrid, which carries certain characteristics of the two governance mechanisms between these two forms of governance (Williamson 1996, pp. 101–109). Hybrid forms mostly manifest themselves in long-term contracts, cooperation, partnership, and franchising (Williamson 1985, p. 104).

The main issue of the theory is to decide on the governance structure that will keep transaction costs at optimum. Although the organization of economic activities depends on balancing the economy of production with transaction costs, the paradigmatic problem of the theory is the "make or buy" decision (outsourcing decision), where transaction costs are central to economization (Williamson 1985).

The transaction cost is a determining factor in outsourcing decisions. According to the theory, transaction costs include pre-contract and post-contract costs. Pre-contract costs cover the costs of drafting a contract, negotiation processes, and maintenance of the contract. Post-contract costs, on the other hand, include performance control costs that arise during the execution of the contracts, adjustment costs, and correction of errors (Williamson 1985, p. 388, 1996, p. 379).

According to the transaction cost theory, the transaction has three characteristics: asset specificity, uncertainty surrounding the transaction, and the frequency of the transaction. According to the theory, transactions with different characteristics will have different transaction cost results. The high asset specificity means that the asset will lose its productive value in alternative channels (Williamson 1985, p. 96). According to Williamson (1996, p. 60), asset specificity appears in different ways such as physical asset specificity, site-specificity, human asset specificity, dedicated assets, brand name specificity, and temporal specificity. An example of the location feature is the positioning of fixed assets close to production to reduce inventory and transportation costs. Transactions with high asset specificity require the parties to make unilateral or mutual relationship-specific investments. In the literature, the concept of relationship-specific investments is used as an equivalent to asset specificity. Relationship-specific investments in an outsourcing relationship are defined as investments that are not suitable for alternative uses or lose their value when the exchange relationship ends (Anderson and Weitz 1992, p. 20). For example, the computer reservation system developed specifically for an airline company is generally the processes that require unique design features, unique human capital and skills, and have high asset specificity. The asset specificity of transactions is an important determinant in deciding to outsource because entity specificity locks the parties in the relationship. In an outsourcing relationship where asset-specific investments are high, both parties are locked in the relationship because the supplier invests in the relationship and the buyer cannot easily find the supplier to make such an investment (Williamson 1981, p. 555).

Uncertainty is another characteristic of the transaction. Generally, environmental uncertainty, internal uncertainty, and behavioral uncertainty were distinguished in studies based on transaction costs and these uncertainty types were operationalized. Environmental uncertainty arises from the dynamism of the environment. For example, the rapid change of technology, rapid price change shows the difficulty of predicting possible future situations (Cannon et al. 2000). The problem of adapting contracts may arise in transactions with high environmental uncertainty. Internal uncertainty indicates the ability of a firm to clarify its needs regarding exactly the type of service or product it needs. Behavioral uncertainty is related to difficulties in predicting the actions of exchange parties in light of the potential for opportunistic behavior (Rindfleisch and Heide 1997). Behavioral uncertainty is seen as a performance evaluation problem regarding whether the performance of the seller is in accordance with the contract (Geyskens et al. 2006). Increasing uncertainty of transactions is a transaction feature that creates transaction dangers. As changing conditions increase, it will be difficult to adapt to them and therefore transaction costs will increase.

Another characteristic of the transaction is its frequency. The frequency of transactions indicates the frequency of repetition of transactions between exchange parties (Williamson 1985, p. 62). Hierarchy is more preferable than the outsource (market) in cases where the frequency of transactions is high due to the recurring cost.

The transaction cost theory has two types of behavioral assumptions: limited rationality and opportunistic behavior. Williamson (1985, p. 45) defines opportunism as the pursuit of self-interest through fraud. Opportunism in an exchange relationship which mostly involves subtle forms of deception. Opportunism appears as an incomplete or distorted transfer of information between the parties, or efforts made to deceive the other party and generate benefits for themselves. The other assumption of the theory is limited rationality. According to Simon (1957, p. 198), people have limited rationality, and people with limited rationality do not have enough mental capacity to predict all possible situations. Therefore, even if decision-makers want to make rational decisions, they can't make rational decisions with limited information processing and limited communication skills (Williamson 1985). According to the transaction cost theory, all contracts are incomplete as it is not possible for the contracts to predict all possible situations due to limited rationality.

10.3.2 Resource Dependence Theory

The resource dependency theory accepts the view that organizations are not fully self-sufficient, cannot provide the needed resources on their own, and therefore have to procure their needs from the elements around them. This means organizations need to engage with other organizations and elements in the environment (Pfeffer and Salancik 1978). Organizations have to establish relationships with other organizations in the external environment in order to obtain resources that they cannot or do not want to produce internally. Organizations become dependent on their environment while providing the resources they need through outsourcing. In this context, resource dependency theory is used to manage the dependency relationships that arise in the process of obtaining the resources that organizations need from their external environment, and the actions and behaviors they exhibit. Understanding the behaviors of organizations requires understanding the context of the behaviors (Pfeffer and Salancik 1978). Accordingly, it is useful to examine the assumptions of the resource dependency theory regarding the outsourcing process.

According to the theory, outsourcing the resources needed by organizations does not constitute a problem for organizations on its own. The main problem for organizations is that when outsourcing, there is a dependency in relationships with suppliers in the market. Dependency is defined as the importance attached to a certain resource in terms of the inputs and outputs of the organization and the degree to which this resource is controlled by a relatively small number of suppliers (Pfeffer and Salancik 1978).

According to the resource dependence theory, power occurs as a result of mutual dependence between two parties. Both parties involved in the exchange relationship have scarce resources and it is important which of these resources is more valuable and critical. In other words, if the resources of one party are more valuable than the other, the relationship between them is asymmetrical. Power plays an important role here because the non-dependent party holds power. Asymmetric relationships occur as a result of power influencing the behavior of the other party by the non-dependent party. The fact that one of the parties is more dependent than the other causes the power not to be distributed equally and evenly and reveals power differences (Pfeffer and Salancik 1978). The power of actor A over B in the relationship shows B's dependence on A. Depending on the type of outsourcing in outsourcing relationships, the dependence of the buyer on the supplier differs. For example, in the case of total outsourcing, the supplier will be the dominant party in the relationship, while selective outsourcing will be more balanced between the power parties. The strong party in a relationship can use its power in different ways as oppressive on issues such as decision controls, price, quality, and service standards (Rokkan and Haugland 2002).

In case organizations outsource the resources they need, the level of dependency (external dependency) on their suppliers is affected by three important structural features of the environment. These are concentration, munificence, and interconnectedness (Pfeffer and Salancik 1978). The low or high number of suppliers in the surrounding where the organization can supply any resource it needs affects the intensity of control of the resource. If the supply of the source is provided by a small number of suppliers, the concentration in that sector is high. Resource shortages can arise when procuring products or services from the same suppliers as competitors. A shortage of resources determines the degree of dependence of an organization on its. In this context, as the scarcity level of a resource increases, the level of dependency on that supplier increases. Interconnectedness refers to the number of connections between organizations and ways of making connections (Pfeffer and Salancik 1978). These three structural features of the environment in which organizations exchange resources affect the interdependence between them.

10.4 Outsourcing of Aircraft Maintenance

Before the Deregulation Act, which took place in the USA in 1978 and was an important development in the airline transportation sector worldwide, airline companies were carrying out most of their maintenance activities with their resources. After liberalization, maintenance activities began to be outsourced and it spread throughout the world. It is well known that for an airline business, aircraft maintenance is not considered the core business of the airline business, but the quality, compliance, and cost-effectiveness of aircraft maintenance have a significant impact on the operations and profitability of the airline business. Aircraft maintenance costs of an airline account for approximately 10% to 15% of its total costs (Al-Kaabi et al. 2007). As airlines are looking for ways to reduce this cost, maintenance cost reduction is seen as the primary factor in deciding to outsource aircraft maintenance (McFadden and Worrels 2012).

Industry trends show that maintenance outsourcing by airlines is increasing. Although full-service airline companies perform maintenance services in-house compared to low-cost airlines, there are also those who outsource maintenance among full-service airlines (Barrett 2004). For example, continental airlines outsource around 60% of aircraft maintenance. The aim of the airline company here is to enable its experienced team to focus on core business activities and to outsource the remaining activities (Rieple and Helm 2008). Another example, United Airlines, which controlled all maintenance services in-house in the past, stops its non-core maintenance divisions and started outsourcing for these services. Delta and Lufthansa prefer to develop themselves more in aircraft maintenance activities, both providing their maintenance services and selling out maintenance services.

The reasons for outsourcing the aircraft maintenance activities of airline companies are summarized below (Al-Kaabi et al. 2007; Czepiel 2003; Rutner and Brown 1999).

- **Reducing costs.** Since aircraft maintenance costs have a significant share in airline costs, the ability of airline operators to gain a cost advantage in maintenance costs depends on their effectiveness and efficiency in these services. It is quite difficult for the airline to be effective and efficient while performing maintenance activities that require expertise by establishing maintenance facilities with extremely high investment costs, using up-to-date technology, providing a qualified workforce, providing continuous training of maintenance personnel. However, a supplier with core business aircraft maintenance activity will use all its resources in this area, so it will be more efficient and effective. Since these supplier businesses will be able to provide maintenance service at a lower cost, they will be able to offer airline businesses at lower prices. Besides, the supplier has many customers and a high business density, allowing it to perform services at lower costs.
- **To improve service quality.** Safe flight time, high reliability for flight, and on-time departure performance are the quality indicators of maintenance service. Together with the service quality, it must be controlled in costs. If the maintenance department of the airline company is not working effectively and efficiently, outsourcing may be necessary to improve quality. Aircraft maintenance companies have more expertise in this area than the airline company because their core business is just maintenance activity.
- **Focus on core business.** It is rare for maintenance activities to be one of the main services of an airline operator carrying passengers. In order for the airline to be successful in maintenance services, it is necessary to allocate very important resources such as qualified human resources, financing, and time for aircraft maintenance. In this case, there will be a decrease in airline transportation, which

should be the main field of activity in the resources to be allocated. If maintenance services are outsourced, some internal resources will be free, they will be allocated to airline transportation, which is its main field of activity, and thus the opportunity to gain a competitive advantage by focusing on core business.

- **Risk sharing**. Airlines are responsible for performing maintenance activities in a way that increases flight safety. Regardless of whether the maintenance activities are carried out with their resources or taken from outside, in case the safety is in danger, the airline company is kept legally responsible. On the other hand, the airline and the maintenance supplier share risks in terms of factors such as technological developments in the field of aircraft maintenance activities and efficiency in investment decisions.
- **Adaptation to changing technology**. Technologies associated with aircraft maintenance are constantly evolving and changing. Failure to follow technological innovations in a timely manner in this industry, where the technology turnover rate is very high, may result in loss of efficiency and productivity and consequently loss of competitive advantages. On the other hand, the existence of a business with self-capability in aircraft maintenance depends on its ability to follow these innovations. For this reason, it is an option to provide maintenance services from suppliers who can closely follow the technology to adapt to technology.
- **Reorganization**. If the airline business wants to reorganize, it must be able to downsize and increase its flexibility. This may be possible by outsourcing highly complex, costly, highly specialized activities such as maintenance. In addition, the flexibility will be increased thanks to the downsizing. If outsourcing agreements are designed flexibly, flexibility will be provided in decisions regarding maintenance.
- **Achieving a Competitive Advantage**. Competition in the air transport industry is increasing and globalizing. Airlines can see the way to gain a competitive advantage by focusing on core business. This will be an important reason to provide outsourcing aircraft maintenance.

Considering routine maintenance tasks, limited rationality and asset specificity is low. Because routine maintenance does not require special expertise, it is possible to find many suppliers on the market with these capabilities. For routine maintenance work, environmental uncertainty is low because most of the maintenance is standard and maintenance steps are specified in the manuals provided by the aircraft manufacturer. The situation is different for heavy maintenance activities because heavy maintenance is done less frequently and involves major repairs. In the heavy maintenance industry, barriers to entry are high because maintenance companies require special skills, and equipment costs are high. As there are few suppliers in the heavy maintenance industry, there is a high risk of hold-up for the buyer, but still many airlines choose to outsource heavy care services to focus on their core capabilities. For example, Continental gets its engine maintenance from General Electric. Southwest Airlines outsources more than half of its heavy maintenance activities (Rieple and Helm 2008).

10.5   Outsourcing of Ground Handling Services

Airline companies can provide ground services required for ground operations with their internal resources, or they can get them from an independent ground handling company through outsourcing (Fuhr 2007). Transactions subject to exchange between the airline operator and the ground handling company are called ground handling services. Ground handling is defined by the International Civil Aviation Organization (ICAO) as "the services required from the arrival of an aircraft to its departure from the airport" (ICAO Conference Paper 2000). The ground handling industry is regulated by national and international aviation authorities. For example, ground services activities in Turkey Airports Ground Services Regulation (SHY 22) are regulated by the (SHY-22 2016). According to the Airport Ground Handling Regulation, ground handling services are grouped under ten service groups. These are representation, passenger traffic, load control and communication, ramp, aircraft line maintenance, flight operation, transportation, catering service, surveillance and management, and aircraft special includes security services. Airline operators can outsource one or more of these service groups from the ground handling operator according to the agreement between them. The reasons for airline companies to outsource ground handling services are similar to the reasons for outsourcing the maintenance we mentioned above because these are the main reasons for outsourcing. The main motivation factors that lead airline companies to use outsourcing for ground handling services are to reduce costs, increase operational quality, reach new technology, focus on the core business, and increase flexibility.

From the point of view of transaction cost theory, asset specificity, uncertainty, and frequency of services are determinants in the outsourcing decision of the airline company. Environmental uncertainty (for example, contract renegotiation) and demand uncertainty are limited by safety and security regulations. Behavioral uncertainty is related to measuring the performance of the service provided by the ground handling company. Monitoring and enforcement of the outsourced services are affected by the complexity of the ground handling process at the respective airport. For example, when a delay occurs at the hub airport, it is difficult to determine the responsible or source of the delay because ground handling is complex and may result from many different factors (fuel, air traffic control, etc.). The poor performance of suppliers at central airports can have consequences that will affect all other flights of the airline company. Therefore, more detailed service agreements and contract negotiations are needed to monitor the performance of the supplier at hub airports. However, as the ground services received at small airports are more limited and clear, it will be relatively easy to monitor the performance of the ground handling company at this airport, and behavior uncertainty will be low. The poor performance of suppliers at central airports can have consequences that will affect all other of the airline companies. Therefore, more detailed service agreements and contract negotiations are needed to monitor the performance of the supplier at hub airports. However, as the ground services received at small airports are more limited and clear, it will be relatively easy to monitor the performance of the ground handling company at this airport, and behavior uncertainty will be low. The decision to use ground service outsourcing for airline companies, especially at the hup airport, is a strategic decision.

Network carriers have high turnaround triggers at hup airports and low at spoke airports. The turnaround number indicates the frequency of ground handling operations. Transaction frequency is secondary importance to the outsourcing decision, but it is an important determinant of the emergence of hybrid governance structures (partnership, shareholding, etc.) (Fuhr 2009).

When we evaluate the airline-ground service relationship, the ground handling equipment used varies according to each aircraft type, and the information technologies (reservation, check-in systems) used for each airline business can differ. For example, if the fleet of the supplier company's (ground handling) customers (airlines) mainly consists of A-320 model aircraft, the ground handling company makes the majority of its ground service equipment investment in accordance with this aircraft type. Again, if the buyer company (airline) uses a different reservation/check-in program from the supplier's other customers (other airline companies), the ground handling operator must also train its personnel for this reservation system. In this context, if the number of flights and/or market share of the airline operator is high at the airport, the assets dedicated to this buyer by the ground handling operator will also be high (Fuhr 2007). In other words, if, for example, an airline company with a high market share wanted to change its ground handling provider, the new supplier would require safeguards before investing in equipment, hiring new personnel, and expanding local maintenance facilities. This will cause the parties to be locked in the relationship because the airline will not be able to easily find a relationship-specific supplier to invest in, and the ground handling business will not want to waste relationship-specific investments (Fuhr 2009).

The outsourcing decision and supplier relationship will differ depending on the type of ground service to be provided with outsourcing. Let's explain this over the example of flight catering, which is one of the ground service types. It is known that airlines provide catering services internally or use hybrid arrangements. Major airlines rarely outsource this service. For example, Turkish Airlines gets its catering service from Turkish Do & Co, and Turkish Airlines owns 50% of this company's share. Although the catering services may not seem uncomplicated and routine at first glance, the risk of logistical delays in this service is high. The airline catering industry actually specializes in logistics and minimizing costs. Since logistics is a critical element in catering services, catering suppliers have developed competence in the design of IT planning systems and delivery of products to aircraft. However, outsourcing this service is risky for airline companies and therefore a strategic decision. For example, in August 2005, British Airway's operations at Heathrow airport were interrupted by catering provider Gate Gourmet, who laid off 350 staff. As a result, British Airways had to cancel 900 flights and lost £45 million. This dispute between Gate Gourmet and British Airways has shown how vulnerable the airline business is to its caterer (Rieple and Helm 2008).

10.6 Outsourcing of Information Technologies and Reservation Systems

The aviation sector is one of the leading sectors in the development and change of information technologies. The use of information technologies is intensive, and managing them is critical for the following reasons:

- The airline industry is a dynamic industry.
- IT-supported operations must be maintained without interruption.
- It is a service sector where customer satisfaction is important.
- The airline industry is pioneering new information technologies.
- IT investments are long term.
- IT projects are extensive and complex.
- National and international regulations have an impact on IT management.

Although information technologies are of critical importance in the airline industry, outsourcing is a frequently used management tool by them. There are financial, operational, and strategic reasons for airlines to outsource information technologies. The main financial reasons are to reduce the costs of owning information technologies, personnel costs, and technology investment costs. The main operational reasons are to reach modern technology in a short time, to adapt to changes in technology, to increase service quality, to ensure the integration of IT systems, and to insufficient internal resources. Strategic reasons are to focus on core business, the demand of the top management, to benefit from the knowledge of the supplier, and to reduce the risk of being dependent on IT employees (Tokgoz and Erdogan 2016).

Airlines outsource a wide range of IT functions. The main IT services subject to outsource are application services, system integration, data center management, training and consultancy, network management, disaster recovery, server management and maintenance, company-specific application development, end-user support, and cloud services. Airline companies often provide application services, system integration, and company-specific applications by outsourcing. However, it is outsourced relatively rarely to disaster recovery and cloud services (Erdogan and Tokgöz 2020).

In terms of information technologies, asset specificity is the uniqueness of the firm's software and hardware and/or products or services that require special expertise (Barthelemy 2003). In the outsourcing of IT products/services with high asset specificity, the existence of detailed contracts is more important for buyer businesses because such transactions have certain characteristics specific to the buyer, so the risk of the supplier business to act opportunistically is high. It can be inferred that while outsourcing IT functions with low routine and asset specificity is not a strategic decision, it is a strategic decision to outsource IT functions with high asset specificity and more closely related to the airline's core business.

The power symmetry or imbalance between the buyer and supplier is important in IT outsourcing because power results in interdependence. Depending on the type of outsourcing, the dependency of the buyer on the supplier will differ. For example, while the supplier will be the dominant party in the case of total outsourcing, it will be more balanced in selective outsourcing. In addition, the amount or criticality of the outsourced IT assets will affect the balance of power between the parties (Kern and Willcocks 2000). The strong party in a relationship can use its power in different ways as oppressive on issues such as decision controls, price, quality, and service standards. The airline's being a large and reputable company in the sector, having sufficient internal resources, and the existence of alternative suppliers providing the same service in the market can make the buyers strong against the supplier (Rokkan and Haugland 2002). However, if there are vendors that market-specific IT products for the industry and create a monopoly, they will be stronger against the buyer and this increases the buyer's dependence on the supplier. There are IT software specific to the aviation sector, such as SITA, which has a monopoly position in the aviation industry and must be used by companies as required by international regulations. The airline does not have bargaining power against such suppliers. In some cases, although there are alternative suppliers providing the same service on the market, due to the high cost of changing IT products, airlines cannot easily change their supplier, which increases the dependency on the supplier.

As mentioned above, IT outsourcing takes place in a very wide range and the dependency relationship between the outsourcing decision and the buyer–supplier differs according to the transaction characteristics of the function (asset specificity, frequency, uncertainty) and the importance level of the function for the airline business. Let's examine outsourcing for reservation systems, a critical IT function close to the airline's core business. Reservation systems used by airlines are usually adaptations of ready-made packages (off the shelf) purchased from IT suppliers. Ready systems cannot meet the needs of every airline, so airline companies want these systems to be customized for them. Reservation systems must be compatible with the distributors' and agents' systems. For this reason, reservation systems have high asset specificity and result in increased dependency between buyer and supplier. Due to the high asset specificity of reservation systems, airlines prefer keeping at least part of the system in a hierarchy or semi-hierarchical arrangement (Rieple and Helm 2008). For example, Pegasus Airlines is a partner of Hitit IT Solutions, the supplier company that uses its customized reservation system.

It has been recognized that effective use of reservation systems not only provides cost advantage but also provides a source of competitive advantage by collecting sophisticated and valuable information about customers. When the information kept by the reservation systems is captured by the supplier, this will cause a security risk and the supplier will act opportunistically and increase the risk of locking the buyer in the relationship. According to the transaction cost theory, if opportunism and hold-up are high, it suggests that the ownership and processing of these data should be kept in-house (Rieple and Helm 2008). Because reservation systems store critical information for the airline, it is a strategic decision to outsource it. Reservation systems consist of many subsystems such as sales and ticketing. Therefore, it is a way for airlines to outsource more routine functions such as ticketing to keep critical functions in-house, to avoid the opportunistic risk of the supplier.

10.7 Conclusion

In this section, the issue of outsourcing in airline companies is explained within the framework of transaction cost theory and research dependence theory. Airlines typically outsource a broad range of business processes such as aircraft maintenance, ground handling, IT systems, crew training, maintenance training, catering services, and aircraft leasing. Some of the airline business outsourcing decisions are tactical and some are strategic. Tactical outsourcing is that involves more cost-oriented decisions and ignores other benefits and risks that may occur. Strategic outsourcing arises with the decision to outsource other activities that are very close to the core business in order to focus on key resources and distinctive skills.

Airline companies outsource in many different functions such as ticket sales and distribution, aircraft leasing, maintenance training, information system and technology, pilot training, advertising, engine overhaul, and ground handling operation. It is not possible to discuss the outsourcing decisions of the airline business in all different areas within the scope of this chapter, but aircraft maintenance, ground handling services, and IT services, which are very common to be outsourced by airline companies, are discussed. It can be said that routine aircraft maintenance outsourcing is a tactical outsourcing decision, while heavy maintenance outsourcing is a strategic decision. Routine maintenance requires less expertise and has low asset specificity, while intensive care jobs require special skill. There are very few heavy maintenance suppliers in the industry. This is an indication that the supplier will be in a stronger position than the airline and the airline will be more dependent on the supplier. Ground handling services cover representation, passenger traffic, load control and communication, ramp, aircraft line maintenance, flight operation, transportation, catering service, surveillance and management, and aircraft special includes security services. Airline companies prefer to outsource one or more of these services at certain stations where they operate. The decision to use ground service outsourcing for airline companies, especially at the hup airport, is a strategic decision. Operation intensity is high at the hub airport and poor performance of suppliers at hub airports can have consequences that will adversely affect all other flights of the airline company. Airlines prefer to outsource some IT functions such as application services, system integration, data center management, training and consultancy, network management, disaster recovery, server management and maintenance, company-specific application development, end-user support, and cloud services. Services such as network management, which do not require special skill and have low relationship-specific investments, are a tactical outsourcing decision. However, outsourcing the reservation system, which affects the main business of the airline, is a strategic decision. Because the reservation system is customized according to the company, that is, it is a process with high asset specificity, which increases the dependency of the buyer on the supplier.

References

Al-kaabi H, Potter A, Naim M (2007) An outsourcing decision model for airlines' MRO activities. J Qual Maint Eng 13(3):217–227

Anderson E, Weitz B (1992) The use of pledges to build and sustain commitment in distribution channels. J Mark Res 29(1):18–34

Barrar P, Gervais R (2006) Global outsourcing strategies: an International reference on effective outsourcing relationships. Gower Publishing Ltd.

Barrett SD (2004) The sustainability of the Ryanair model. Int J Transp Manag 2(2):89–98

Barthelemy J (2003) The hard and soft sides of IT outsourcing management. Eur Manag J 21(5):539–548

Brown D, Wilson S (2005) The black book of outsourcing: how to manage the changes, challenges and opportunities. Wiley, Inc.

Cannon JP, Achrol RS, Gundlach GT (2000) Contracts, norms, and plural form governance. J Acad Mark Sci 28(2):180–194

Czepiel E (2003) Practices and perspectives in outsourcing aircraft maintenance. Federal Aviation Administration Office of aviation research. FAA Publications, pp 1–25

Doganis R (2006) The airline business, 2nd ed. Routledge

Erdogan D, Tokgöz N (2020) Bilgi teknolojileri dış kaynak kullanımı başarısında biçimsel ve ilişkisel yönetişimin rolü: Havacılık sektöründe bir araştırma. İzmir İktisat Dergisi 35(2):221–239

Fink RC, Edelman LF, Hatten KJ, James WL (2006) Transaction cost economics, resource dependence theory, and customer–supplier relationships. Ind Corp Chang 15(3):497–529

Fuhr J (2007) Contractual design and functions–evidence from service contracts in the European air transport industry. Working paper series no 2007–03. Center for Network Industries and Infrastructure Publications

Fuhr J (2009) Liberalization of the European ramp-handling market: a transaction cost assessment. JTEP 43(1):105–122

Geyskens I, Steenkamp JBE, Kumar N (2006) Make, buy, or ally: a transaction cost theory meta-analysis. Acad Manag J 49(3):519–543

Ghobrial A (2005) Outsourcing in the airline industry: policy implications. J Transp Law Logist Policy 72(4):457–473

ICAO (2000) Ground handling at airports. Conference on the economics of airports and air navigation services. Montreal: International Civil Aviation Organisation. https://www.icao.int/Mee tings/ansconf2000/Documents/wp10e.pdf

Kern T, Willcocks L (2000) Exploring information technology outsourcing relationships: theory and practice. J Strat Inf Syst 9(4):321–350

McFadden M, Worrels DS (2012) Global outsourcing of aircraft maintenance. J Aviat Technol Eng 1(2):63–73

Morrell P (2005) Airlines within airlines: an analysis of US network airline responses to low cost carriers. J Air Transp Manag 11(5):303–312

Pfeffer J, Salancik G (1978. The external control of organizations: a resource dependence perspective. Harper and Row

Rieple A, Helm C (2008) Outsourcing for competitive advantage: an examination of seven legacy airlines. J Air Transp Manag 14(5):280–285

Rindfleisch A, Heide JB (1997) Transaction cost analysis: Past, present, and future applications. J Mark 61(4):30–54

Rokkan AI, Haugland SA (2002) Developing relational exchange: effectiveness and power. Eur J Mark 36(2):211–230

Rutner SM, Brown JH (1999) Outsourcing as an airline strategy. J Air Transp World Wide 4(2):24–27

SHY-22 (2016) Yer hizmetleri yönetmeliği. http://www.resmigazete.gov.tr/eskiler/2016/08/201 60823-3.htm

Simon HA (1957) Models of man: social and rational: mathematical essays on rational human behavior in a social setting. Wiley, New York

Tokgoz N, Erdogan D (2016) IT outsourcing reasons in the aviation industry. In: Proceedings of 39th international business research conference, Tokyo, Japan

Willcocks LP, and Lacity MC (1999), IT outsourcing in insurance services: risk, creative contracting and business advantage. Inf Syst J 9(3):97–242

Williamson OE (1981) The economics of organization: the transaction cost approach. Am J Sociol 87(3):548–577

Williamson OE (1985) The economic institutions of capitalism. Simon & Schuster Inc.

Williamson OE (1991) Comparative economic organization: the analysis of discrete structural alternatives. Adm Sci Q 36(June):269–296

Williamson OE (1996) The mechanism of governance. Oxford University Press Inc.

Dilek Erdoğan Assistant Professor in the Department of Aviation Management at Tarsus University, Turkey. She received her B.Sc., M.Sc., and Ph.D. all in Civil Aviation Management at Anadolu University, Turkey. She teaches courses on aviation management and research methods. She focused on the IT outsourcing relationship in Turkey and the aviation industry. She pursues her research in the areas of airport management, airline management, information technology management, strategic management, transaction cost theory. She has recently been working on the impact of the covid epidemic on the aviation industry. She has articles about aviation, information technology and covid in well-known journals

Chapter 11
The Importance of Organizational Behavior Model Applications in Air Transportation Industry

Metin Reyhanoğlu and **Harun Yılmaz**

Abstract Because the air transportation is a labor-intensive industry, human resource is one of the most crucial elements in order to survive and achieve the sustainable growth in the increasing competitive environment. As the industry has two types of staffs, flight and ground staff, and with the spread of new technologies and globalization, all human resources processes, from personnel selection to motivation, should implement new organizational behavior models to justify labor costs. The aim of this chapter is to emphasize the importance of applying organizational behavior models in the air transportation industry. To the goal of keeping the organizational commitment of the employees and reducing the intention to leave the job employees, who have undergone intensive training, implementation of organizational behavior models will provide significant gains to the organization. Unlike many labor-intensive industries, the aviation industry requires experienced and qualified employees to run operations safely and effectively. However, due to intense competition, low profit margin pressure, seasonal effects on demand, and intense working pace, turnover and turnover rates in the aviation industry are high. The most effective use of human resources in the aviation industry is through understanding the human nature. The term "human factor" has become increasingly popular as the commercial aviation industry notices human error that underlies most aviation accidents and incidents, rather than mechanical failure. The human phenomenon in the aviation sector differs from other sectors on eight essentials: punctuality and simultaneity, service quality, risk factor, interdisciplinary cooperation, high rules and standardization, command-based activities, jargons and world language, being subject to different laws. The success of businesses operating in the air transport sector depends on effective human resources policies and practices. In this respect,

M. Reyhanoğlu (✉)
Faculty of Economics and Administrative Sciences, Hatay Mustafa Kemal University, Hatay, Turkey
e-mail: mreyhan@mku.edu.tr

H. Yılmaz
Faculty of Aeronautics and Astronautics, Iskenderun Technical University, Hatay, Turkey
e-mail: harun.yilmaz@iste.edu.tr

K. Kiracı and K. T. Çalıyurt (eds.), *Corporate Governance, Sustainability, and Information Systems in the Aviation Sector, Volume I*, Accounting, Finance, Sustainability, Governance & Fraud: Theory and Application, https://doi.org/10.1007/978-981-16-9276-5_11

one of the most important parts of the airline business is the human resources department. The impact of recruitment and training on employees increases the psychological state, loyalty, and commitment of employees to the organization. The public and private aviation companies applying organizational behavior models will be one-step ahead in achieving sustainable growth besides gaining advantage in competition.

Keywords Human resources · Organizational behavior model · Human factors · Organizational commitment · Labor-intensive industry · Labor costs · Flight and ground staff · Safety and security

11.1 Introduction

Competition in the air transportation industry has been increased due to the enhancement in human needs and demands, developments in global economies, facilitation of liberalization worldwide, and market entry conditions. Human resources is one of the most important elements in order to survive and achieve the sustainable growth in this increasing competitive environment. In the air transportation industry, which is a labor-intensive industry, there are flight and ground staff serving in operations and administrative departments. It is an obligation to select the suitable staff who are competent in public and private aviation companies. Therefore, significant costs are incurred for many tangible and intangible activities. According to IATA data (2019, 2020), besides oil prices and interest rates, it is seen that there are significant increases in the costs incurred for the labor force. Among these costs for human resources include items that are expensed during initial and recurrent training activities and other costs such as transportation, equipment, documents, and protective clothing. These expenses, which are incurred by the companies in order to be competent for their staff, rank first in some companies and second in others. As a result of all these efforts, a high level of performance and efficiency is expected from the staff.

Moreover, it is crucial to ensure the organizational commitment of employees and to decrease their intention to quit who have high performance and efficiency, which is the most important factor in the achieving goals of company. The importance of organizational behavior practices that will align employees with their organizations and commit them to their organizations is increasing day by day in the air transportation industry. Organizational behavior practices that ensure the interaction of the employees with other employees by adopting the history, norms and values, goals, policy, and language of the organizations without any uncertainty can reduce the intention to leave by ensuring organizational commitment (Yilmaz 2020a). In the air transportation industry, the holding cost of the labor force has become more important than the recruitment and placement costs in ensuring the sustainability of the organizations.

Today, with the spread of new technologies and globalization in the transition from industrial societies to information societies, it is emphasized that managers need to implement new organizational behavior models in order to keep sustainable competitive advantage of their organizations. Because of humans being at the center of organizational behavior theories, in the literature, it has been emphasized that it is very difficult to understand management issues without specifying the expectations and desires of staff, and without going into the details of human and group behaviors. In this context, the importance of organizational behavior models is increasing in companies operating in the air transportation industry. Giving importance to the human resources and planning human resources can directly affect the success of the company. In this respect, human resources management is accepted as a tool used for the organizations to achieve its strategic goals. It is desired that employees benefit their organizations in all the phases—from their employment to leaving the organization. Due to the great importance of trained and skilled human resources, which have contributed to the rapid growth of the civil air transportation industry in recent years, great attention is paid to the legal and social rights of the employees across the industry. With the aim of keeping the organizational commitment of the employees and reducing the intention to leave the job employees, who have undergone intensive training, implementation of organizational behavior models will provide significant gains to the organization. It is shown that researches on organizational behavior model applications in the air transport industry are insufficient.

The aim of the study is to emphasize the importance of applying organizational behavior models in the air transportation industry. Thus, it will be possible to identify the problems experienced by individuals working in the aviation industry and to produce solutions. In addition, it is another goal to encourage business managers to apply the new organizational behavior models, and researchers to investigate organizational behavior model researches in the air transport industry. In this context, studies conducted in this field in the literature are examined. Companies applying organizational behavior models will be one-step ahead in achieving sustainable growth besides gaining advantage in competition.

11.2 The Importance of Human Resources in Aviation Industry

Air transport is a sector that provides significant economic and social benefits to cities and regions with airports. Air transport, which has a time advantage in passenger and cargo transportation, is constantly growing. But epidemics (for example, the COVID-19 pandemic in 2020 has come to a standstill and some companies have gone bankrupt all over the world, for the effects, see ICAO 2020), political crises (Bükeç and Erdoğan 2010; for the effects of the tension with the Russians on the Turkish aviation industry in 2015, see Demir 2015), war, terrorist attacks and illegal acts, stagnation in the economy, etc. factors may slow this growth or cause a decline.

Despite these factors that adversely affect air transport, the growth in the sector becomes sustainable with the support of advanced technology and qualified personnel as well as the legislative work and government incentives needed in the sector. Among the characteristics of the airline industry are economies of scale, growth through merger, interdependence, price rigidity, and non-price competition (Wensveen 2007, p. 176). In the airline industry, which has an oligopolistic market structure, there are high barriers such as significant capital requirements, the need for technical and technological know-how, and control of patent rights in the market entry.

The aviation industry is an industry in which research and development of various air- and ground-based equipment required for the testing, use, and maintenance of aviation systems, including manned and unmanned aerial vehicles, and aircraft (Wensveen 2007, p. 4). Even if there is a period of stagnation and regression, the development in the industry is always upward.

Air transport is at the center of world tourism and trade due to its effect. According to IATA 2019 annual evaluation report data, tourists traveling internationally by airline spent around $ 900 billion in 2019. Again according to the data in this report; in 2019, approximately 4.5 billion passengers and 61.2 million tons of cargo were transported on 46.8 million flights between 22,000 city pairs. Approximately, 65 million people work in this sector (IATA 2020, p. 39).

It is seen that the effect of qualified personnel on the growth of the sector is great. There are expert personnel working in the operational and administrative departments that have a license or certificate. Especially with the liberalization in the civil aviation sector in the beginning of 1980, the increasing competition has made labor an important factor in achieving sustainable competitive advantage. Therefore, flexible working models have started to be widely applied in the air transport sector all over the world (Koç 2018, p. 1060). It has required increasing the safety, effectiveness, and efficiency of air navigation in the face of the increase in the density of flight traffic and operations globally; in order to reduce flight costs and environmental impact, it is ensured that the correct information is shared between various systems at the right place and time by cooperating with international institutions and organizations (SHGM 2020).

The importance of human resources is increasing day by day in the growth and development of the sector. Intensive training is provided to employees from the recruitment phase to leaving the organization. In this training process, the employee can gain some important competencies. These competencies can be used nationally and internationally with the licenses or certificates they have. In this context, organizations want to employ and retain trained, competent, qualified personnel who are experts in their fields. Therefore, analyzing organizational behavior models such as job satisfaction, organizational commitment, and well-being of employees is the subject of academic research.

11.3 Human Nature in Aviation Industry

The most effective use of human resources in the aviation industry is through understanding the human nature. Human nature refers to general psychological traits, emotions, and behavioral traits shared by others. The reason for these feelings and behaviors is a certain need and motivation to meet this need. Motivation is the reason why a person behaves in a certain way and is at the center of the goals. The goal is the object of a person's effort. The study of human behavior is an attempt to explain how and why people behave this way. Human behavior, which is a complex subject, is a product of both innate human nature and individual experience and environment. Although there are many definitions of human behavior depending on the field of study, human behavior in the scientific world is seen as the product of the factors that cause people to act in predictable ways.

The term "human factor" has become increasingly popular as the commercial aviation industry notices human error that underlies most aviation accidents and incidents, rather than mechanical failure. According to Virovac et al. (2017, p. 257), the human factor seen in the aviation industry in general describes the human errors that are conducted involuntarily while the employee is doing his/her job and resulting damage. The human factor includes collecting information about human abilities, limitations, and other characteristics to ensure safe, comfortable, and effective use and applying them to vehicles, machines, systems, tasks, jobs, and environments. However, technological changes and innovations may cause human-induced errors, as aviation industry employees require new skills and knowledge (Reason and Maddox 1996).

The human factor in aviation focuses on better understanding how people can be integrated into technology in the safest and most efficient way. This understanding then entails being translated into design, training, policies, or procedures to help people perform better. Since technology continues to evolve faster than the ability to predict how people will interact with it, the industry should no longer rely on experience and intuition to make decisions about human performance. Instead, a solid scientific foundation is required to evaluate human performance effects in design, training, and procedures (Graeber 2020). Therefore, it is necessary to focus on the human factor in the aviation industry.

The human phenomenon in the aviation sector differs from other sectors on eight essentials: punctuality and simultaneity, service quality, risk factor, interdisciplinary cooperation, high rules and standardization, command-based activities, jargons and world language, being subject to different laws (Fig. 11.1).

11.3.1 Punctuality and Simultaneity

In addition to the processes that follow each other, the complexity of the works with each other but their dependence requires the timely execution of the works (Uyar

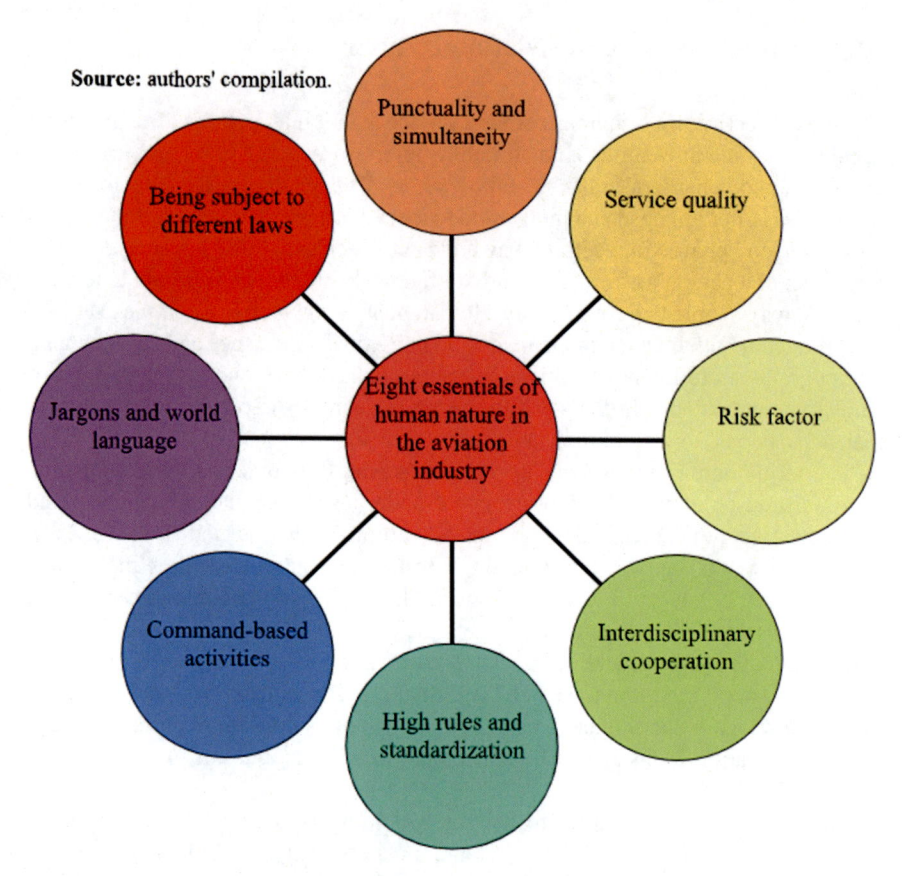

Fig. 11.1 Eight essentials of working life in the aviation industry

2018, p. 2). In many industries, such as the manufacturing industry, jobs are one after the other. Things are interdependent and one cannot begin before one job is finished. However, in the aviation industry, besides the sequencing of the works one after the other, the complexity with each other requires that the works be carried out simultaneously. Therefore, employees in the aviation industry require them to have versatile thinking ability and to be resistant to stress due to time pressure. Time pressure is an important source of stress that mostly negatively affects the decision-making process and work performance of individuals (Saptari et al. 2015). In this sector, where the service lasts almost 24 h without interruption, aviation employees have to carry out their work with great devotion, despite the shift and long working order, work stress, excessive attention, time pressure, and heavy responsibilities (Koç 2018, p. 1070).

Transportation, which is one of the basic needs of human beings, is a sector that never stops except in extraordinary situations. Therefore, the working hours of individuals working in the transportation sector are generally irregular. The same is true in the air transport industry. The working hours of the individuals involved in the operation may differ, except for those working in the administrative department. Flight personnel start their duty before the flight and continue their duty until they leave the plane after the flight is over. The working hours of the ground personnel may vary according to the airport traffic density and plane arrival and departure flights. Therefore, the working hours of ground staff can be more irregular.

11.3.2 Service Quality

The human factor plays a key role in the success of organizations operating in the aviation industry as a service sector, especially due to the principle of inseparability of service (the service cannot be considered separately from the service providers) (Doğan et al. 2020). Survival and having a lasting competitive advantage has always been an important organizational strategy for the airline industry. Service quality, which is one of these strategies, is usually the first priority to be considered. Because ensuring the loyalty and satisfaction of passengers who intend to revisit and buy depends on it. As is widely known, employees play an important key role in the process of delivering high-quality service. However, how to motivate employees to do more to deliver superior service quality is still a difficult process for the airline industry (Ho and Wu 2019, p. 17).

As a global sector, air transport has a technology and labor-intensive feature. Airline companies strive to achieve flight and ground safety, passenger comfort, and economic goals in their operations. In this sector where the risk factor is very important, businesses also pay attention to the high level of service quality to ensure customer satisfaction. Quality service is one of the factors determining airline demand. This means achieving a high level of income. In addition to socio-economic variables, there are studies in the literature that income is in a strong relationship with airline demand (Eren et al. 2020, p. 240). This feature of the airline transport sector makes it important and seems to contribute to the economic development of societies.

11.3.3 Risk Factor

Compared to other transportation sub-industries, the aviation industry is the sector with the lowest risk per km traveled due to the strict implementation of international standards (Uyar 2018, p. 1). Despite this, the risk impact of very small mistakes in this sector is very high (Uyar 2018, p. 2). Human error has been documented as a primary contributor to more than 70% of aircraft crashes (Graeber 2020, p. 23). The safety of both the aircraft crew and the people on the ground is extremely important, as transportation is done by air. Because it is not possible to save people in the air and on the ground if the plane crashes. In addition, the possibility of having explosive and flammable materials on the ground can cause great risks. Therefore, the aviation industry is in the high risk category. People at all levels of safety management to reduce or even eliminate risk, and those who manage risk and make risk decisions are a source of risk (Uyar 2018, p. 2).

11.3.4 Interdisciplinary Cooperation

The aviation industry leads many related industries and can move them forward (Türk and Şener 2018, p. 5). This requires working with different disciplines. Besides, it includes different professions and disciplines in aviation. This industry, where the costs are high due to the high rate of use of labor and the necessity of qualifying a certain part of this workforce, requires an intensive employee–employee interaction (Chen 2006, p. 74) to be carried out in cooperation with different professions and disciplines. Safe and efficient execution of air transport activities depends on the efficiency of all remaining aviation activities. Because civil aviation is a system and all elements of this system are in interaction with each other and should be viewed as a whole (Gerede 2003, p. 20).

Flight and ground personnel work in operational (line) and administrative (support) departments in the air transport sector. Operationally, there are flight crew, air traffic controller, flight operations specialists (dispatchers), aircraft maintenance technicians, ground services officers (passenger, operations, ramp, and cargo services personnel), security, and fuel personnel. Apart from this, departments such as human resources, purchasing, cost control unit, legal consultancy, business development, quality, accounting-finance, and public relations also provide administrative support services in the business. People from many different disciplines and professions have to work together in harmony.

11.3.5 High Rules and Standardization

Aviation industry is a sector where rules and regulations of especially international organizations are intense (Hirst 2008). In each section, there is an obligation to comply with international rules and standards established in order to ensure flight and ground safety. In this context, licensing is required in some occupational groups in the air transport sector. International Civil Aviation Organization (ICAO 2006) has created international standards and recommendations for licensing, for example, the following personnel:

- As flight personnel: private, commercial, and airline transport pilot (airplane and helicopter); glider pilot; balloon pilot; flight navigation specialist; and flight engineer.

- As ground staff: aircraft maintenance personnel (technician and engineer), air traffic controller, flight operations specialist (dispatcher), and aviation information management officer (AIM). AIM officer is an occupational group that implements flight safety, efficiency, effectiveness, order, aviation information, and data within the framework of established standards (SHGM 2020).

11.3.6 Command-Based Activities

Together with high rules and standardization, employees act according to previously determined commands. This is especially true for pilots. Leading the team members is the pilot-in-command or commander, who is assigned as command and responsible for the operation of the aircraft and the safe conduct of the flight during the flight duty period. Apart from the captain in charge, there is a co-pilot or first officer who will take flight instructions and perform other pilot duties (ICAO 2006, pp. 1–3).

Individuals working in the air transport sector work on a command basis. Especially flight personnel (cockpit and cabin) and maintenance personnel have to use checklists during their duties. This requirement was created to reduce the rate of human error, which has the largest share in aircraft accidents and incidents. Human error has been documented as the primary cause of more than 70% of aircraft crashes. According to Türen et al. (2015, p. 8), controlling complex devices in a certain order during the flight; updating these devices and possible malfunctions in pilots; and installing, maintaining, controlling, and updating the electronic software of high-tech aircraft in maintenance technicians can create technostress. Although generally thought to be related to flight operations, human error has recently become a major problem in maintenance practices and air traffic management. Aircraft and flight manual (handbooks) documents created by the manufacturers and operators guide the personnel working in the operations and help in terms of the correctness of the applications.

11.3.7 Jargons and World Language

Every industry and profession has its own jargon, which is normal for it to formulate; like the use of Latin-predominantly medical terms in hospitals, and the use of chemical terms in a chemical factory. In air transportation, the terms aviation and engineering are used in English. These terms are common all over the world and are bound by rules. Especially pilots, flight personnel, and air traffic personnel use aviation jargon at a high level. In terms of job descriptions, there is an obligation to comply with the different qualifications of the employees and also the international rules and standards established to ensure flight and ground safety. In order to understand and apply these rules and standards correctly, the language factor always comes first in the air transport industry. Because the individual who knows a foreign language can show a successful performance in both oral and written communication. Therefore, businesses employ individuals with foreign language knowledge.

11.3.8 Being Subject to Different Laws

Due to the aforementioned features, the legal status of those working in aviation transportation works differently than other sectors. In air transportation, all flight personnel such as pilots, flight engineers, stewardesses, and cabin crew are out of the scope of the Labor Law, and their rights are guaranteed in the Code of Obligations within the framework of the employment contract they have signed, while the rights of employees working in all ground facilities are covered by Labor Law No. 4857 (Koç 2018, p. 1063).

11.4 Researches and Models of Organizational Behavior in Aviation

Flight and ground personnel have an intense working tempo in air transport operations where speed factor comes first. Emotions and behaviors during their duties can affect the performance of individuals positively and negatively. In this context, it has become important to examine, research, and analyze organizational behavior issues in the air transport sector. After identifying the problems experienced by the personnel, producing appropriate solutions for them will contribute to the development of the sector. In this context, organizational issues such as human resources management, career planning, leadership styles, or individual issues that measure job satisfaction of technicians, pilots, and cabin crew were addressed in studies conducted in the air transport sector (Koç 2018, p. 1055).

Due to the fact that the aviation industry is affected by technological (Türk 2020, p. 1257) and political (Bükeç and Erdoğan 2010) change much more than other sectors, it directly affects the way of doing business and the working styles of employees in this industry. In addition, due to the high unit costs and the regulations changes on the employees to reduce these costs also lead the working styles and motivation of the employees either positively or negatively. The change in the aviation industry inevitably causes employees to resist this change. Generally speaking, although older ones have higher resistance to change, on the contrary, young ones may have higher resistance to change in the aviation industry (Türk 2020, p. 1257). The biggest reason for this situation the older people is the acceptance of change and have experienced this more in the aviation sector.

The aviation industry is a complex set of services as well as technical. Fulfillment of these services is possible first of all with people trained in the field of aviation, having knowledge and skills (Küçükönal and Korul 2002: 68). The performance, motivation, efficiency, job satisfaction, desire to stay at work, and, accordingly, life and job satisfaction of the employees in the aviation sector depend on the working conditions in the workplace and the weight of the work done with the material and moral benefits obtained (Koç 2018, p. 1068).

The level of job satisfaction of employees in the aviation industry affects their attitudes and behaviors. Therefore, job satisfaction is of critical importance (Doğan et al. 2020, p. 28) in this sector as it affects the perceptions of customers in the service provision of employees (Yeh 2014, p. 94). Unlike many labor-intensive industries, the aviation industry requires experienced and qualified employees to run operations safely and effectively. However, due to intense competition, low profit margin pressure, seasonal effects on demand and intense working pace, turnover, and turnover rates in the aviation industry are high. The existence of unfair practices reduces job satisfaction, and this can increase employees' intention to quit. In a study conducted in the aviation industry in Turkey (Doğan et al. 2020), job satisfaction has a fully mediation role in the effect of organizational justice dimensions on the intention to leave, showed so job satisfaction is an important factor in terms of aviation organizations. Therefore, it is suggested that organizations in the sector should focus on practices aimed at increasing the job satisfaction of their employees.

Since employees in the aviation industry take action against time, proper time management will also affect their individual performance. Because even the slightest mistake has to be costly and therefore time must be managed correctly. Performance can also cause stress due to time pressure. This is especially seen in the pilot profession. According to a study (Yiğitel and Ertemsir 2020), it was found that pilots' time management skills had a positive effect on their individual performance.

In a study conducted in the aviation sector (Çoban and Aydoğdu), it was determined that time pressure and technostress are positively related and that time pressure is a determinant of the change in technostress perception of maintenance technicians. It is possible to say that time pressure and technostress may cause human-induced errors in the aircraft maintenance industry. According to Tarafdar et al. (2011, p. 304), increasing workload due to the use of technology, constant updates, technical problems, technological complexity, and uncertainty causes employees to experience technostress. In this respect, time pressure creates technostress and this can cause human-induced errors.

In another study (Akalın 2019), the work performance of pilots is adversely affected by stress factors varieties (task conflict, overwork, task uncertainty, etc.). Cabin crew members, on the other hand, may have a negative working performance due to their feelings of burnout (Tuna 2019).

The attitudes and behaviors of employees in the aviation industry depend on many factors. It can be affected not only by organizational structure and management policies, but also by relationships within the sector. In the aviation industry, service contracts are frequently signed between the airline operator and ground handling companies. This airline company frequently procures ground handlers from a different business within the framework of outsourcing (Erdoğan 2019). The duration of the contracts signed between the two parties will affect the nature of the relationships of not only the managers but also the lower level employees of both parties.

11.5   Designs of an Aviation Company

Organizational structures and charts differ due to the diversity of activities in the air transport sector. An organizational structure can be considered as a set of entities that cooperate collectively and contribute to a common goal (Machado et al. 2015, p. 103). The organizational chart shows the formal authority relationships between superiors and subordinates at various hierarchical levels, as well as formal communication channels within the company. Organizational chart, which helps managers to apply organizational principles such as range of control and unity of objectives, plays an important role in identifying organizational deficiencies, such as a person reporting to more than one manager or one manager with a very wide range of control. It helps the members of the organization to understand more clearly where they stand in the company compared to others and how and where managers and employees fit into the overall organizational structure. The organizational chart is a static model of the company as it shows how the company is organized at a particular point. Airline carriers must constantly adapt to changing conditions, as they operate in a dynamic environment. Some old positions in the hierarchical structure may no longer be required or new positions may need to be created to achieve new goals. Therefore, the organizational chart should be periodically revised and updated to reflect these changing conditions.

Airlines have grown so rapidly in the past 25 years that it is difficult to say that it is typical of any organizational chart, or that the chart of a company at a particular time is still in effect even after a few months. However, all airlines have divisions where airline activities are similarly divided. Understandably, the larger the carrier, the greater the specialization of the tasks and the greater the departmentalization.

Departments in airline businesses are divided into three different activities: operation, service, and sales. Here, some inputs such as employees are shared, and different airline companies have different numbers of employees for each field of activity. Airline businesses have to reach the optimum number of personnel in order to achieve high efficiency. However, different airlines may have different numbers of employees for the same departments (Li and Cui 2018, p. 150).

Ownership and management structures of organizations doing business around the world may differ significantly from each other due to their financial (tax) obligations. Ownership structure varies depending on a number of factors such as the number of employees, the overall size of the business, the scope of its responsibility, the number of owners, and their type and obligations (Vasigh et al. 2014, p. 52). According to Wells (1989), personnel working in an airline business is gathered under two main sections, Line Personnel and Staff Personnel, in an organizational structure where the management can achieve its goals (as cited in Küçükönal and Korul 2002, p. 78).

Figure 11.2 shows the organization chart of a large airline company. As seen in the figure, there are two main sections under the executive Vice-President, namely, staff administration and line administration. Within the scope of support activities in Staff Administrations, subsections have been created for finance and property, information services, personnel, corporate communications economic planning, and legal and medical activities. In line administration, there are subsections where the activities related to aviation, which is the main business of the enterprise, are carried out: flight operations, engineering and maintenance, marketing.

Due to its feature of being a service organization, research, development, and marketing activities for service development and diversification are heavily involved in airline companies (Küçükönal and Korul 2002, p. 77). Airline companies aiming to fulfill aviation services, to be ahead of the rapid growth and increasing competition

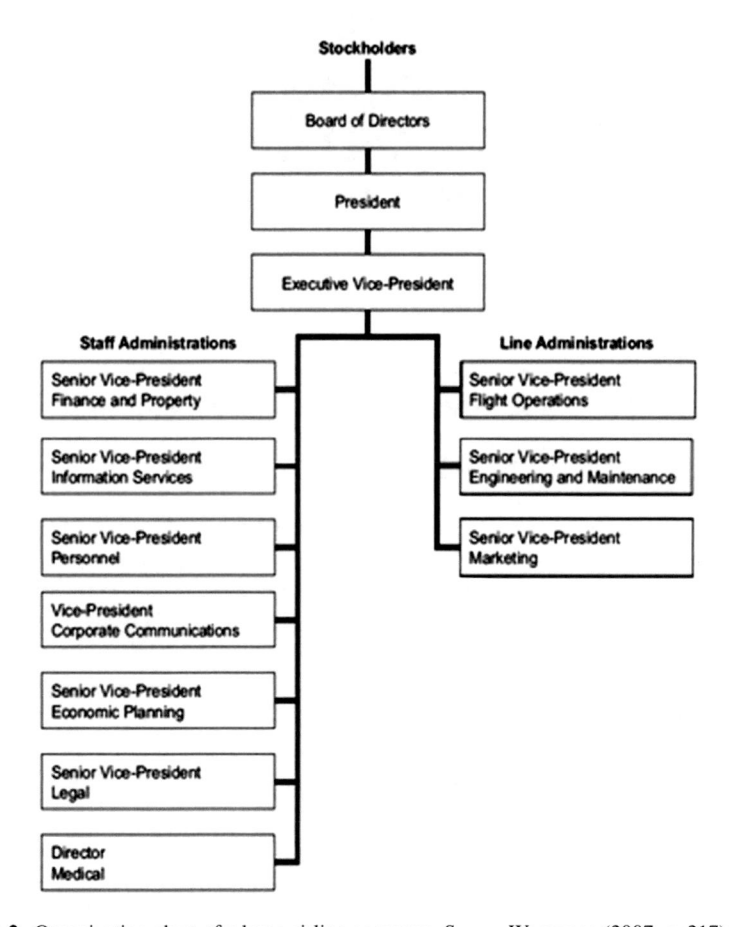

Fig. 11.2 Organization chart of a large airline company. *Source* Wensveen (2007, p. 217)

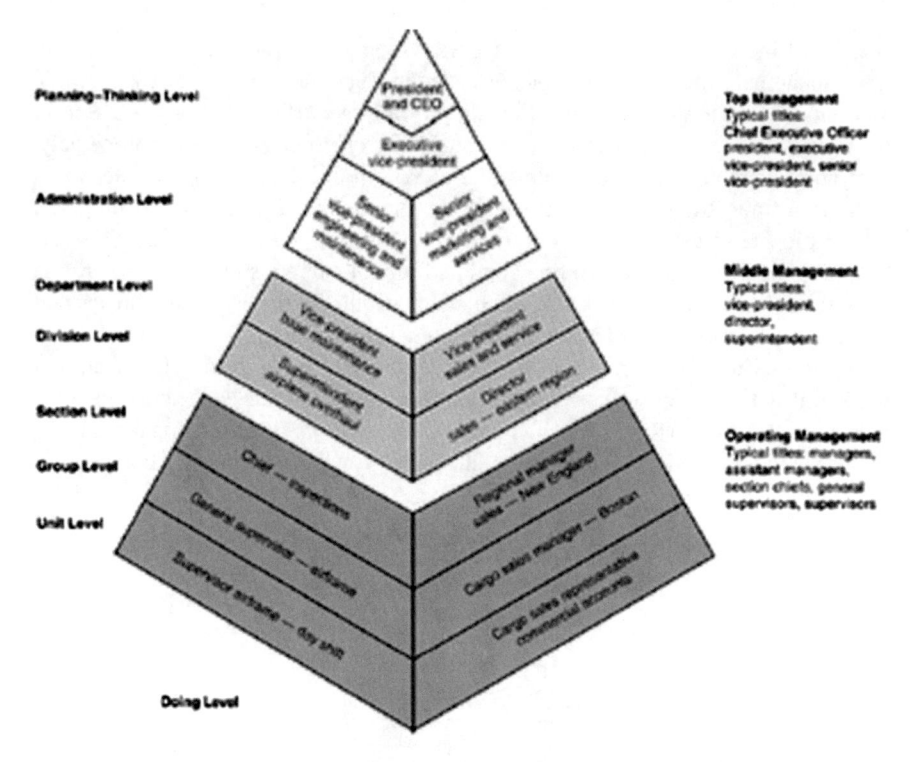

Fig. 11.3 Authority pyramid in an airline business. *Source* Wensveen (2007, p. 204)

in parallel with this, and to realize a safe flight activity by providing better service, should analyze and manage human errors and their root causes well (Kanbur and Gökalp 2014, p. 1).

Figure 11.3 shows the authority pyramid in any airline business. The part here constitutes the Line Administration part in Fig. 11.2. Strategic activities based on coordination and decision-making are carried out on the higher level and things are carried out on the lower level. In the figure, the darker parts of the pyramid show where jobs are performed (e.g., obtaining data, receiving reservations, aircraft maintenance). Lighter parts show activities such as planning, consulting, and policy-making.

Terms such as top management, middle management, and lower management are commonly used in the business world to distinguish levels of management within an organization. Unfortunately, not every level has a clear definition and the meanings added to the terms sometimes differ from one company to another. However, the top management of a business is generally considered to be the policy-making group responsible for the overall management of the company. The middle management is responsible for the execution and interpretation of policies throughout the organization, and the lower management is directly responsible for the final implementation of policies by employees under their supervision (Wensveen 2007, p. 202).

From a socially sustainable perspective, the general belief among top executives is to develop the right employees to embrace the right culture. The culture of an organization ensures that its employees maintain ethical, egalitarian, and transparent values while making decisions that affect stakeholders (Philips et al. 2019, p. 32).

11.6 Implications of Human Resources in Aviation

The success of businesses operating in the air transport sector depends on effective human resources policies and practices. In this respect, one of the most important parts of the airline business is the human resources department. Although recent technological developments, globalization, social trends, and changes bring new challenges for recruiting and selection, recruiting and selection is an important human resource management function as it covers all organizational practices and decisions (Talukder 2016, p. 38).

The impact of recruitment and training on employees increases the psychological state, loyalty, and commitment of employees to the organization (Alola and Alafeshat 2020, p. 1). Employees are key inputs to ensure service excellence and productivity, which can be important sources of competitive advantage. Still, among the most challenging jobs in service organizations, employees are expected to be friendly and helpful in their dealings with customers, as well as being quick and efficient in performing operational tasks (Wirtz et al. 2008, p. 4).

The interdependencies of the fleet and other investment plans of the airline companies are closely related to the planning of human resources, as the success of the enterprises in the medium and long term is in integrity (Küçükönal and Korul 2002, p. 68). Strategic human resources try to show that positive employee attitudes and behaviors increase organizational performance based on airline industry data and experience (Pate and Beaumont 2006, p. 327).

Many studies in the literature have drawn attention to the important impact of job design and acknowledged that interpersonal relationships can cause employees to experience their duties significantly and meaningfully (Ho and Wu 2019, p. 17). Moreover, the experience of important and meaningful tasks plays a key role in increasing employees' resources and motivation at work.

Human behavior is also defined as a result of efforts to meet specific needs. These needs, such as the need for food and water, can be easy to understand and define. They can also be complex, such as the need for respect and acceptance. Human performance refers to human capabilities and limitations that have an impact on the safety, security, and efficiency of aviation operations (ICAO 2006, pp. 1–2).

Human factors are often considered to be synonymous with team resource management or maintenance resource management. Human resources management is quite complicated due to the requirement of personnel that can fulfill a wide range of activities in the airline sector and can undertake multiple tasks (Küçükönal and Korul 2002: 79). It includes collecting information about human factors, human capabilities, limitations, and other characteristics to ensure safety, comfort, and effectiveness, and applying them to vehicles, machines, systems, tasks, jobs, and environments. Despite rapid gains in technology, the industry continues to invest heavily in training, equipment, and systems with long-term implications, as workers are ultimately responsible for ensuring the success and safety of the aviation industry (Graeber 2020).

Generally, personnel costs representing more than 35% of the operating cost of an airline company are definitely one of the most important areas of concern for management (Wensveen 2007, p. XVI). Therefore, when there is a decrease in flight and passenger traffic, airlines choose the way to reduce capacity. Because it is seen that the cost per unit started to increase. In such a situation, airlines may have to lay off some of their personnel. As a matter of fact, it is estimated that the average turnover rate in this industry is 20–30% (Miles and Mangold 2005, p. 543). The remaining personnel, on the other hand, expand their responsibilities by undertaking new assignments. Airlines often try to persuade their unionized staff to take more responsibility (Wensveen 2007, p. 180). When airline companies want to do the right distribution of personnel, they can adjust by hiring or firing employees for optimum efficiency rather than asking employees to do the jobs at every stage (Li and Cui 2018, p. 151). It has shown that low-cost carriers who switch to a hybrid business model can improve the attitudes and behaviors of employees in the short term, and there is a way to increase job performance in this sector (Pate and Beaumont 2006, pp. 322–325).

In addition to wages, personnel working in aviation enterprises expect some other payments, supports, and facilities that will increase their living standards. These opportunities not only increase job applications but also increase the commitment of existing staff to the business (Küçükönal and Korul 2002, p. 85). Employees are provided with training and certification to ensure that the existing personnel in the aviation sector and the personnel to be employed do the job correctly, comply with the law, feel safe, and trust themselves when making decisions (Phillips et al. 2019, p. 36). With regard to training and development of employees, many companies make the mistake of viewing training as a cost rather than an investment, and many who see it as an investment limit training to the technical aspects of the business rather than aiming to develop employees in a more holistic way (Wirtz et al. 2008, p. 15).

For example, the world's leading Singapore Airlines uses a variety of forms of reward and recognition, including diverse business model and scope, symbolic behavior, performance-based sharing options, individual contributions, and a significant percentage of variable compensation components linked to the company's financial performance (Wirtz et al. 2008, p. 13). Based on the interviews with Singapore Airlines' senior management and experienced flight crew, five elements that form the cornerstones of Singapore Airlines' human resources management and strengthen its service excellence strategy have been identified. These are (Wirtz et al. 2008, p. 7).

- strict selection and recruitment processes,
- comprehensive training and retraining of employees,
- forming successful service delivery teams,

- empowering operational staff, and
- the motivation of the employees.

Therefore, it is seen that all of the routine human resource activities in the aviation industry are carried out as other industries. However, these activities should be considered within the framework of eight essentials given in the title of "Human Nature in Aviation Industry". For example, applicants to be recruited to an aviation company; it is expected to have the ability of using the jargon of the aviation industry, a general knowledge and ability to work with different disciplines, the ability to work under high stress, and the ability to think fast and to put those into practice. In this direction, it is required applicants to be subjected to various tests.

11.7 Recommendations and Conclusion

Civil aviation industry is one of the sectors with the highest growth rates in Turkey. Developments in the Turkish Civil Aviation Sector have made significant progress not only with an increase in the number of passengers, but also with applications used to ensure flight safety and aviation security at airports, ground handling services, and other different areas (Yilmaz 2020b). As a result of the opening-up policy in Turkey, in order to manage the increasing tourism and trade potential in recent years, a significant meaning has been attributed to the civil aviation, and it is considered as a strategic element. Bakır, Bal and Akan (2017) emphasized in their study on the civil aviation sector that important criteria in Turkey are terrorist threats, tourism potential, geopolitical, economic, and political developments. For this reason, it is recommended to consider these criteria in strategies and planning to be implemented in the civil aviation industry. One way to ensure the development of the aviation industry in Turkey is to minimize the problem of human resources. Accordingly, some findings and suggestions are given below.

Although the aviation industry is affected by negative factors periodically, it shows a sustainable development. The key to this development is human resources. The contribution of human resources is great in the development of technology and innovation studies and the emergence of useful inventions. Specialization is the top priority of this industry. With the specialization given importance in the aviation sector, the time loss that occurs when employees move from one job to another tends to be eliminated (Wensveen 2007, p. 179).

Despite this, the employee turnover rate in the aviation sector is high. This situation is particularly intense in ground handling services. The most important reason for this is that airline companies outsource ground handling services from subcontractors. This situation shows itself even more especially in low-cost oriented airline companies. The high rate of workforce turnover in the aviation sector means the loss of qualified employees who have gained years of experience and experience, and as a result, the conduct of activities with inexperienced employees (Chung et al. 2017,

p. 89). It is not possible to expect job satisfaction and productivity from employees who feel the danger of losing their jobs.

In service businesses, it is a difficult process to do human resource management properly, and most successful service organizations have a strict commitment to effective human resource management, including recruitment, selection, training, motivation, and employee retention (Wirtz et al. 2008, p. 5). Strategic human resource management, which has an impact on both human and social outcomes, should contribute to the sustainability of employment (Santana et al. 2019, p. 924). Organizational success is largely determined by how successfully employee talents are defined, developed, and put into use in critical positions. One of them is a suitable performance evaluation.

Airline businesses should be able to design business and work systems to achieve organizational goals; employ people who have the desire and ability to perform effectively; and train, motivate, and reward employees to increase performance and be more productive (Talukder 2016, p. 43). According to the research conducted by Alola and Alafeshat (2020), it was concluded that selection and recruitment and training and development of practitioners and human resources managers have an effect on organizational performance and employee engagement.

In order to combat time pressure, the quality and quantity of resources such as the number of technicians in the workplace, necessary tools, and technical documents should be increased. Daily, weekly, and monthly workload plans should be rearranged. Improvements should be made on organizational factors that cause time loss. Employees should be given time management training. Technological workload should be distributed equally to employees, who experience intense time pressure such as pilots and technicians, to be able to cope with technostress. Basic and refresher trainings should be given in order to adapt to new technologies. Technical problems and updates should be done in a timely manner. Working hours should be reorganized and social activities should be given importance in order to prevent the physical and psychological negative effects of technostress (Çoban and Aydoğdu 2020).

All processes, from personnel selection to training, should be designed in accordance with time planning in the aviation industry, where time must be planned very well by nature, where even seconds are of critical importance, and where the slightest delay or the smallest mistake can cause disasters. In personnel selection, emphasis should be placed on analytical thinking, teamwork predisposition, communication ability, ability to work under stress and pressure, attention to detail, technology predisposition, agility and dynamic, rapid decision-making, problem-solving ability, self-discipline, and technical skills. It is considered appropriate to implement training programs on time management and to encourage employees to improve themselves on the subject (Yiğitel and Ertemsir 2020).

Considering the relationship of psychological ownership with positive organizational outcomes with income and position; career management, manager development, performance management, and remuneration policies should be prepared. It should not be forgotten that opportunities such as premiums and promotions may affect psychological ownership (Şenol and Üzüm 2020). Another important factor in creating psychological ownership is to provide an environment of trust within the organization. The environment of trust not only makes a positive contribution to the economy and entrepreneurship (Akın et al. 2017), but also can shape human resources practices within the organization. Studies show how trust can facilitate effective and productive outcomes within the organization on the basis of individual, group, and organizational criteria (Akın et al. 2018, p. 263). Human resources practices (job security, teamwork, performance-based payment, training and development, equality and information sharing, etc.) that will ensure high commitment in organizations should be implemented. In this sense, trust will improve social communication between employees, provide willingness to work for the benefit of the organization without fear of exploitation, and create a psychological contract.

Social and environmental activities carried out within the scope of corporate social responsibility in a business focus on improving the relations of the company with its stakeholders, charities and community organizations, employees, suppliers, customers, and stakeholders (Phillips et al. 2019, p. 30). It is believed that businesses will achieve healthier and lasting success in the field of civil aviation if they implement both wage and legal practices equally and fairly with these self-sacrifices of their employees (Koç 2018, p. 1070).

References

Akalın ZD (2019) Stress factors on pilots and relationship between business performance. Unpublished Master Dissertation, Beykent University, Institue of Socail Sciences, Istanbul

Akın Ö, Erdost Çolak HE (2018) İnsan kaynaklari uygulamalari çerçevesinde güven [Trust within the framework of human resources practices]. In: Reyhanoğlu M (ed) Araştırmalar Işığında Yönetimde Güncel Konular [Current issues in management in the light of research]. Nobel Yayıncılık, Ankara, pp. 263–285

Akın Ö, Reyhanoğlu M, Güzel AE, Arslan Ü (2017) To determine the relationship between entrepreneurship and trust: the case of Turkey. In: Aydın R, Yıldız H (eds) Conference proceedings of mediterranean international conference on social sciences, May 19–22, 2017, University of Donja Gorica, Podgorica, Montenegro, pp 411–419

Alola UV, Alafeshat R (2020) The impact of human resource practices on employee engagement in the airline industry. J Pub Aff e2135. https://doi.org/10.1002/pa.2135

Bakır M, Bal HT, Akan Ş (2017) Integrated SWOT-AHP approach in the assessment of Turkish Civil Aviation Sector. J Aviat 1(2):154–169

Bükeç CM, Erdoğan D (2010) 11 Eylül sonrası dönemde havacılıkta güvenlik ve işbirliği anlayışındaki değişim [The change in the understanding of security and cooperation in aviation after September 11]. 3. In: Ulusal Havacılık ve Uzay Konferansı [3rd National Aviation and Space Conference]. September 2020, Anadolu University, Eskişehir, Turkey. https://www.researchg ate.net/publication/319826050_11_EYLUL_SONRASI_DONEMDE_HAVACILIKTA_GUV ENLIK_VE_ISBIRLIGI_ANLAYISINDAKI_KURESEL_DEGISIM. Accessed September 29, 2020

Chen CF (2006) Job satisfaction, organizational commitment, and flight attendants' turnover intentions: a note. J Air Transp Manag 12(5):274–276

Chung EK, Jung Y, Sohn YW (2017) A moderated mediation model of job stress, job satisfaction, and turnover intention for airport security screeners. Saf Sci 98:89–97

Çoban R, Aydoğdu T (2020) The effect of time pressure on technostress in aviation industry: research on aircraft maintenance technicians. J Bus Res Turk 12(3):2442–2460

Demir E (2015) Rusya'nın yaptırımlarının Türkiye ekonomisine olası etkileri [the possible impact of Russia's economic sanctions on Turkey]. Bilgi Notu, Türkiye İş Bankası, İktisadi Araştırmalar Bölümü, İstanbul [Information Note, Turkey Business Bank, Economic Research Department, Istanbul]. https://ekonomi.isbank.com.tr/ContentManagement/Documents/ar_15_2015.pdf. Accessed september 29, 2020

Doğan Ü, Aktemur Ş, Uzgör M, ve Yeloğlu HA (2020) The mediating role of job satisfaction on the effect of organizational justice to the intention to leave among employees in the aviation sector in Turkey. J Aviat Res 2(1):26–44

Erdoğan D (2019) Determination of contract duration: an empirical study on the ground handling agreements with the view of transaction cost theory. Gaziantep Univ J Soc Sci 18(2):623–637

Eren AS, Eryer A, Eryer S (2020) Examining the relationship between air transportation and economic growth: the Turkey case empirical analysis. Int J Soc Educ Sci 2(3):236–257. https://dergipark.org.tr/en/pub/usbed/issue/57491/815735. Accessed September 29, 2020

Gerede E (2003) Türk Sivil Havacılık sisteminin sorunları [Problems of the Turkish Civil Aviation system]. 78. Yılda Türk Hava Kurumu ve Türk Havacılığının Geleceği Paneli [Turkish Aeronautical Association and the Future of Turkish Aviation in 78th Anniversary Panel], February 19, 2003, Ankara, pp 19–39

Graeber C (2020) The role of human factors in in improving aviation safety. Aero Mag (8):23–31. http://www.boeing.com/commercial/aeromagazine/aero_08/human_textonly.html. Accessed September 29, 2020

Hirst M (2008) The air transport system. Woodhead Publishing Limited, Cambridge

Ho CW, Wu CC (2019) Using job design to motivate employees to improve high-quality service in the airline industry. J Air Transp Manag 77(2019):17–23

IATA (2019) IATA annual review 2019. 75th annual general meeting, Seoul, June. https://www.iata.org/contentassets/c81222d96c9a4e0bb4ff6ced0126f0bb/iata-annual-review-2019.pdf. Accessed September 29, 2020

IATA (2020) IATA annual review 2020. 76th annual general meeting, Amsterdam, November. https://www.iata.org/contentassets/c81222d96c9a4e0bb4ff6ced0126f0bb/iata-annual-review-2020.pdf. Accessed December 1, 2020

ICAO (2006) Annex 1 personnel licensing (Tenth edition). https://www.theairlinepilots.com/forumarchive/quickref/icao/annex1.pdf. Accessed December 1, 2020

ICAO (2020) Economic impacts of COVID-19 on civil aviation. https://www.icao.int/sustainability/Pages/Economic-Impacts-of-COVID-19.aspx. Accessed September 29, 2020

Kanbur E, Gökalp Ç (2014) Havacılıkta ekip kaynak yönetimi (CRM): Türkiye ve dünyada yapılan araştırmalardan seçmeler Aviation crew resource management (CRM): Turkey and the selection of researches in the World]. V. Ulusal Havacılık ve Uzay Konferansı [VII. National Aviation and Space Conference], 8–10 September 2014, Erciyes University, Kayseri. http://uhuk.org.tr/bildiri.php?No=UHUK-2014-108. Accessed December 1, 2020

Koç N (2018) An elasticized of the working time practices in air transport: split working model. Gaziantep Univ J Soc Sci 17(1):1054–1073

Küçükönal H, Korul V (2002) Human resources management in airline companies. Afyon Kocatepe Univ J Soc Sci IV(2):67–90

Li Y, Cui Q (2018) Airline efficiency with optimal employee allocation: an input-shared network range adjusted measure. J Air Transp Manag 73:150–162

Machado N, Castro AJM, Oliveira E (2015) Impact of the organizational structure on airline operations. In: Proceedings of the ITSC 2010 workshop on artificial transportation systems and simulation, Madeira Island, Portugal, September 19, pp 103–124. https://paginas.fe.up.pt/~niadr/PUBLICATIONS/2010/ITSC2010_nm_ajmc_eco.pdf. Accessed September 29, 2020

Miles SJ, Mangold WG (2005) Positioning Southwest Airlines through employee branding. Bus Horiz 48(6):535–545

Pate JM, Beaumont PB (2006) The European low cost airline industry: the interplay of business strategy and human resources. Eur Manag J 24(5):322–329

Phillips S, Thai VV, Halim Z (2019) Airline value chain capabilities and CSR performance: the connection between CSR leadership and CSR culture with CSR performance, customer satisfaction and financial performance. Asian J Shipp Logist 35(1):30–40

Reason J, Maddox ME (1996) Human error. In: Human factors guide for aviation maintenance. Federal Aviation Administration/Office of Aviation Medicine, Washington, DC, pp 14/1–14/45

Santana M, Valle R, Galan JL (2019) How national institutions limit turnaround strategies and human resource management: a comparative study in the airline industry. Eur Manag Rev 16(4):923–935

Saptari A, Leau JX, Mohamad NA (2015) The effect of time pressure, working position, component bin position and gender on human error in manual assembly line. In: International conference on industrial engineering and operations management, 3–5 March 2015, Dubai, United Arab Emirates, pp 1–6

Şenol L, Üzüm B (2020) Demographic features and psychological ownership: a research in the aviation sector. Afyon Kocatepe Univ J Soc Sci 22(3):760–770

SHGM (2020) AIM memuru. http://web.shgm.gov.tr/tr/havacilik-personeli/4143-aim-memuru. Accessed December 2, 2020

Talukder MH (2016) Human resource practices predicting manager performance appraisal: evidence from an airline company in Bangladesh. Bus Rev 11(1):37–47

Tarafdar M, Tu Q, Ragu-Nathan TS, Ragu-Nathan BS (2011) Crossing to the dark side: examining creators, outcomes, and inhibitors of technostress. Commun ACM 54(9):113–120

Tuna G (2019) A study on the effect of the burnout levels of the flight attendants on organizational commitment and work performance. Unpublished Master Dissertation, Beykent University, Institue of Socail Sciences, Istanbul

Türen U, Erdem H, Kalkın G (2015) Techno-stress at work scale: a research in aviation and banking sectors. J Labour Relat 6(1):1–19

Türk A (2020) Organizational analysis of attitude perception of employees against change with demographic difference: an application in the aviation sector. J Bus Res Turk 12(2):1256–1266

Türk A, Şener A (2018) Analysis of the relationship between service quality and competitive strategies: an application in aviation area. J Nişantası Soc Sci 6(1):1–21

Uyar T (2018) Uluslararası Sivil Havacılık Örgütü (ICAO) emniyet yönetim sisteminde risk kavramı [Risk concept in International Civil Aviation Organization (ICAO) safety management system]. VII. Ulusal Havacılık ve Uzay Konferansı [VII. National Aviation and Space Conference], September 12–14, 2018, Ondokuz Mayıs University, Samsun. http://www.uhuk.org.tr/bildiri. php?No=UHUK-2018-095. Accessed September 29, 2020

Vasigh B, Fleming K, Humphreys B (2014) Foundations of airline finance methodology and practice, 2nd edn. Routledge, USA

Virovac D, Domitrović A, Bazijanac E (2017) The influence of human factor in aircraft maintenance. Promet Traffic Transp 29(3):257–266

Wensveen JG (2007) Air transportation: a management perspective, 6th ed. Ashgate Publishing Limited

Wirtz J, Heracleous L, Pangarkar N (2008) Managing human resources for service excellence and cost effectiveness at Singapore Airlines. Manag Ser Qual Int J 18(1):4–19

Yeh YP (2014) Exploring the impacts of employee advocacy on job satisfaction and organizational commitment: case of Taiwanese airlines. J Air Transp Manag 36:94–100

Yiğitel S, Ertemsir E (2020) Investigation of the relationship between time management skills and individual performance: a research on the pilots working in civil aviation sector. Bus Manag Stud Int J 8(2):1546–1575. https://doi.org/10.15295/bmij.v8i2.1447

Yilmaz F (2020a) Evaluation of industry and historical development between 2003 and 2018 years of civil aviation sector in Turkey. Eurasian J Res Soc Econ 7(1) 113–129

Yılmaz H (2020b) The mediation role of organizational commitment and organizational citizenship in the effect of organizational socialization on intention of leave: a research in the aviation industry. Unpublished Ph.D. Dissertation, Hatay Mustafa Kemal University, Institute of Social Sciences, Hatay, Turkey.

Dr. Metin Reyhanoğlu is an Associated Professor in the Department of Business Administration at the Hatay Mustafa Kemal University, Hatay, south of Turkey where he has been a faculty member since 2008. Reyhanoglu completed his Ph.D. at the Faculty of Political Sciences, Ankara University, graduated from Hatay Mustafa Kemal University and his undergraduate studies at the Inonu University. He teaches lectures mainly in the area of management and organization since 2008, examples of his lectures are entrepreneurship and SMEs, organization theories, cross-cultural management to undergraduate and graduate level. His research interests lie in the mean area of organizational behavior and organizational theory. In recent years, he has focused on job embeddedness, burnout, organizational cynicism and trust, mobbing, ambidexterity, and work-family conflict.

Dr. Harun Yılmaz is a lecturer in the Department of Aviation Management at the Iskenderun Technic University, Hatay, south of Turkey where he has been a faculty member since 2019. Yılmaz completed his Ph.D. at the Faculty of Economics and Administrative Sciences Hatay Mustafa Kemal University, graduated from Anadolu University and his undergraduate studies at the Anadolu University. He teaches lectures mainly in the area of aviation management since 2012, examples of his lectures are Introduction to Civil Aviation, Airport Equipment, Flight Operations, Cabin Crew Training, Air Traffic Rules and Services, Aviation Security, Operation and Performance, Dangerous Goods Regulations and Air Cargo Transportation to undergraduate level. His research interests lie in the mean area of organizational behavior. In recent years, he has focused on organizational socialization, organizational commitment, organizational trust, organizational citizenship and turnover intention.

Chapter 12
A Systematic and Bibliometric Review on the Role of Strategic Alliances in Achieving Sustainable Competitive Advantage in the Airline Industry: From Resource Dependence Theory Perspective

Gökhan Tanrıverdi and Ümit Doğan

Abstract Airlines depend on various resources to provide air transportation service to passengers in a competitive, dynamic, and turbulent environment. According to Resource Dependency Theory (RDT), strategic alliances are one of the political actions and behaviors that organizations operating in intense competitive environments resort to gain access to necessary resources and avoid resource uncertainty. In line with, airlines use strategic alliances such as code share alliances and global airline alliances to access to the resources they need more easily for their sustainability. We aim to argue how strategic alliances provide sustainable competitive advantage to airlines from resource dependence theory perspective in this systematic and bibliometric review. The study reviewed the top 30 most cited studies out of 156 studies from the WoS database. In addition, all the studies in the dataset were subjected to citation network and co-word analysis, supporting the findings of the review of the most cited studies. The findings confirm that strategic alliances are seen as a network and that airlines achieve a sustainable competitive advantage through access to network resources. The study contributes to the literature by determining the conditions to be considered in the success of strategic alliances.

Keywords Airline industry · Strategic alliances · Sustainable competitive advantage · Resource dependence theory · Systematic and bibliometric review · PRISMA

G. Tanrıverdi (✉)
Department of Aviation Management, Ali Cavit Çelebioğlu School of Civil Aviation, Erzincan Binali Yıldırım University, Yalnızbağ Campus, 24000 Erzincan, Turkey
e-mail: gtanriverdi@erzincan.edu.tr

Ü. Doğan
Department of Civil Aviation Management, Anadolu University, PO Box 26170, Eskişehir, Turkey
e-mail: umit_dogan@anadolu.edu.tr

12.1 Introduction

How organizations provide resources for their survival and how they manage these resources effectively and productively are among main problems of organizational theory and strategic management literature (Wernerfelt 1984; Williamson 1981; Barney 1991; Pfeffer and Salancik 2003). Organizations can take several actions to gain access to critical resources that can enable them to achieve sustainable competitive advantage in the long term. The actions and behaviors mentioned are a strategic response to changes and uncertainty in the external environment of organizations (Pfeffer and Salancik 2003). Being a member of strategic alliances is expressed as one of these strategic actions of organizations. Strategic alliances are arrangements among two or more organizations based on a series of cooperation agreements. These agreements can range from a very formal agreement that can be seen in the establishment of a multilateral limited company to an informal arrangement of a short-term project (Campbell, Stonehouse, and Houston 2002, p. 220). The main reason for participating in strategic alliances is that alliances facilitate access to the critical resources needed. Organizations that access critical resources, on the other hand, aim to gain and maintain a competitive advantage (Albers et al. 2005; Chen and Chen 2003, p. 1). The literature supports the idea that resources obtained through strategic alliances have positive effects on the success of the organization (Afuah 2000; Lee et al. 2001; Rothaermel 2001; Lavie 2006). Competition and economic uncertainty in the environments in which organizations operate are increasing and becoming more complex. Strategic alliances, on the other hand, are important mechanisms that provide hedging through risk sharing, especially in competitive industries (Liou 2012). Therefore, risk sharing is another important factor for creating a strategic alliance. The rapid and constant change in customer requirements is another driving force that leads organizations to create strategic alliances. The prerequisites for organizations to gain sustainable competitive advantage through strategic alliances are that the new resources accessed are strategically valuable, costly to imitate, rare, and non-substitutability. From the perspective of resource dependency theory, strategic alliances significantly contribute to the success of the organization by enabling share of intangible resources such as reputation, legitimacy, experience, and social capital as well as tangible resources among members (Wang et al. 2020). For example, a relatively small organization in terms of market share can gain legitimacy and reach critical resources by forming a strategic alliance with an organization that attains a place in the market. Thus, organizations take the opportunity to strengthen their position and further grow in the markets in which they operate (Hillman et al. 2009; Orhan and Tasci 2019, p. 71). Airlines depend on various resources to provide air transportation service to passengers in a competitive, dynamic, and turbulent environment (Aldrich 1976, p. 421). Pfeffer and Salancik (2003, 1978) claims that it is impossible for organizations without interacting with other organizations to have all of their resources they needed to perform their activities in the context of resource-based theory. In line with, airlines often use strategic alliances such as code share alliances or global airline alliances to access the resources they need more easily for their sustainability in the airline industry where competition is intense (Gerede and Yalçınkaya 2015, p. 131; Garg 2016). Thus, airlines could offer more value to their customers in the international market.

In the literature, there is a limited number of systematic literature review, bibliometric review, and bibliometric visualization studies on air transport management. In addition to one systematic literature review on air transport literature (Ginieis et al. 2012), there are two bibliometric reviews, one focusing on the fundementals of air transport research and the other one on the relationship between air transport and business model, and airport performance (Aldemir and Sengur 2018; Bergiante et al. 2015; Loos et al. 2016). Moreover, two bibliometric visualization analysis studies, analyzing studies of airport capacity management, which has critical importance for airlines to prevent disruptions of their flights and of aviation's future after COVID-19 based on JATM's knowledge body have differed from previous reviews (Dixit and KumarJakhar 2021; Tanrıverdi et al. 2020). Tanrıverdi et al. (2020) stated that airline alliance is one of the main topics in the aviation management literature. In this study, we aim to argue how strategic alliances provide sustainable competitive advantage to airlines from resource dependence theory perspective by integrating systematic literature review (SLR) and bibliometric visualization analysis. Previous studies have examined strategic alliances in various industries based on RDT in the context of organizations' responses to environmental uncertainty. This study also focuses on the role of strategic alliances in providing sustainable competitive advantage in contradistinction to extent literature. The study includes five sections. Remains of the study are organized as follows: Sect. 12.2 clarifies foundations and significance of strategic alliances in the airline industry. Section 12.3 describes methodology. Section 12.4 reveals result of this systematic and bibliometric review conducted on strategic alliances in the airline industry. The last section presents discussion and conclusion.

12.2 Strategic Alliances in the Airline Industry

Since the deregulation of the airline industry in 1978 and the deregulation of the European airline industry in 1986, airlines have been engaged in bilateral airline agreements to gain access to international markets. Thereafter, the privatization of carriers in Europe and East Asia and more airlines entering international markets further intensified competition among airlines. Due to competition among airlines, many flag carrier airlines have begun to attempt to merge their operations and create economies of scale. Accordingly, strategic alliances became significant, and the foundation of alliances gained momentum (Wassmer et al. 2017; Min and Joo 2016). Domestic market and foreign ownership restrictions have also led airlines to create strategic alliances for providing passenger/cargo services (Zhang et al. 2004). In addition to this, open sky agreements, enabling full competition at airports and international routes, pave the way for strategic alliances in many countries which restrict foreign ownership in domestic airlines and airports.

Strategic alliances are important approach that allows airlines to serve international markets (Lazzarini 2007). According to Oum et al. (2000) and Delbari et al. (2016), strategic alliance is an agreement in which two or more airlines come together for a common goal in the medium or long term, increasing their competitiveness. The first examples of strategic alliances in the airline industry included bilateral agreements between two airlines. These bilateral agreements, common among airlines from different countries, are named as code-sharing (Chen and Chen 2003, p. 19; Upham et al. 2003, p. 26). These agreements are the most popular and the simplest form of airline alliances. Code-sharing is defined as a commercial agreement that allows an airline to place its designator code, assigned by the International Civil Aviation Organization (ICAO) on a flight operated by another airline as it appears on computer booking systems (Min and Joo 2016; Oum et al. 1996). In code-sharing agreements, the routes of the two carriers which are presented as a single product to customers, passenger, and baggage flow are coordinated jointly, and airport resources (check-in, ground handling staff, passenger lounge, etc.) are shared (Lazzarini 2007). For example, Turkish Airlines might have an agreement to operate flights between Istanbul Airport and Houari Boumediene Airport for Algeria's flag carrier airline, Air Algérie. This flight would be grounded under Air Algérie's ICAO code (DAH) in that operated by Turkish Airlines. Both airlines have been able to provide services with their passengers by using each other's flight codes mutually without additional investment through their code-sharing agreement (Oum et al. 1996, p. 190). Thus, both airlines can increase their market shares, providing their customers with a wider flight network and improve passenger service through one-stop booking. However, code-sharing has also some disadvantages. First, code-sharing might cause confusion among passengers who do not know with which airline to operate their flights. Second, code-sharing may increase the chance of monopoly for a certain route and leave no alternative option for passengers. Third, it can negatively affect and complicate airlines' pricing strategy, frequent flyer programs, flight plans, services, and branding strategies. This may be a potential source of inefficiency for airlines (Min and Joo 2016).

Strategic alliances between airlines, which were firstly implemented as code-sharing, evolved into a larger and more comprehensive form in the late 1990s. Although some of the airline alliances have failed such as Wings alliance, there are global airline alliances that still have continued their activities today, such as Star Alliance, Oneworld, and SkyTeam. Star Alliance, the first of these alliances was found by five leading airlines in 1997. Thereafter, One World was established in 1999 and Skyteam in 2000, respectively. Of these three global airline alliances, Star Alliance has 26 members; One World has 15 members; and Sky Team has 19 members. In terms of number of passengers carried and destinations served, Star Alliance is relatively the largest compared to other global alliances (Fan et al. 2001). Collaborative marketing efforts, including shared lounges, frequent flyer program recognition, and brand investments, are among the activities of global alliances (Lazzarini 2007). Global airline alliances are becoming increasingly larger structures day by day. The fact that approximately 70% of airline passenger market and turnover of airlines globally are dominated by global airline alliance members such as Star Alliance, Sky Team, and One World is stated as a substantial proof of afore-mentioned development (Garg 2016).

Airlines are involved in numerous cooperative relationships due to the nature of the airline industry that involves high fixed costs and fierce competition that exists, especially in international markets. Membership of code share alliances and global alliances are strategic actions taken by airlines in this direction. Airlines aim to be involved in these alliances for two main reasons. The first reason is to respond to changes in economic, market, and regulatory conditions (Albers et al. 2005). Achieving and effectively managing critical resources which determines the survival and sustainability of an airline is vitally important (Bazargan 2010, p. 1; Doğan 2018, p. 1). Another main reason for underlining airlines' membership of strategic alliances is also that these strategic alliances give legitimacy to airlines (Eisenhardt and Schoonhoven 1996). Organizations' fundraising efforts are not sufficient for the sustainability of the organization. In addition to these efforts, organizations should have political, institutional, social and economic legitimacy (Carroll and Delacroix 1982; DiMaggio and Powell 1983). In this regard, and airlines which have the resources they need to facilitate the access to strategic alliances need to be included to guarantee the sustainability of these resources and gaining legitimacy by Dyer and Singh (1998), Lavie (2006), Wassmer et al. (2017). In another words, airlines get a chance to benefit from attractive route network, frequent flyer recognition, joint purchasing, and joint development of operational planning systems, enhancing their competitive advantages and sustainability through alliances by gaining legitimacy (Casanueva et al. 2013, p. 442). The other benefits of code-sharing agreements in the airline industry include risk-sharing, attracting more passengers, offering customers a standard product, integrating networks, sharing airport facilities, and seamless travel, all resulting in lower cost operations (Garg 2016; Zhang et al. 2004; Min and Joo 2016; Tanrıverdi and Küçük Yılmaz 2018). As a result of such joint use of resources within strategic alliances, member airlines can boost their load factors and revenues by increasing their passenger traffic. However, resources shared within the alliance are not firms' own internal resources. However, as just stated, they are stated as potential sources that can improve performance of the firm (Lavie 2006).

12.3 Data and Method

This systematic and bibliometric review analyzes the effects of strategic alliances on sustainable competitive advantage focusing on knowledge bodies of global airline alliances and code share alliances. In the study, on the one hand, PRISMA approach was used to systematize the literature review. On the other hand, VOSviewer visualization software was used to reveal the clusters and bibliometric mappings of knowledge body and relationship between most cited documents by analyzing citation network. Concurrently, outstanding words in the airline strategic alliance literature were identified by performing co-word analysis through VOSviewer visualization software. It was observed that the results obtained with citation network analysis and co-word analysis supported the results obtained from systematic review.

Data were collected from WoS database. WoS database includes all of the best document sources in one we used to collect documents in this study, such as the Social Science Citation Index (SSCI), Science Citation Index (SCI), and Emerging Science Citation Index (ESCI). On November 11, 2020, "airline strategic alliances" was searched in the WoS search box for data collection. From all documents published up to the relevant date, only published "articles"were selected and the first dataset was created. Out of 180 documents, the number of articles was 161. Then, 156 studies were analyzed by eliminating unrelated studies through Preferred Reporting Items for Systematic Reviews and Meta-Analyses (PRISMA) approach. The identification and application process of this systematic and bibliometric review is shown in Fig. 12.1.

According to Fig. 12.2, indicating publication trend by years on airline strategic alliances, although by year of 1994, the number of studies fluctuated until 2010, then the number of studies on strategic airline alliances increased significantly.

Table 12.1 provides information on the characteristics of journals in which relevant studies are published. The journal published the most studies so far is Journal of Air Transport Management, which is prominent journal in the field of air transport management and has an h-index value of 62. Transportation Research Part E-Logistics and Transportation Review, which has 97 h-index, ranks second with nine studies it has published. The most prestigious journals in which studies were published are Tourism Management (h-index: 181) and Transportation Research Part B-Methodology (h-index: 138), respectively.

12.4 Results

To demonstrate the impact of strategic airline alliances on airline competitiveness more clearly, a citation network analysis of 156 documents obtained from the Web of Science was performed. In line with, the following steps were followed:

(1) Articles are defined as the unit of analysis.
(2) To focus on studies better, the dataset of 156 documents is limited to most cited 30 documents (with at least 23 citations).

Most frequently cited primary documents including Gimeno (2004) (170 citations), Park (1997), Lazzarini (2007) are presented in Table 12.3 accord with their cluster. As a result of the performed citation analysis within the scope of the study, five documents, which were independent of the network, were excluded from the analysis. On the one hand, the document network was created with citation network analysis, on the other hand, the remaining 25 documents were grouped into 6 different clusters, as in Table 12.2. The total number of citations of these documents is 1188.

In the study, each cluster obtained through document citation network analysis stands for the different effects of strategic alliances on the airline industry. Studies grouped into six clusters are summarized below under specific themes. Of

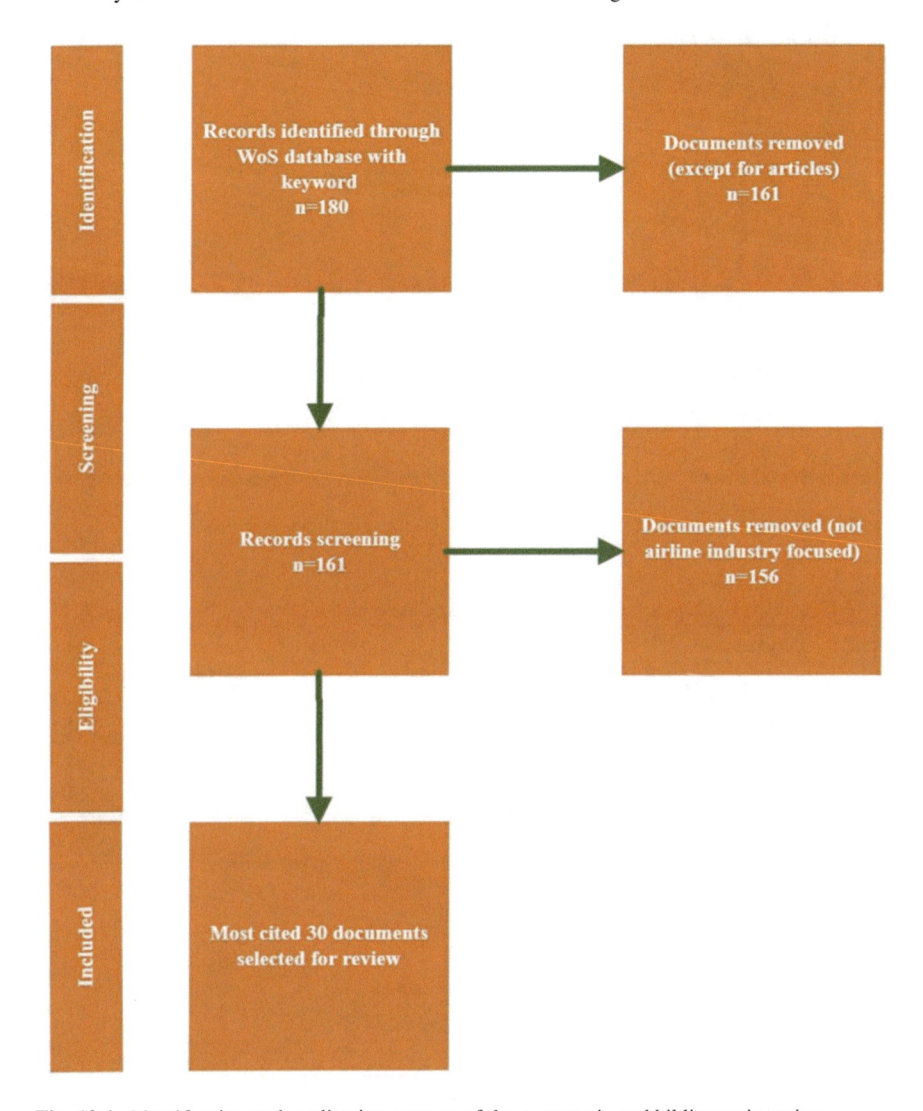

Fig. 12.1 Identification and application process of the systematic and bibliometric review

these, cluster 1 stands for the partner selection, cluster 2 focuses on productivity, cluster 3 identifies competitive advantages, cluster 4 underlines competitive rules, cluster 5 illustrates network complementarity, and finally cluster 6 calls attention to market concentration. Although each study in clusters provides important insight into strategic alliances that play an important role in the development of the airline industry, this study only includes the most cited studies in each cluster due to page restrictions and time constraint. Figure 12.3 shows the graphic mapping of document citation network analysis, which includes all the studies.

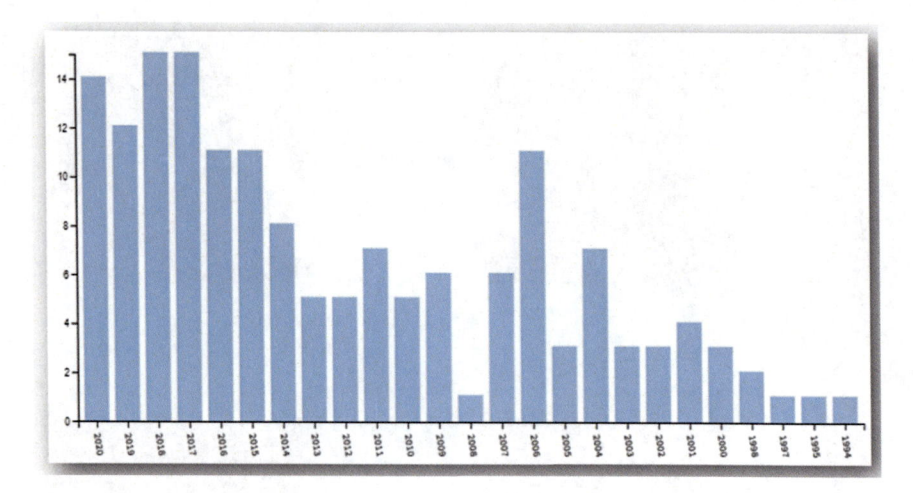

Fig. 12.2 The number of studies published on airline strategic alliance by years

Table 12.1 Top 10 journals publishing the most articles on airline strategic alliance

Rank	Journal	Publication number	H-index
1	Journal of Air Transport Management	30	62
2	Transportation Research Part E-Logistics and Transportation Review	9	97
3	Research in Strategic Alliances	5	7
4	Transportation Research Part A-Policy and Practice	5	124
5	International Journal of Industrial Organization	4	84
6	Journal of Transport Economics and Policy	4	60
7	Tourism Management	4	181
8	Transportation Research Part B-Methodological	4	138
9	Economics of Transportation	3	18
10	Sustainability	3	68

Cluster 1 includes six studies. Three of the studies mentioned focus on partner selection to strategic alliances, and the others focus on alliance efficiency in general (Garg 2016; Liou 2012; Liou et al. 2011). Garg (2016) and Liou (2012) have developed a model for partner selection with multiple-criteria decision-making analyses in the context of the Indian airline industry. Liou (2012) considered good relationships between a firm and its partners as the basis for a successful strategic alliance. Liou, Tzeng, Tsai, and Hsu (2011) suggested that the choice of a particular partner is critical to the alliance's strategic goals and success. In this context, a weak alliance created by the wrong choice of partner leads to losses and unexpected risks on core competencies. As for the other three studies, Fan et al., (2001) examined the

Table 12.2 Top 25 most cited papers published on strategic airline alliances

Clusters	Title	Author(s)	Year	Journal published	TC	AC/EC
#1	A hybrid ANP model in fuzzy environments for strategic alliance partner selection in the airline industry	Liou et al.	2011	Applied Soft Computing	52	5,2
	Developing an integrated model for the selection of strategic alliance partners in the airline industry	Liou	2012	Knowledge-Based Systems	45	5
	Analytical models of international alliances in the airline industry	Park et al.	2001	Transportation Research Part B-Methodological	42	2,1
	Evolution of global airline strategic alliance and consolidation in the twenty-first century	Fan et al.	2001	Journal of Air Transport Management	42	2,1
	A robust hybrid decision model for evaluation and selection of the strategic alliance partner in the airline industry	Garg	2016	Journal of Air Transport Management	29	5,8
	Air cargo alliances and competition in passenger markets	Zhang et al.	2004	Transportation Research Part E-Logistics and Transportation Review	29	1,71
#2	The effect of horizontal alliances on firm productivity and profitability: Evidence from the global airline industry	Oum et al.	2004	Journal of Business Research	63	3,71
	The empirical analysis of the impact of alliances on airline operations	Iatrou and Alamdari	2005	Journal of Air Transport Management	42	2,63
	A comparative performance analysis of airline strategic alliances using data envelopment analysis	Min and Joo	2016	Journal of Air Transport Management	26	5,25

(continued)

Table 12.2 (continued)

Clusters	Title	Author(s)	Year	Journal published	TC	AC/EC
	The evolution of coopetitive and collaborative alliances in an alliance portfolio: The Air France case	Chiambaretto and Fernandez	2016	Industrial Marketing Management	24	4,8
	Productivity of airline carriers and its relation to deregulation, privatization, and membership in strategic alliances	Sjögren and Söderberg	2011	Transportation Research Part E-Logistics and Transportation Review	24	2,4
#3	Competition within and between networks: The contingent effect of competitive embeddedness on alliance formation	Gimeno	2004	Academy of Management Journal	170	10
	The impact of membership in competing alliance constellations: Evidence on the operational performance of global airlines	Lazzarini	2007	Strategic Management Journal	80	5,71
	An investigation of key competitiveness indicators and drivers of full-service airlines using Delphi and AHP techniques	Delbari et al.	2016	Journal of Air Transport Management	37	7,4
	Resource Ambidexterity through Alliance Portfolios and Firm Performance	Wassmer et al.	2017	Strategic Management Journal	34	8,5
	Network resources and social capital in airline alliance portfolios	Casanueva et al.	2013	Tourism Management	28	3,5
#4	Strategic alliances, shared facilities, and entry deterrence	Chen and Ross	2000	Rand Journal of Economics	36	1,71

(continued)

Table 12.2 (continued)

Clusters	Title	Author(s)	Year	Journal published	TC	AC/EC
	Strategic formation of airline alliances	Flores-Fillol and Moner-Colonques	2007	Journal of Transport Economics and Policy	42	3
	Strategic airline alliances and endogenous Stackelberg equilibria	Lin	2004	Transportation Research Part E-Logistics and Transportation Review	26	1,53
	Airline alliances and partner firms' outputs	Park and Zhang	1998	Transportation Research Part E-Logistics and Transportation Review	42	1,83
	Rivalry between strategic alliances	Zhang and Zhang	2006	International Journal of Industrial Organization	38	2,53
#5	Multilateral airline alliances: Balancing strategic constraints and opportunities	Gudmundson and Lchner	2006	Journal of Air Transport Management	23	1,53
	The effects of airline alliances on markets and economic welfare	Park	1997	Transportation Research Part E-Logistics and Transportation Review	113	4,71
#6	Airline alliances-who benefits?	Morrish and Hamilton	2002	Journal of Air Transport Management	51	2,68
	Consequences of Strategic Alliances between International Airlines-The Case of Swissair and SAS	Youssef and Hansen	1994	Transportation Research Part A-Policy and Practice	50	1,85

Notes TC = Total Citations, and AC/EY = Average citations per each year

economic factors and industry forces influencing the macroscopic development of strategic airline alliances. The study suggests that airlines will head toward larger mergers, mega carrier systems, and alliance structures due to the economic power inherent in the sector, the increasing pace of liberalization, and anti-trust concerns. In another study in the cluster, a model was developed to examine the effects of the air cargo alliance of two carriers on competition in passenger markets, as opposed to studies focusing on global alliances on passenger service (Zhang et al. 2004). In the established model, an alliance is proposed for partners to provide an integrated cargo service using airliners, while continuing to offer passenger services. It is argued that with the mentioned alliance, members (two separate alliances formed by four airlines) do not only improve their own performance, but also have an edge over competitors by reducing the output of their competitors in both cargo and passenger markets. In this context, the joint profit of the alliance partners will increase, and the economic output of competitors will decrease. This is because the alliance reduces the marginal cost of passenger services and helps each member increase its marginal

Table 12.3 Keywords' occurrence and total link strengths

Keyword	Occurrence	Total link strength
Strategic Alliances	19	67
Airlines	15	58
Airline Alliances	13	38
Strategic Alliance	12	36
Airline	9	38
Alliances	9	34
Airline Industry	8	27
Alliance Portfolio	5	23
Competition	5	13
Aviation	4	18
Network Resource	4	18
Networks	4	15
Alliance	4	15
Data Envelopment Analysis	4	10
Airline Competition	3	10

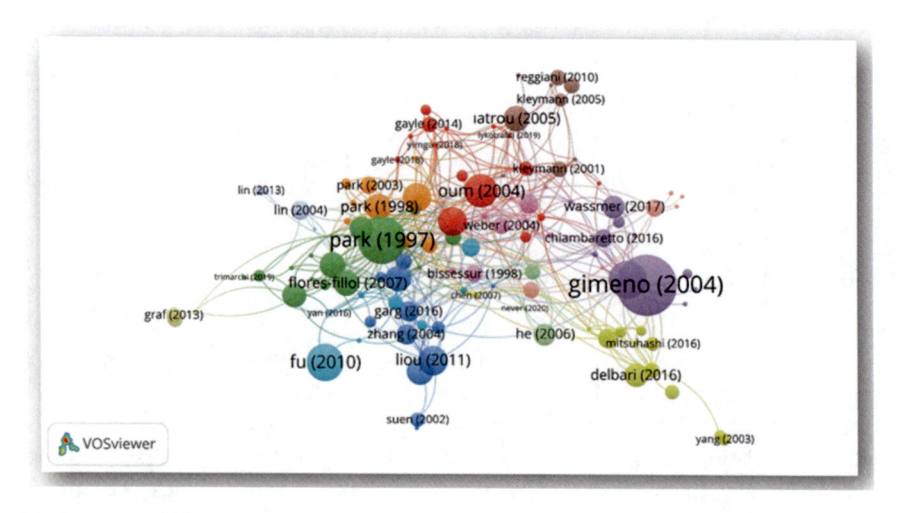

Fig. 12.3 Mapping document citation analysis

profit due to complementary cargo and passenger services. Park et al. (2001) examined the effects of alliances on demand and costs. In the study, an alliance model was developed in which each firm maximized not only its own profit, but also its partner's profit to a certain degree. The study reveals that complementary alliances positively affect overall economic outputs, and that competitors' outputs decrease in contrast to complementary alliance partners. In addition, it is noted that the low prices of partner airlines in local markets will create new incentives.

Cluster 2 combines productivity-oriented studies. Chiambaretto and Fernandez (2016) discussed the evolution of strategic alliance portfolio composition from the perspective of coopetition within Air France's alliance portfolio. The authors argued that in situations where market uncertainty is high, airlines re-create their portfolios rather than increasing the number of deals, prefer to coopetitive alliances, and form horizontal alliances rather than vertical ones. Iatrou and Alamdari (2005) revealed the effects of alliances on airline traffic and performance in their study on members of four global airline alliance in 2002. Alliances provide an increase in passenger numbers and load factors, while reducing costs. Airlines can experience these positive effects 2 years after the foundation of alliances, while the increase in traffic mainly takes place on hub-to-hub routes. In addition, the authors emphasize that the word "strategic" remains on paper for most alliances, and alliances cannot achieve deeper collaborations by going beyond frequent flyer program coordination and code share cooperation (Min and Joo 2016). The authors also explored the comparative effectiveness performance of alliances members, focusing on strategic airline alliances that have experienced one of the major changes in the last 20 years with the "open sky" agreements. Although airline alliance reduces costs and provides brand recognition, it does not provide an adequate improvement in the comparative operating activities of member airlines owning to customer bases and shared resources. A more interesting finding is that smaller alliances tend to outperform larger ones. According to the authors, airlines can outperform their competitors by prioritizing organizational learning and resources that are hard to replicate in saturated markets. In the study conducted on the impact of alliances on airline efficiency and profitability, Oum et al. (2004) concluded that while alliances affected airline efficiency at a substantive level, they did not have a significant and positive impact on profitability. In addition, it was stated that the important factor was the level of cooperation (low–high). Accordingly, the increase in the level of cooperation indicates a stronger significant and positive impact on productivity and profitability (Sjögren and Söderberg 2011). Oum et al. (2004) focused on the question of how deregulation, privatization, and the occurrence of strategic alliance affect airline efficiency. It has been determined that alliance membership has an uncertain effect, and the ownership structure (state or private ownership) has no significant effect while deregulation increases efficiency.

The third cluster includes five studies focusing on "increasing competitive advantage" and "alliance portfolio and network resources" of strategic alliances. The first study by Casanueva, Gallego, and Sancho (2013) examined effects of network resources obtained through cooperation and the different dimensions of social capital owned by each alliance member on firm performance (operational and financial performance). Access to partner resources is influenced by the structural and relational factors of the network. In the study examining a total of 214 airlines, it was concluded that access to network resources improved company performance (Casanueva et al. 2013). Delbari et al. (2016) examined full service airlines' competitiveness indicators and key drivers that reflected competitiveness. They found that strategic alliances are important determinant for meeting customer demands and increasing airline competitiveness. Wassmer et al. (2017) suggested that partner resources obtained through alliances can increase the success of airlines when evaluated as a whole portfolio. The study draws attention to the balance mechanism in resource access, that is, the importance of strategic ambidexterity which is related to prevent the exploitation of competent resources. It is emphasized that airlines must have a certain level of strategic ambidexterity both for how to achieve their strategic goals by diversifying products (increasing their revenue) and for how to increase their efficiency through their partners' resources (reducing costs). The research suggests that the more airlines increase revenues and reduce costs, the more firm success will increase. Another study examining how competitive dynamics affect the formation of alliance, partner selection and network development claims that parties' access to capabilities and information from direct and indirect partners is influenced by intra-network and inter-network competition. In intra-network competition, airlines can try to get the same network advantages by connecting to their competitors' networks. In inter-network competition, airlines try to form alliances in complementary networks which they can achieve similar benefits with their competitors. Airlines create value by leveraging mutual joint expertise and efficiency provided through these alliances (Gimeno 2004). Lazzarini (2007) argued that membership in strategic alliances serves as an intermediary in obtaining resources from membership of other firms. In addition, it is emphasized that the ties of a non-member airline with key airlines which are central to strategic alliances facilitate its access to network resources. Joint operations and activities carried out through strategic alliances are key to reducing costs and improving service quality due to the economics of scale and scope (Lazzarini 2007; Wassmer et al. 2017).

Cluster 4 mostly focuses on competitive rules. Chen and Ross (2000), focused on the possible anticompetitive effects that could have as a result of joint use of ground facilities by member airlines. The authors underline that alliances prevent competition and lower prices by adding new capacities to the industry, which have implications for reducing social welfare. However, they also point out that strategic alliances improve social efficiency, enabling airlines to share high fixed costs. Flores-Fillol and Moner-Colonques (2007) discussed the formation of strategic alliances for a specific city-pair market as a two-stage game and analyze the effects of two complementary strategic alliances. They reveal that alliance members have an edge over their competitors when low economies of density and competition are important. In another study, Lin (2004) focused on the economic impact of code-sharing alliances on international and national airlines. Noting that airlines have two options in terms of price (price leader or follower), the authors argue that this condition, which they call the Stackelberg leadership model, helps airlines choose whether to be a price leader or a follower. Park and Zhang (1998) explored the effects of alliances on airline outcomes by comparing changes in airline traffic on routes in which airlines operate both within and outside the alliance. The results indicate that there has been a significant increase in traffic on the routes where airlines operate. Zhang and Zhang (2006) underlined that although complementary alliances provide strategic advantages to airlines, alliances can also face contradictory outcomes as a result of their competition with each other. Furthermore, the study emphasizes that although alliances contribute to economic welfare in general, alliances formed due to the threat of market entry have a negative impact on economic welfare.

Cluster 5 emphasizes network complementarity. Gudmundsson and Lechner (2006) investigated structural holes and network closure in global airline alliances. The results indicate that alliance closure is highly beneficial. Sufficient opportunities and structural holes can only be maintained through adding partners with many non-redundant ties and shedding partners with many redundant ties. Therefore, highly distributed multi-member multilateral alliances such as STAR should be better positioned to exploit opportunities. Another study, focusing on the economic impact of complementary and parallel alliances on member airlines, highlights the competitive advantage that alliance members achieve through alliances. The study claims that complementary alliances increase economic welfare, while parallel alliances have a negative impact on it. They also discovered that the traffic increases as alliances forced non-members to incur expenditures. According to this study, with the formation of alliances, the profit and competitive advantages of partner airlines increase, while other airlines have difficulty competing against partner airlines with a decrease in profits (Park 1997).

Cluster 6 as the latter part of clusters focuses on the impact of strategic alliances on market concentration: Morrish and Hamilton (2002) analyzed the last 15 years of strategic alliances in the airline industry. The results indicate that airlines with alliance membership do not monopolize the market and settle for modest gains due to similar-sized fare reductions, despite increased levels of productivity. Another study examined the effects of the alliance on quality of service, market concentration, and ticket prices, noting the redistributive nature of alliance effects. While a slight reduction in concentration in markets in which the partners offer connecting service and fare increases were determined, partners with the strongest service increases also had the lowest fare increases (Youssef and Hansen 1994).

Co-word analysis is performed on the basis of common keywords used in the dataset including studies examined. Most frequently repeated words in the dataset appear partially larger than other words in terms of shape size in the center of Fig. 12.4.

Table 12.3 shows figures of most frequently repeated keywords' occurrence and total link strengths. It is revealed that "Strategic Alliances", "Airlines", and "Airline Alliances" are the most frequently repeated words.

12.5 Discussion and Conclusions

This study examines the role of strategic alliances in achieving sustainable competitive advantage in organizations from the perspective of resource dependence theory. Accordingly, a number of findings were obtained as a result of systematic and bibliometric analysis of 157 articles collected from the WoS database. It is observed that studies on strategic airline alliances first appeared in 1994 and have been on the rise, especially since 2015. This indicates the special interest of researchers caused by more airlines participating in strategic alliances due to the impact of financial

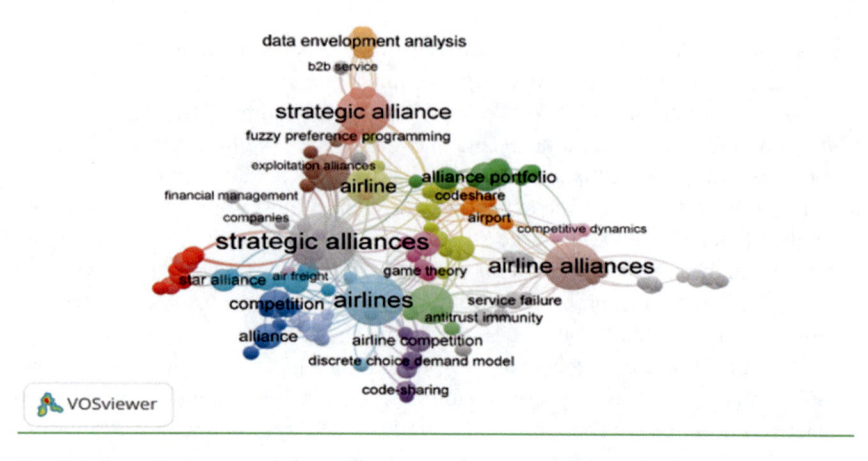

Fig. 12.4 Mapping of co-word analysis

crises around the world. The main source of this study is 30 articles collected from the Journal of Air Transport Management (h-index: 62). This situation is closely related to the journal being a leading journal in aviation management domain (Tanrıverdi et al. 2020). In addition, Transportation Research Part E-Logistics and Transportation Review, Transportation Research Part A-Policy and Practice, Tourism Management, and Transportation Research Part B-Methodological are important journals contributing to the study. Two systematic analyses were performed as citation network analysis and co-word analysis for bibliometric analysis of documents. As a result of citation network analysis, documents were grouped under the following six clusters: partner selection, productivity, competitive advantages, competitive rules, network complementarity, and market concentration. The prominent articles in each cluster were Liou et al. (2011), Oum et al. (2004), Gimeno (2004), Flores-Fillol and Moner-Colonques (2007), Park (1997), and Morrish and Hamilton (2002), respectively. These studies lead to a more effective explanation of the themes that represent the cluster in which they are grouped. Cluster 3 directly emphasizes the positive impact of strategic alliances on sustainable competitive advantage. In this sense, it is an important evidence that Delbari et al. (2016) stated strategic alliances are one of the drivers that increases airlines' competitiveness. Although other clusters do not directly focus on sustainable competitive advantage, they clarify the foundations that provide it.

RDT claims that organizations operating in highly competitive environments have political actions and behaviors to minimize environmental uncertainty. It is asserted that one way of minimizing environmental uncertainty is to decrease resource uncertainty (Fink et al. 2006). Accordingly, it is stated that organizations form strategic alliances or are members of existing alliances to reduce uncertainty. Organizations provide access to the resources they need and ensure resource sustainability in this way. Critical knowledge, skills, and resources gained through strategic alliances provide a sustainable competitive advantage in the long term. Airlines also aim to achieve these potential benefits by participating in strategic alliances (code-sharing alliances or global airline alliances), especially in competitive international markets. Accordingly, the results of this systematic and bibliometric review also reveal strong evidence that strategic alliances provide airlines with a sustainable competitive advantage. These results correspond to the results of studies which reveal that strategic alliances in other industries contribute to sustainable competitive advantage (Culpan 2008; Ireland et al. 2002).

It has been determined that obtaining competitive advantages through strategic alliances depends on a number of conditions. First, strategic alliances are seen as a network, and resources shared through the network are also characterized as network resources. Considering resource heterogeneity in the selection of network members is another important issue (Ireland et al. 2002). By this way, it is aimed to maximize the advantage that members will gain from the network in terms of alliance performance and efficiency by ensuring that network members have resources that complement each other. Second, it is emphasized that access to network resources is based on the location of the alliance member on the network. Strong relationships with key members have an effect on facilitating access to network resources. Moreover, a non-member can reach network resources through its relationship with key members in the alliance. This is among the notable results of the study. This result indicates that network position obtained through organizational networks and strong relationships is key to achieving sustainable competitive advantage. Finally, it has been concluded that alliances can provide their members with sustainable competitive advantage in two ways. Alliances with high network closure offer its members a competitive advantage by ensuring effective and productive use of network resources through strong relationships among members. Alliances with members being widely geographically distributed provide sustainable competitive advantage by taking advantage of the network opportunities offered by brokers and weak ties that exist in structural gaps.

The findings of co-word analysis also support the above-mentioned results. "Strategic alliances", "Airlines", and "Airline Alliances", which are the most repeated words as a result of the analysis, do not provide any significant data by themselves. These words should be taken into consideration holistically with the relatively less frequent words of "Competition", "Alliance portfolio", "Network resource", "Networks", and "Airline competition". This proves airlines' efforts to achieve sustainable competitive advantage by accessing to the network resources mentioned above through strategic alliances.

On the other hand, according to another result, the effect of alliance membership on the efficiency of airlines and their sustainable competitive advantages is unclear (Oum et al. 2004; Morrish and Hamilton 2002). Tanrıverdi and Küçük Yılmaz (2021) assert that although strategic airline alliances deployed in airline industry still benefit member airlines, they are criticized by member airlines due to their firm position preventing cooperative cost reduction.

In a nutshell, according to this systematic and bibliometric review focusing on strategic airline alliances, apart from opinions questioning the efficiency of the airlines, there is a consensus that alliances have considerable direct and indirect positive effects on sustainable competitive advantage of airlines. However, it should be taken into account that strategic alliances succeed if certain conditions are met. Accordingly, it is believed that studies that empirically test the relationship of the above-mentioned conditions with strategic alliances will contribute to the literature. In this context, it is suggested that network position, network closure, and structural holes variables should be addressed holistically by integrating the resource dependency theory and network theory in further studies. In parallel with the development of scientific studies, newly published studies trail relatively behind previously published studies in terms of the number of citations. This is the limitation of bibliometric analysis which leads studies to play second fiddle. However, it does not mean that low-cited studies are less important. Another limitation of this study is the number of documents (157). In this context, another suggestion is to examine studies on strategic alliances in different industrial areas with bibliometric analysis and to reveal the role of strategic alliances in terms of sustainable competitive advantage.

References

Afuah A (2000) How much do your co-opetitors capabilities matter in the face of technological change? Strateg Manag J. https://doi.org/10.1002/(sici)1097-0266(200003)21:3%3c397::aid-smj88%3e3.0.co;2-1

Albers S, Koch B, Ruff C (2005) Strategic alliances between airlines and airports-theoretical assessment and practical evidence. J Air Transp Manag. https://doi.org/10.1016/j.jairtraman.2004.08.001

Aldemir HO, Sengur FK (2018) Academic foundations of air transportation research in an emerging country. Int J Aviat Syst Oper Train. https://doi.org/10.4018/ijasot.2017010102

Aldrich H (1976) Resource dependence and in terorganiza tional relations: local employment service offices and social services sector organizations. Adm Soc. https://doi.org/10.1177/009539977600700402

Barney J (1991) Firm resources and sustained competitive advantage. J Manag. https://doi.org/10.1177/014920639101700108

Bazargan M (2010) Airline operations and scheduling. In: Airline operations and scheduling, 2 ed

Bergiante NCR, Santos MPS, Espírito Santo RA (2015) Bibliometric study of the relationship between business model and air transport. Scientometrics. https://doi.org/10.1007/s11192-015-1711-6

Campbell D, Stonehouse G, Houston B (2002) Business strategy. Butterworth-Heinemann, Oxford

Carroll GR, Delacroix J (1982) Organizational mortality in the newspaper industries of Argentina and Ireland: an ecological approach. Adm Sci Q. https://doi.org/10.2307/2392299

Casanueva C, Gallego Á, Sancho M (2013) Network resources and social capital in airline alliance portfolios. Tour Manage 36:441–453. https://doi.org/10.1016/j.tourman.2012.09.014

Chen FCY, Chen C (2003) The effects of strategic alliances and risk pooling on the load factors of international airline operations. Transp Res Part E Logist Transp Rev 39(1):19–34. https://doi.org/10.1016/S1366-5545(02)00025-X

Chen Z, Ross TW (2000) Strategic alliances, shared facilities, and entry deterrence. Rand J Econ 31(2):326. https://doi.org/10.2307/2601043

Chiambaretto P, Fernandez AS (2016) The evolution of coopetitive and collaborative alliances in an alliance portfolio: the air France case. Ind Mark Manage 57:75–85. https://doi.org/10.1016/j.indmarman.2016.05.005

Culpan R (2008) The role of strategic alliances in gaining sustainable competitive advantage for firms. Manag Revu. https://doi.org/10.5771/0935-9915-2008-1-2-94

Delbari SA, Ng SI, Aziz YA, Ho JA (2016) An investigation of key competitiveness indicators and drivers of full-service airlines using Delphi and AHP techniques. J Air Transp Manag 52:23–34. https://doi.org/10.1016/j.jairtraman.2015.12.004

DiMaggio PJ, Powell WW (1983) The iron cage revisited: institutional isomorphism and collective rationality in organizational fields. Am Sociol Rev. https://doi.org/10.2307/2095101

Dixit A, KumarJakhar S (2021) Airport capacity management: a review and bibliometric analysis. J Air Transp Manag 91

Doğan Ü (2018) Management of dependency relations and factors effecting the outsourcing decision in their ground handling activities of airlines in Turkey. Anadolu University Institute of Social Science

Dyer JH, Singh H (1998) The relational view: Cooperative strategy and sources of interorganizational competitive advantage. Acad Manag Rev. https://doi.org/10.5465/AMR.1998.1255632

Eisenhardt KM, Schoonhoven CB (1996) Resource-based view of strategic alliance formation: strategic and social effects in entrepreneurial firms. Organ Sci. https://doi.org/10.1287/orsc.7.2.136

Fan T, Vigeant-Langlois L, Geissler C, Bosler B, Wilmking J (2001) Evolution of global airline strategic alliance and consolidation in the twenty-first century. J Air Transp Manag 7(6):349–360. https://doi.org/10.1016/S0969-6997(01)00027-8

Fink RC, Edelman LF, Hatten KJ, James WL (2006) Transaction cost economics, resource dependence theory, and customer-supplier relationships. Ind Corp Chang. https://doi.org/10.1093/icc/dtl008

Flores-Fillol R, Moner-Colonques R (2007) Strategic formation of airline alliances. J Transp Econ Policy 41(3):427–449

Garg CP (2016) A robust hybrid decision model for evaluation and selection of the strategic alliance partner in the airline industry. J Air Transp Manag 52:55–66. https://doi.org/10.1016/j.jairtraman.2015.12.009

Gerede E, Yalçınkaya A (2015) Ekonomik Düzenlemelerin Havayolu Yönetimine Etkisi. In: Havayolu Taşımacılığı ve Ekonomik Düzenlemeler Teori ve Türkiye Uygulaması. Ankara: Sivil Havacılık Genel Müdürlüğü Yayınları

Gimeno J (2004) Competition within and between networks: the contingent effect of competitive embeddedness on alliance formation. Acad Manag J 47:820–842. https://doi.org/10.2307/20159625

Ginieis M, Sánchez-Rebull MV, Campa-Planas F (2012) The academic journal literature on air transport: analysis using systematic literature review methodology. J Air Transp Manag. https://doi.org/10.1016/j.jairtraman.2011.12.005

Gudmundsson SV, Lechner C (2006) Multilateral airline alliances: balancing strategic constraints and opportunities. J Air Transp Manag. https://doi.org/10.1016/j.jairtraman.2005.11.007

Hillman AJ, Withers MC, Collins BJ (2009) Resource dependence theory: a review. J Manag. https://doi.org/10.1177/0149206309343469

Iatrou K, Alamdari F (2005) The empirical analysis of the impact of alliances on airline operations. J Air Transp Manag 11(3):127–134. https://doi.org/10.1016/j.jairtraman.2004.07.005

Ireland RD, Hitt MA, Vaidyanath D (2002) Alliance management as a source of competitive advantage. J Manag. https://doi.org/10.1016/S0149-2063(02)00134-4

Lavie D (2006) The competitive advantage of interconnected firms: an extension of the resource-based view. Acad Manag Rev. https://doi.org/10.5465/AMR.2006.21318922

Lazzarini SG (2007) The impact of membership in competing alliance constellations: evidence on the operational performance of global airlines. Strateg Manag J 28(4):345–367. https://doi.org/10.1002/smj.587

Lee C, Lee K, Pennings JM (2001) Internal capabilities, external networks, and performance: a study on technology-based ventures. Strateg Manag J. https://doi.org/10.1002/smj.181

Lin MH (2004) Strategic airline alliances and endogenous Stackelberg equilibria. Transp Res Part E: Logist Transp Rev 40(5):357–384. https://doi.org/10.1016/j.tre.2003.09.001

Liou JJH (2012) Developing an integrated model for the selection of strategic alliance partners in the airline industry. Knowl Based Syst 28:59–67. https://doi.org/10.1016/j.knosys.2011.11.019

Liou JJH, Tzeng GH, Tsai CY, Hsu CC (2011) A hybrid ANP model in fuzzy environments for strategic alliance partner selection in the airline industry. Appl Soft Comput J 11(4):3515–3524. https://doi.org/10.1016/j.asoc.2011.01.024

Loos MJ, Taboada Rodriguez CM, Petri SM, dos Santos Matos L (2016) Mapping the state of the art of airport performance measurement. Espacios

Min H, Joo SJ (2016) A comparative performance analysis of airline strategic alliances using data envelopment analysis. J Air Transp Manag 52:99–110. https://doi.org/10.1016/j.jairtraman.2015.12.003

Morrish SC, Hamilton RT (2002) Airline alliances-who benefits? J Air Transp Manag 8(6):401–407. https://doi.org/10.1016/S0969-6997(02)00041-8

Orhan G, Tasci D (2019) The role of integration in achieving sustainable benefits from airline alliances. Int J Sustain Aviat 5(1):70. https://doi.org/10.1504/ijsa.2019.10021482

Oum TH, Park JH, Kim K, Yu C (2004) The effect of horizontal alliances on firm productivity and profitability: evidence from the global airline industry. J Bus Res. https://doi.org/10.1016/S0148-2963(02)00484-8

Oum TH, Park J-H, Zhang A (2000) Globalization and strategic alliances: the case of the airline industry. Pergamon Press, Oxford

Oum TH, Park JH, Zhang A (1996) The effects of airline codesharing agreements on firm conduct and international air fares. J Transp Econ Policy

Park JH (1997) The effects of airline alliances on markets and economic welfare. Transp Res Part E Logist Transp Rev 33(3):181–195. https://doi.org/10.1016/S1366-5545(97)00013-6

Park JH, Zhang A, Zhang Y (2001) Analytical models of international alliances in the airline industry. Transp Res B: Methodol 35(9):865–886. https://doi.org/10.1016/S0191-2615(00)00027-8

Park JH, Zhang A (1998) Airline alliances and partner firms' outputs. Transp Res Part E Logist Transp Rev 34E(4):245–255. https://doi.org/10.1016/S1366-5545(98)00018-0

Pfeffer J, Salancik G (2003) The external control of organizations: a resource dependence perspective (Stanford Business Classics), New York.

Pfeffer J, Salancik GR (1978) The external control of organizations: a resource dependence approach. Harper and Row Publishers, NY

Rothaermel FT (2001) Incumbent's advantage through exploiting complementary assets via interfirm cooperation. Strateg Manag J. https://doi.org/10.1002/smj.180

Sjögren S, Söderberg M (2011) Productivity of airline carriers and its relation to deregulation, privatisation and membership in strategic alliances. Transp Res Part E Logist Transp Rev 47(2):228–237. https://doi.org/10.1016/j.tre.2010.09.001

Tanrıverdi G, Bakır M, Merkert R (2020) What can we learn from the JATM literature for the future of aviation post Covid-19?-a bibliometric and visualization analysis. J Air Transp Manag. https://doi.org/10.1016/j.jairtraman.2020.101916

Tanrıverdi G, Küçük Yılmaz A (2018) Coopetition strategy: a research on traditional airlines. Gaziantep Univ J Soc Sci 17(1):317–333. https://doi.org/10.21547/jss.333589

Tanrıverdi G, Küçük Yılmaz A (2021) The transformation of strategic airline alliances and airline joint ventures: where are they heading? Indep J Manag Prod 12(1)

Upham P, Thomas C, Gillingwater D, Raper D (2003) Environmental capacity and airport operations: current issues and future prospects. J Air Transp Manag. https://doi.org/10.1016/S0969-6997(02)00078-9

Wang T, Wu J, Gu J, Hu L (2020) Impact of open innovation on organizational performance in different conflict management styles: based on resource dependence theory. Int J Confl Manag. https://doi.org/10.1108/IJCMA-09-2019-0165

Wassmer U, Li S, Madhok A (2017) Resource ambidexterity through alliance portfolios and firm performance. Strateg Manag J 38(2):384–394. https://doi.org/10.1002/smj.2488

Wernerfelt B (1984) A resource-based view of the firm. Strateg Manag J. https://doi.org/10.1002/smj.4250050207

Williamson OE (1981) The economics of organization: the transaction cost approach. Am J Sociol. https://doi.org/10.1086/227496

Youssef W, Hansen M (1994) Consequences of strategic alliances between international airlines: the case of Swissair and SAS. Transp Res Part A 28(5):415–431. https://doi.org/10.1016/0965-8564(94)90024-8

Zhang A, Hui YV, Leung L (2004) Air cargo alliances and competition in passenger markets. Transp Res Part E Logist Transp Rev. https://doi.org/10.1016/S1366-5545(03)00034-6

Zhang A, Zhang Y (2006) Rivalry between strategic alliances. Int J Ind Organ 24(2):287–301. https://doi.org/10.1016/j.ijindorg.2005.04.005

Gökhan Tanrıverdi works at Department of Aviation Management, Ali Cavit Çelebioğlu School of Civil Aviation in Erzincan Binali Yıldırım University. He graduated from the department of civil air transport management at Erciyes University in 2014. He earned his master's and Ph.D. degrees in the civil aviation management field at Anadolu University in 2016 and 2021, respectively. His studies are mainly on strategic management in the airline business. He was also a part of works on various topics in the field of aviation management. He uses qualitative methods, multi-criteria decision-making methods, social network analysis, and quantitative methods in his studies.

Ümit Doğan was born in 1990. He received his Bachelor's Degree in Aviation Management at Erciyes University in 2014 and then his Master's Degree in Civil Aviation Management of Anadolu University in 2018. He has been working as a research assistant in the Faculty of Aeronautics and Astronautics, Eskisehir Technical University, (formerly Anadolu University), and continuing his Ph.D. in civil aviation management. His research interests include mainly strategic management, organizational theories, organizational behavior and aviation management. He carried out qualitative and quantitative studies related to ground handling services, outsourcing, transaction cost theory, resource dependence theory, job satisfaction and organizational justice.

Chapter 13
The Political Economy of the Board of Directors: The Case of Turkish Airlines (1956–2020)

Leyla Adiloğlu-Yalçınkaya and Akansel Yalçınkaya

Abstract Turkish Airlines, the first civil airline organization in Turkey, has positioned itself as a flag carrier. The aim of this chapter is to examine the board of directors of THY, which has changed over time. For this purpose, a single case study approach was designed to allow in-depth analysis of the composition of the board. The findings of this study indicate that Turkish Airlines has changed its board members to comply with the changes in the environment. This study contributes to resource dependence theory literature by revealing the factors influencing the board of directors of Turkish Airlines.

Keywords Airline governance · Board of directors · Resource dependence theory

13.1 Introduction

The environment surrounding the organization is important to understand the activities of the organization. Resource dependence theory argues that organizations need the resources from the environment for their sustainability. However, these needs make organizations to be dependent on the environment which provides the scarce resources for their operations (Casciaro and Piskorski 2005). The resource dependence theory argues that organizations try to manage these dependencies toward the environment in which they are (Pfeffer and Salancik (2003), Sargut and Özen (2010), Üsdiken and Leblebici (2009)). One of these management strategies is changing of the board members to gain the power due to compliance with the environment (Pfeffer and Salancik 2003). These changes make the organization control the resources and survive in a highly competitive environment.

L. Adiloğlu-Yalçınkaya (✉)
Faculty of Aviation and Aeronautical Sciences, Özyeğin University, Istanbul, Turkey
e-mail: leyla.adiloglu@ozyegin.edu.tr

A. Yalçınkaya
Faculty of Tourism, Istanbul Medeniyet University, Istanbul, Turkey
e-mail: akansel.yalcinkaya@medeniyet.edu.tr

This chapter points out that Turkish Airlines is under the influence of the environment and this can be observed in the composition of its board. In the aviation industry, the board of directors plays a critical role for the decisions of the organizations and reflects the changes in the environment. The question of whether the change of the board of directors is derived by the needs of the resources has necessitated an analysis within the framework of the resource dependence theory.

The first part presents the theoretical framework of this chapter. In the second part of this chapter, related literature will be presented, followed by the third part focusing on the research method. The fourth part discusses the results of changes in the board member of Turkish Airlines, which is the first airline in aviation industry in Turkey.

13.2 Theoretical Framework

Resource dependence theory is one of the theories which focus on the effects of the surrounding environment on the organization in order to understand and explain the practices of the organizations (Aldrich and Pfeffer 1976; Pfeffer and Salancik 2003). It emerged in the 1970s same as other organizational theories that examine the factors at macro-level instead of the micro-level factors. The most comprehensive explanation of the resource dependence theory has come out with the book entitled "*The External Control of Organizations: A Resource Dependence Perspective*" published in 1978 by Jeffrey Pfeffer and Gerald R. Salancik (Aldrich and Pfeffer 1976; Hatch and Cunliffe 2013). Resource dependence theory examines the organizational management of dependences through scarce resources (Pfeffer 1982; Pfeffer and Salancik 2003; Üsdiken and Leblebici 2009).

There are three fundamental assumptions of the theory. The first assumption highlights the importance of the environment and the context to understand and explain the organization. According to the theory, organizations are not self-sufficient and need the resources provided by the environment. This situation creates uncertainty that should be managed by the organizations. For this reason, organizations try to manage these dependencies in different ways. The second assumption of the theory points that organizations are able to react to the environment and manage these dependencies toward their interests. The third assumption of the theory is about the power imbalance originating from owning the resources. The theory argues that organizations are more powerful if they own the resources. This unbalanced power may also affect the organizations (Aldrich and Pfeffer 1976; Pfeffer 1982; Pfeffer and Salancik 2003; Üsdiken 2010).

Resource dependence theory states that organizations try to manage the dependencies through different strategies such as cooptation, interlocking, and executive succession. Resource dependence theory addresses that organization might prefer cooptation which means the appointment of the representative(s) to the decision-making positions such as the board of directors, advisory committee, interlocking directorates, and/or selection or removal of top managers (executive succession) in order to comply with the environment. Thus, cooptation, interlocking, and executive succession are seen as a way of reducing uncertainty in light of theory (Pfeffer and Salancik 2003).

This study analyzes the effect of the environment on the changes in the board of directors of Turkish Airlines.

13.3 Literature Review

Considering the studies in aviation literature, few studies explores the relation between directors and the environmental change based on resource dependence theory. For instance, Hermann and Rammal (2010) advocate that one of the reasons causing bankruptcy of Swissair was failure to change the board in accordance with the environment. Authors highlight that company has been collapsed for this reason. On the other hand, Hillman et al. (2000) examined the changes in corporate directors in airline companies after the deregulation. The researchers found that, airlines respond to environmental change with the change of directors. They conclude that airline companies comply with environment and decrease their dependences in this way. Similarly, Lang and Lockhart (1990) found that airlines focused on their indirect interlocking more than direct competitors after the deregulation of the airline industry. Kole and Lehn (1999), who investigated the governance structure during 22-year period, found that after deregulation act of 1978 board size was decreased. Considering the studies regarding the governance structure in airline organizations, Staniland (2003) examined the role of the state in the airline governance and comparised governance structures of the big four European Airlines (Air France, KLM, Lufthansa, BEA/BOAC). Staniland (2003) stated that considerable variations in both the role government and distribution of power between executives and boards.

Furthermore, Bogers et al. (2015) argue that although bringing external people to the board can contribute to innovation and entrepreneurship it can also create complexity. Authors show that Cimber Airlines preferred external members for a while then they changed approach again and decided to have just internal managers on the board. Contrary to this study, Hiatt et al. (2018) who analyzed the effect of stakeholder relations on new venture across ten countries found that organizational ties to high-ranking military officers are beneficial for new ventures in air transport industry. Similarly Daft (2018) highlights in his book that retired generals and executives from defense industry usually are hired in aerospace industry. More recently, Ozdemir (2020) found a positive relation between board diversity and financial performance in his study in which airlines are included in the sample.

13.4 Method

The aim of this chapter is to explore how the board member of Turkish Airlines has changed over time and explain the change of the board members in light of the resource dependence theory assumptions. For this purpose, a single case study approach was designed by following the guidelines of Yin (2016) to allow in-depth analysis of the changes in board of directors. To examine the chronological change of the board members within the scope of the study, Turkish Airlines was selected by applying purposeful sampling because the organization has a long history. The data were obtained from the secondary sources (e.g., archival documents, news clippings, internal company documentation, websites, industry reports).

13.5 Findings

Turkish Airlines, the first civil airline company, has been found in 1933 under the name State Airlines Administration (THY 2001). Turkish Airlines was %100 public company. Later, THY was consolidated as a partnership named Turkish Airlines in 1956.

13.5.1 Composition of the Board Between 1956 and 1982

As it is mentioned above, THY has been transformed from the State Airlines Administration to Turkish Airlines Inc. with the Statute No. 6623, enacted in 1956, and published in 1955 (THY 2001). The change of the regulation allowed THY to operate under private law (Milliyet 1956) and change the composition of board of directors. Thus, external members were allowed to become members of the board. On the other hand, it has been found that key actors in the Turkish press in that period have become the shareholders of THY (Table 13.1).

As it might be seen from Table 13.1, the representatives of almost all newspapers became the shareholder of THY. This critical step of transformation in the decision process might be explained in light of resource dependence theory. Resource dependence theory addresses that if there is mutual dependencies between organizations, this leads organizations to be able to negotiate easily (Casciaro and Piskorski 2005). In this case, the reason behind involvement the owners or representatives of press into the decision-making process is management of the mutual dependencies. While the press companies need to distribute the daily newspapers rapidly and extend their networks, THY also needs to promote the air transportation and it's position to the society because in that period air transportation was not so common among citizens. Thus, THY allowed the representatives of press to be in the decision-making process in order to promote itself, gain legitimacy among society, and increase the number of

Table 13.1 Representatives of Turkish press as a shareholder of THY

Newspaper	Shareholder
Cumhuriyet Matbaacılık ve Gazetecilik T.A.Ş.	
Haldun ve Erol Simavî Kollektif	Sedat Simavî successors
Yeni Sabah Gazetesi-owner	Sefa Kılıçlıoğlu
Milliyet Gazetesi-owner	Ercüment Karacan
Vatan Gazetecilik ve Matbaacılık T.A.Ş.	
Hergün Gazetesi-owner	Faruk Gürtunca
Kazım Şinasi Dersan ve Ortakları Akşam Komandit Şirketi	
Yeni İstanbul Neşriyat Limited Şirketi	
Tan Gazetecilik ve Neşriyat Evi	Halil Lütfi Dördüncü-owner
Nebioğlu Yayınevi	Dr. Osman Nebioğlu-owner
Doğan Kardeş Yayınları Anonim Şirketi	
Hayat Mecmuası Tifdruk Matbaacılık Sanayı A.Ş.	
Yelpaze Mecmuası-owner	Alâeddin Kıral
İstanbul Ekspres Gazetesi-owner	Mithat Perin
Dünya Basın, Yayın ve Ticaret Ltd. Ortaklığı	
Akbaba Yayınevi Sahibi	Yusuf Ziya Ortaç
Son Telgraf ve Gece Postası-owner	Ethem İzzet Benice
İleri Matbaacılık ve Neşriyat T.A.Ş.	
Tercüman Gazetesi-owner	Selim Baban
Son Posta Gazetesi	Selim Ragıp Emeç

THY (1956

the airline passenger. As Pfeffer and Salancik (2003) highlight legitimacy is one of the important resources for the organizations. According to the published memories of airline entrepreneurs such as Hürkuş (2000) and Altunbay (1989), private airline companies named Hürkuş and Göktur were also mainly operated for the distribution of the newspapers in that period. This highlights the demand among press companies.

In the beginning of 1950s, Turkish government of that period decided to renew the fleet of THY. While the government was searching for the aircraft to modernize the fleet, the close relation with Britain was established in that period. Afterward, British Overseas Airways Corporation (BOAC) became the shareholder of THY with %6,5 partnership ratio and this was the first and last foreign partnership in the history of THY. For this reason, George Cribbert, General Manager of BOAC became the member of the Board of Directors of THY in 1957(Yalçınkaya and Nergiz 2020).

The reason behind this milestone was clearly explained in the report of THY that this was because the need of the aircraft in that time (THY 2008b). While THY needs this relation in order to allow capital inflows, ensure the supply of credit for foreign consultants and modernize the fleet (Nergiz 2008), on the other hand, British government cooperates with Turkish government in order to prevent THY from buying aircraft from the US Convair and supports the British aircraft industry (Higham 2013). As a result of this cooperation, THY owns first turboprop aircraft (Viscount type) for long-haul flights (THY 2008b).

Another important milestone in the history of THY was the coup d'état in 1960. This coup d'état, which increased uncertainty in the environment, created a radical change for THY. This radical change played a role in the militarization of the board of director and operational department of THY. This is also the starting point of militarization phenomena which characterize the aviation industry in Turkey, lasts until the 2000s.

It should be mentioned that July 1960 was a breaking point in THY's administrative structure (Nergiz 2008). People with the military background became the board member and managers of the different departments except Enver Akoğlu, Retired General of the Turkish Air Force who was appointed in 1957. For instance, Air Force Colonel Halit Elgin became head of the board of directors, Major İhsan Göksaran and Colonel Şerafettin Kırmızı in the Turkish Air Force (Albayrak 1983). New board members in that time questioned the law fundamentals of having a foreign member, Sir George Cribbert in the board which is strategic position for Turkish Airlines, flag carrier in Turkey. Their hesitations were reflected in the meeting minutes of THY as stated by Nergiz (2008). This paves the way for replacing Sir George Cribbert with Selahattin Beyazıt, who was Turkish citizen and representative of BOAC in Turkey in that time.

Later, Retired General of the Turkish Air Force Şahap Metel became a member of the board of directors in 1962. Another important example is Ağasi Şen, Air Force Colonel, who became general manager in THY between 1965 and 1968. Mr. Şen was the team member who captured Former Prime Minister Adnan Menderes during the coup d'état in 1960. After this coup d'état, Mr. Şen became an aide-de-camp to the new President Cemal Gürsel.

During the 1970s, new members with the military background became the board of directors. It has been found that presidential decree (no.7/2385) was published for the appointment of Brigadier General of Air Force A. İhsan Göksaran to be a member of the board of directors of THY in 1971 ("Presidential decree", 1971). Furthermore, similar appointments might be observed until the 1980s such as General in Turkish Army Kemallettin Gökakın (1971), Lieutenant General Remzi Yelman (1972), and Colonel Ahmet Öztürk (1974) (Albayrak 1983).

The data obtained in this study show that a change in the environment increases the uncertainty for THY. Thus, the members with military background are the reflection of this change. Although appointment of these members is mandatory because of the coup d'état, it is seen that THY maintained the close relations with the military in order to balance the power. At that time, the majority of airports, important sources for the operations of the airlines were just opened to military aviation until 1980s (Korul and Küçükönal 2003). Accordingly, THY needed strong relations for expanding the network easily. On the other hand, THY provided the another important resources-airline pilots from the military services for a long time until the foundation of the university for pilot training (Yalçınkaya and Adiloğlu 2012).

Last important milestone for THY was another coup d'état in 1980. It is observed that the coup d'état affected the board of directors of THY because most of the members were replaced with people with military backgrounds. This was a radical change for the composition of the board. After the coup d'état, Major General Cengiz Sakaryalı (1980) became a general manager of THY. Also, retired Major General Nuri Gök became a chairman of the board and Major General of Air Force Hayri Gülşeni became a vice chairman of the board. On the other hand, retired Military Judge Alaattin Aksoy and retired General Emcet Edizel became the members of the board of directors (Albayrak 1983; Milliyet 1980).

13.5.2 Composition of the Board Between 1983 and 2003

After all, Turgut Özal (the Prime Minister) came to power with his Homeland Party in 1983. After the change of the government, new appointments were made in THY in 1984 through decision number 84/30759. Accordingly, the positions of a general manager and a chairman of the board were merged and former military pilot Yılmaz Oral was appointed to this position. On the other hand, former expert of State Planning Organization Yüksel Dinçer and former military pilot Naci Doğan Dinçer became the members of the board (Arşivi 1984). Furthermore, public finance expert Fikret Gümüşdere and retired General of Air Force Osman Ural became Assistant Managers of THY. Although Mr. Özal supports the liberal economy, it is observed that the members with the military background have existed until the end of the 1980s. Then, as it is mentioned by Yalçınkaya (2020) the role of the state started to reduce in air transportation industry in the 1980s. For instance, although Cem Kozlu didn't have experience in the public and/or military service he was appointed as a General Manager and Chairman of the Board of Directors of THY in 1988 by Turgut Özal. More interestingly, requirements for this appointment were changed for this position which were later criticized (Adiloğlu-Yalçınkaya 2019). The following article in the official gazette No. 19996 published in 1988 harmonized for Mr. Kozlu and requirement like experience in public service was changed. Thus, although Cem Kozlu had no background in public sector he became manager to be appointed to THY's management (Article 12(2)):

> Having served at least ten years, with a minimum of four years in the public sector, and six years in the private sector (for those who have less than four years of public service, two years of private sector service shall be considered to equal a year of service in the public).

Mr. Kozlu states the following in his book for Özal's explanation for this appointment (Kozlu 2007, p. 13):

> What I want from you are modernization and demilitarization of the company, building of a commercial-driven logic in THY and organizing of THY for the privatization.

A close look at the environment surrounding THY, it is observed that there were coalition governments in Turkey during Mr. Kozlu's two terms on the General Manager and Chairman of the Board of Directors position. Our analysis revealed that the coalition government was reflected in the composition of board of directors. Mr. Kozlu states the following in his memoir:

> I believe in the positive effect of the members with different political connection except behaving politically in the organization and in the board. Explaining the strategies and practices to the political parties provides transparency and increases trustfulness and support. For instance, Prof. Murat Demircioğlu was consultant to Democratic Left Party (DSP), our auditor İbrahim Pektaş has connection with the conservative political parties and Alaaddin Kuday may explain THY to Nationalist Movement Party (MHP).

Such political diversified approach within the decision-making process worked as a catalyst especially after the privatization period. In 1994, Law No. 4046 was published in 1994 and Turkish Airlines became a State Economic Enterprise and placed under the jurisdiction of the Privatization Administration, Articles of Association (THY 2001). This means THY is outside of the scope of the Ministry of Transportation anymore and this change explains the importance of the aforementioned composition of the board of directors during the coalition governments where ministries are in different political parties.

13.5.3 Composition of the Board Between 2003 and 2020

Among aforementioned important developments, 2003 is another important year for the composition of the board of directors of THY. When the Justice and Development Party (AKP) came to power in 2002, new government changed all of the members of the board (THY 2008b). Most of the newly appointed members had worked in different affiliates of the Istanbul Metropolitan Municipality, where the prime minister previously chaired. For instance, newly appointed General Manager of THY, Abdurrahman Gündoğdu, was working as a General Manager of Transportation Inc. (Ulaşım A.Ş.,) which is an affiliate of Istanbul Metropolitan Municipality (Genc 2016). Another member who were appointed from Istanbul Metropolitan Municipality is Candan Karlıtekin. Mr. Karlıtekin was appointed as vice chairman of the board of THY. Newly appointed board member Hamdi Topçu was close to Prime Minister Tayyip Erdoğan and he served as a financial advisor to Mr. Erdoğan (Topçu 2019). Another member Prof. Dr. Ömer Dinçer served as an undersecretary of the Prime Minister, Tayyip Erdoğan from 2003 to 2007. One and the only member who had an aviation background was Hüseyin Atilla Öksüz, a helicopter pilot in the military, who became one of the board members. Considering the privatization process of

THY, in 2006, the share of the state was firstly reduced to 46.43% and then increased to 49.12% in the same year. Although decreasing share of the state below 50% paves the ways of having member proposed by the market, concerns of the state effect on the composition of the board still exist (Gerede 2010). After decreasing the state-owned shares of THY, Muzaffer Akpınar became a member of the board for representing the shareholder in 2007.

Another important issue for THY is Orhan Birdal who was working as General Manager and Chairman of the Board in General Directorate of State Airports Authority (DHMI), and he became a member of the board of THY in 2008. DHMI has monopoly power in the country because it operates most of the airports and controls all of the airports as an authority. From the resource dependence perspective specially the role of the controlling the slots at the airports creates dependency for all of the airlines and with the cooptation strategy THY could manage this uncertainty which was later criticized by Competition Authority (*Competition Report* 2012; Şenyücel 2011) and State Supervisory Council (*State Supervisory Board Research and Examination Report* 2011). In 2009, Turan Erol who was member of Capital Markets Board of Turkey and Chief Advisor to the Prime Minister became a member of the board (THY 2009). In 2012, Mehmet Nuri Yazıcı who was a member of Istanbul Metropolitan Municipality Council and consultant to the President became a member of the board. Also, Naci Ağbal who had strong relations with central bureaucracy and experience in the Ministry of Finance became a member of audit board of THY in 2012 (THY 2006; Topçu 2019).

Among these members the findings demonstrate that M. İlker Aycı and Arzu Akalın became the members of the board in 2014 (THY 2014). While Arzu Akalın was legal advisor of prime minister of that time, M. İlker Aycı was working in Prime Ministry Investment Support and Promotion Agency (HaberTürk 2014).

When the board of directors' changes were examined, it is observed that the politicization of the board of directors has been maintained as it is stated in the literature (Gerede 2010). For instance, İsmail Cenk Dilberoğlu who was a Chairman of the Board of Trustees in Ensar Foundation found by the high-level politicians from the AKP (Yabanci 2019) became a member of the board in 2015 (THY 2015); Ogün Şanlıer who was working as a manager in the one of the affiliate (Halk Ekmek A.Ş.) of Istanbul Metropolitan Municipality became a member of the board in 2016 (THY 2016); Mithat Görkem Aksoy who was working as a Presidential Pilot (NTV 2018) and Fatmanur Altun who was member in one of the affiliates (Kültür A.Ş.) of Istanbul Metropolitan Municipality and Chairman of the Board Youth and Education Foundation (TÜRGEV) found by two children of president (Yabanci 2019) became the members of the board of THY in 2018 (THY 2018); Arda Armut who was working in the Presidency of the Republic of Turkey Investment Office same as İlker Aycı became a member of the board of THY in 2019 (THY 2019). Finally, Melih Şükrü Ecertaş who was member of Youth Branches of AKP became a member of board of directors (THY 2020).

13.6 Conclusion

The primary aim of the study is to understand and investigate the historical change of the board composition of Turkish Airlines from 1956 to 2020. For this purpose, the members of the board have been analyzed from the political economy perspective and resource dependence theory and the data obtained from the various secondary data. Research findings showed that there are three periods which can be named as Militarization period (1956–1982), Demilitarization period (1983–2002), and Politicization period (2003–2020).

During the militarization period, it is observed that after coup d'état in 1960, retired high-ranking military officers became executives and members of the board of THY. This militarization has affected the organizational and operational culture of the airline until 2000s. The demilitarization period which starts in the beginning of the 1980s characterized by the attempts of neo-liberalist Turgut Özal. With commercial-driven approach, Mr. Özal tried to demilitarize, commercialize, and privatize THY. Although some of the members of the board were former military officers, majority of the members were professionals. Lastly, after 2003, it is found that all of the members were replaced with the professionals with strong connections with the prime minister. These findings support the assumptions of the resource dependence theory which address that organizations as political actors comply with the environment in order to balance the power and gain the resources for the operations (Pfeffer and Salancik 2003).

Appendix: THY Board of Directors (1956–2020)

1956	1957	1958	1959	1960
Rıza Çerçel	Firuz Kesim	Firuz Kesim	Daniş Koper	Halit Elgin
Firuz Kesim	Sir. George	Dr. Osman	Dr. Osman	Suphi İşcen
Dr. Osman	Cribbett	Nebioğlu	Nebioğlu	İhsan Göksaran
Nebioğlu	Ulvi Yenal	Semih	Enver Mersinoğlu	Ulvi Yenal
Semih	Enver Akoğlu	Sipahioğlu	Seyfi Kurtbek	Şerafettin Kırmızı
Sipahioğlu	Hüseyin Ünsal, Dr.	Sir. George	Si George	
Muhiddin Asral,	Osman Nebioğlu,	Cribbett	Cribbett	
Hüseyin Ünsal,	Semih Sipahioğlu	Enver Akoğlu	Abdullah Parla	
Ulvi Yenal,		Hüsetin Ünsal,	Ulvi Yenal	
Enver Akoğlu,		Dr. Reflat	Hüseyin Ünsal	
Sir. George		Tuncel		
Cribbett				

(continued)

(continued)

1961	1962	1963	1964	1965
Halit Elgin Suphi İşcen İhsan Göksaran Selahattin Beyazıt	Şahap Metel Suphi İşcen Turgut Sayar Selahattin Beyazıt Halit Elgin	Sami Şehbenderler Suphi İşcen Turgut Sayar Bekir Akkan Selahattin Beyazıt Şahap Metel	Arif Demirer Suphi İşcen Şehbenderler, Selahattin Beyazıt	Ağasi Şen, Arif Demirer, Selahattin Beyazıt, Ziya Altınoğlu
1966	1967	1968	1969	1970
Arif Demirer Ziya Altınoğlu Selahattin Beyazıt Ağasi Şen	Muharrem Tuncay Ziya Altınoğlu Tevfik Tığlı Nuri Togay Gilbert Lee Ağasi Şen, Hüseyin Yeğin	Ağasi Şen Hüseyin Yeğin, Muharrem Tuncay Ziya Altınoğlu Tevfik Tığlı Nuri Togay Gilbert Lee	Tevfik Tığlı Ziya Altınoğlu Muharrem Tuncay Nuri Toğay Gilbert Lee Hüseyin Yeğin Orhan Batı	Kemal Aygün Kemallettin Gökakın Muhittin Günaltay Nurettin Ayasun Gilbert Lee Orhan Batı Nurettin Erguvanlı
1971	1972	1973	1974	1975
Kemallettin Gökakın Nurettin Ayasun Kemal Aygün Muhittin Günaltay Gilbert Lee Nurettin Erguvanlı İhsan Göksaran	Kemal Aygün Muhittin Günaltay Feridun Cemal Erkin Adil Sağıroğlu Gilbert Lee Remzi Yelman	Kemal Aygün Muhittin Günaltay Feridun Cemal Erkin Adil Sağıroğlu Gilbert Lee Remzi Yelman	Kemal Aygün Ağasi Şen Erhan Işıl Ahmet Öztürk Prof. Dr. Fahri Terzioğlu Recai Kutan Remzi Yelman	Kemal Zeren Recai Kutan Nihat Kürşat Mithat Perin Fehmi Alpaslan Ağasi Şen Nurettin Erguvanlı
1976	1977	1978	1979	1980
Kemal Zeren Recai Kutan Nihat Kürşat Mithat Perin Fehmi Alpaslan Nurettin Erguvanlı	Nihat Kürşat İsmail Yetiş İbrahim Sıtkı Hatipoğlu Kemal Zeren Mithat Perin Nurettin Erguvanlı	Rıza Çerçel Adnan Erdaş Nazif Arslan Hacim Kamoy Nuri Gök Argun Yelutaş Selahattin Babüroğlu Ertuğrul Alper	Rıza Çerçel Adnan Erdaş Nazif Arslan Hacim Kamoy Nuri Gök Mehmet İsvan Ertuğrul Alper	Ertuğrul Alper Abdullah Uraz Aykut Kuranel Uğur Gümüştekin Gülcemal Dinçkan Erdoğan Akünal Cengiz Sakaryalı

(continued)

(continued)

1981	1982	1983	1984	1985
Nuri Gök Hayri Gülşeni Cengiz Sakaryalı Cevat Yeyman İsmai Arar, Ali Sait Özcivril	Ali Sait Özçivril Hayri Gülşeni Yaşar Yıldırım Cengiz Sakaryalı Cevat Yeyman	Ali Sait Özcivril Hayri Gülşeni, İsmet Dinçel Yaşar Yıldırım Cengiz Sakaryalı Cevat Yeyman, Osman Ural, Yüksel Dinçer	–	Yılmaz Oral Yüksel Dinçer Naci Dinçer Fikret Gümüşdere Osman Ural Aytekin Tece

1986	1987	1988	1989	1990
Yılmaz Oral Yüksel Dinçer Naci Dinçer Fikret Gümüşdere Osman Ural Halil Gündüz	Yılmaz Oral Yüksel Dinçer Naci Dinçer Reşat Erkmen Fikret Gümüşdere Osman Ural Halil Gündüz Vahdettin Gündüz	Dr. Cem M.Kozlu Yılmaz Oral Yüksel Dinçer Dr. Ahmet Ertuğrul Osman Ural Vahdettin Gündüz Bülent Öztürkmen Reşat Erkmen Naci Dinçer	Dr. Cem M. Kozlu, Vahdettin Gündüz Dr. Ahmet Ertuğrul Bülent Öztürkmen İbrahim Pektaş A. Yener Kalyoncu	Dr. Cem M. Kozlu Dr. Ahmet Ertuğrul İbrahim Pektaş Meftun Yurdagül Bülent Öztürkmen Yener Kalyoncu M.Ökkeş Özuygur

1991	1992	1993	1994	1995
Dr. Cem M. Kozlu Yusuf Bolayırlı Dr. Ahmet Ertuğrul Meftun Yurdagül Bülent Öztürkmen Sait Gönenç Sözer Özel İbrahim Pektaş M. Ökkeş Özuygur	Erman Yerdelen Tezcan M. Yaramancı Yusuf Bolayırlı Rahmi Gümrükçüoğlu İhami Yılmazer Üstün Sanver Nail Kurt	Erman Yerdelen Tezcan M.Yaramancı Yusuf Bolayırlı Rahmi Gümrükçüoğlu İlhami Yılmazer Nail Kurt Alev Akiş Ersin Bener	Erman Yerdelen Tezcan M.Yaramancı Burhan Ayhan Atilla Çelebi Yusuf Bolayırlı Ersin Bener Yusuf Baki Aydın İhami Yılmazer Nail Kurt Alev Akiş Rahmi Gümrükçüoğlu	Erman Yerdelen Burhan Ayhan Ersin Bener Atilla Çelebi Yunus Baki Aydın İlhami Yılmazer Erkan Dereli Nail Kurt Alev Akiş İbrahim Yerebakan

(continued)

(continued)

1996	1997	1998	1999	2000
Erman Yerdelen	Dr. Cem M. Kozlu	Dr. Cem M. Kozlu	Dr. Cem M. Kozlu	Dr. Cem M. Kozlu
Dr. Cem Kozlu	Yusuf Bolayırlı	Yusuf Bolayırlı	Yusuf Bolayırlı	Yusuf Bolayırlı
Atilla Çelebi	Sühan Özkan	Sühan Özkan	Murat Demircioğlu	Dr. Ahmet Ertuğrul
Burhan Ayhan	Mehmet Gök	Mehmet Gök	Mehmet Gök	Mehmet Gök
Engin Aras	Vahit Erdem	Dr. Ahmet Ertuğrul	Alaaddin L. Kuday	Önder Doğu
Nail Kurt	Önder Doğu	Önder Doğu	Dr. Ahmet Ertuğrul	Murat Demircioğlu
Yunus Baki	Dr. Ahmet Ertuğrul	Tolga Akgün	Önder Doğu	Uyur Bayar
Aydın				
Sühan Özkan				
Atay				
Şefkatlioğlu				
Nail Kurt				
Sibel Çarmıklı				
Semih Öztürk				
İbrahim				
Yerebakan				
Erkan Dereli				
Dr. Ahmet				
Ertuğrul				
Şeref Has				

2001	2002	2003	2004	2005
Dr. Cem M. Kozlu	Dr. Cem M. Kozlu	Abdurrahman Gündoğdu	Dr. Candan Karlıtekin	Dr. Candan Karlıtekin
Yusuf Bolayırlı	Yusuf Bolayırlı	Dr. Candan Karlıtekin	Abdurrahman Gündoğdu	Hamdi Topçu
Dr. Ahmet Ertuğrul	Dr. Ahmet Ertuğrul	Prof. Dr. Cemal Şanlı	Prof. Dr. Cemal Şanlı	Assoc. Dr. Temel Kotil
Mehmet Gök	Mehmet Gök	Prof. Dr. Ömer Dinçer	Prof. Dr. Oğuz Borat	Atilla Öksüz
Önder Doğu,	Önder Doğu	Hüseyin Atilla Öksüz	Hüseyin Atilla Öksüz	Prof. Dr. Cemal Şanlı
Murat Demircioğlu	Alaaddin Kuday	Hamdi Topçu	Hamdi Topçu	Prof. Dr. Oğuz Borat
Uğur Bayar	Turgut Bozkurt	Mehmet Büyükekşi	Mehmet Büyükekşi	Mehmet Büyükekşi

2006	2007	2008	2009	2010
Dr. Candan Karlıtekin	Dr. Candan Karlıtekin	Dr. Candan Karlıtekin	Hamdi Topçu	Hamdi Topçu
Hamdi Topçu	Hamdi Topçu	Hamdi Topçu	Prof.Dr. Cemal Şanlı	Prof.Dr. Cemal Şanlı
Assoc. Dr. Temel Kotil	Assoc. Dr. Temel Kotil	Assoc. Dr. Temel Kotil	Assoc. Dr. Temel Kotil	Assoc. Dr. Temel Kotil
Atilla Öksüz	Atilla Öksüz	Orhan Birdal	Orhan Birdal	Orhan Birdal
Prof. Dr. Cemal Şanlı	Prof. Dr. Cemal Şanlı	M. Muzaffer Akpınar	M. Muzaffer Akpınar	M. Muzaffer Akpınar
Prof. Dr. Oğuz Borat	M. Muzaffer Akpınar	Mehmet Büyükekşi	Mehmet Büyükekşi	Mehmet Büyükekşi
M. Mehmet Büyükekşi	Mehmet Büyükekşi	Prof.Dr. Cemal Şanlı	Dr.Turan Erol, Dr. Candan Karlıtekin	Dr.Turan Erol

(continued)

(continued)

2011	2012	2013	2014	2015
Hamdi Topçu	Hamdi Topçu	Hamdi Topçu	Hamdi Topçu	M.İlker Aycı
Prof.Dr. Cemal Şanlı	Prof.Dr. Cemal Şanlı	Prof.Dr. Cemal Şanlı	Prof.Dr. Mecit Eş	Assoc. Dr. Temel Kotil
Assoc. Dr. Temel Kotil	Assoc. Dr. Temel Kotil	Assoc. Dr. Temel Kotil	Assoc. Dr. Temel Kotil	İsmail Cenk Dilberoğlu
Mehmet Büyükekşi	M. Muzaffer Akpınar	M. Muzaffer Akpınar	M. Muzaffer Akpınar	İsmail Gerçek
İsmail Gerçek	Mehmet Büyükekşi	Mehmet Büyükekşi	Mehmet Büyükekşi	Prof. Dr. Mecit Eş
Gülsüm Azeri	İsmail Gerçek	İsmail Gerçek	İsmail Gerçek	Mehmet Büyükekşi
M. Muzaffer Akpınar	Gülsüm Azeri	Naci Ağbal	Naci Ağbal	Ogün Şanlıer
	Naci Ağbal	Mehmet Nuri Yazıcı	M.İlker Aycı	M. Muzaffer Akpınar
	Mehmet Nuri Yazıcı	Prof. Dr. Mecit Eş	Arzu Akalın	Arzu Akalın

2016	2017	2018	2019	2020
M.İlker Aycı	M.İlker Aycı	M.İlker Aycı	M.İlker Aycı	M.İlker Aycı
Bilal Ekşi	Bilal Ekşi	Bilal Ekşi	Prof. Dr. Mecit Eş	Prof. Dr. Mecit Eş
İsmail Cenk Dilberoğlu	İsmail Cenk Dilberoğlu	Ogün Şanlıer	Arda Ermut	Arda Ermut
İsmail Gerçek	İsmail Gerçek	Orhan Birdal	Bilal Ekşi	Bilal Ekşi
Prof.Dr.Mecit Eş	Prof.Dr.Mecit Eş	İsmail Cenk Dilberoğlu	Mithat Görkem Aksoy	Mithat Görkem Aksoy
Mehmet Büyükekşi	Ogün Şanlıer	Mithat Görkem Aksoy	Orhan Birdal	Orhan Birdal
Ogün Şanlıer	Arzu Akalın	Prof.Dr.Mecit Eş	Ogün Şanlıer	Melih Şükrü Ecertaş
Muzaffer Akpınar	Orhan Birdal	M. Muzaffer Akpınar	M. Muzaffer Akpınar	M. Muzaffer Akpınar
Arzu Akalın	M. Muzaffer Akpınar	Dr. Fatmanur Altun	Dr. Fatmanur Altun	Dr. Fatmanur Altun

Resource Compiled by authors (Albayrak 1983; THY, 2008a, 2008b, 2009, 2010, 2011, 2012, 2013, 2014, 2015, 2016, 2017, 2018, 2019, 2020)

References

Adiloğlu-Yalçınkaya L (2019) Kurumsal Unsurların ve Kaynak Bağımlılıklarının Havayolu İş Modeli Değişimi Üzerindeki Etkileri: Türk Hava Yolları A.Ş. ve Pegasus Hava Taşımacılığı A. Ş. Örnekleri [Unpublished PhD thesis, Anadolu University]
Albayrak İ (1983) Turkish Airlines' 50th Yearbook (1933–1983). Türk Hava Yolları
Aldrich HE, Pfeffer J (1976) Environments of organizations. Ann Rev Soc 2(1):79–105. https://doi.org/10.1146/annurev.so.02.080176.000455
Altunbay M (1989) Hürriyete uçan Türk Mehmet Altunbay'ın hatıraları. Azerbaycan Kültür Derneği
Arşivi C (1984) Karar/84-30759. Fon No: 30-11-1-0-Kutu No.688-Dosya No:63. Sıra No:19
Bogers M, Boyd B, Hollensen S (2015) Managing turbulence: business model development in a family-owned airline. Calif Manag Rev 58(1):41–64

Casciaro T, Piskorski MJ (2005) Power imbalance, mutual dependence, and constraint absorption: a closer look at resource dependence theory. Adm Sci Q 50(2):167–199. https://doi.org/10.2189/asqu.2005.50.2.167

Competition Report (2012)

Daft RL (2018) Organization theory & design. Cengage learning

Genc A (2016) Yüksek irtifa: yerelden küresele THY'nin başarı öyküsü. Alfa

Gerede E (2010) The evolution of turkish air transport industry: significant developments and the impacts of 1983 liberalization. Manag Econ 17(2)

HaberTürk (2014) THY'ye üst düzey atama! HaberTürk. https://www.haberturk.com/ekonomi/is-yasam/haber/935968-thyye-ust-duzey-atama

Hatch M, Cunliffe A (2013) Organization theory: modern, symbolic and postmodern perspectives. Oxford

Hermann A, Rammal HG (2010) The grounding of the "flying bank". Manag Decis 48(7):1048–1062. https://doi.org/10.1108/00251741011068761

Hiatt SR, Carlos WC, Sine WD (2018) Manu Militari: the institutional contingencies of stakeholder relationships on entrepreneurial performance. Organ Sci 29(4):633–652

Higham R (2013) Speedbird: the complete history of BOAC. Bloomsbury Publishing

Hillman AJ, Cannella AA, Paetzold RL (2000) The resource dependence role of corporate directors: strategic adaptation of board composition in response to environmental change. J Manag Stud 37(2):235–256. https://doi.org/10.1111/1467-6486.00179

Hürkuş V (2000) Bir tayyarecinin anıları: yaşantı. Yapı Kredi

Kole SR, Lehn KM (1999) Deregulation and the adaptation of governance structure: the case of the US airline industry. J Financ Econ 52(1):79–117

Korul V, Küçükönal H (2003) Türk sivil havacılık sisteminin yapisal analizi. Ege Akademik Bakış Dergisi 3(1):24–38

Kozlu C (2007) Bulutların Üstüne Tırmanırken: THY Bir Dönüşüm Öyküsü. Remzi Kitabevi

Lang JR, Lockhart DE (1990) Increased environmental uncertainty and changes in board linkage patterns. Acad Manag J 33(1):106–128

Milliyet (1956). Devlet Hava Yolları şirket halinde çalışmaya başlıyor. Milliyet, 2

Milliyet (1980). THY Genel Müdürlüğüne Emekli Hava Tümgeneral Cengiz Sakaryalı Seçildi. Milliyet, 7

Nergiz A (2008) Türkiye'de sivil havacılığın gelişimi ve THY [MSc, Marmara University]

NTV (2018) Cumhurbaşkanı Erdoğan'ın pilotu Mithat Görkem Aksoy: Yeni havalimanı Türkiye'nin önünü açacak. https://www.ntv.com.tr/video/turkiye/cumhurbaskani-erdoganin-pilotu-mithat-gorkem-aksoy-yeni-havalimani-turkiyenin,rx_G70mQIU-62P6FIv8HfA

Ozdemir O (2020) Board diversity and firm performance in the US tourism sector: the effect of institutional ownership. Int J Hosp Manag 91:102693

Pfeffer J (1982) Organizations and organization theory. Pitman Boston

Pfeffer J, Salancik GR (2003) The external control of organizations: a resource dependence perspective. Stanford University Press

[Record #371 is using a reference type undefined in this output style.]

Sargut S, Özen Ş (2010) Örgüt Kuramlarına Genel Bir Bakış: Karşılaştırmalı Bir Çözümleme. In: Özen ASSŞ (ed), Örgüt Kuramı, 2 ed. İmge Kitabevi

Staniland M (2003) Government birds: air transport and the state in Western Europe. Rowman & Littlefield Publishers

State Supervisory Board Research and Examination Report (2011)

Şenyücel O (2011) Sivil havacılık sektöründe Rekabet Kurumu'nun yeri Ulaştırma Sektöründe Serbestleştirme, Rekabet ve Rekabet Hukuku Sempozyumu, Ankara

THY Regulation & Main Contract| THY Kanunu & Esas Mukavelenamesi (1956)

THY (2001) Annual report. THY. https://investor.turkishairlines.com/documents/ThyInvestorRelations/download/yillik_raporlar/faaliyetRaporu2001_en.pdf

THY (2006) Annual report. https://investor.turkishairlines.com/tr/mali-ve-operasyonel-veriler/faaliyet-raporlari

THY (2008a) Annual report. https://investor.turkishairlines.com/tr/mali-ve-operasyonel-veriler/faa liyet-raporlari
THY (2008b) Turkish Airlines' 75th Yearbook (1933–2008b). THY
THY (2009) Annual report. THY. https://investor.turkishairlines.com/tr/mali-ve-operasyonel-ver iler/faaliyet-raporlari
THY (2010) Annual report. https://investor.turkishairlines.com/tr/mali-ve-operasyonel-veriler/faa liyet-raporlari
THY (2011) Annual report. https://investor.turkishairlines.com/tr/mali-ve-operasyonel-veriler/faa liyet-raporlari
THY (2012) Annual report. https://investor.turkishairlines.com/tr/mali-ve-operasyonel-veriler/faa liyet-raporlari
THY (2013) Annual report. https://investor.turkishairlines.com/tr/mali-ve-operasyonel-veriler/faa liyet-raporlari
THY (2014) Annual report. https://investor.turkishairlines.com/tr/mali-ve-operasyonel-veriler/faa liyet-raporlari
THY (2015) Annual report. https://investor.turkishairlines.com/tr/mali-ve-operasyonel-veriler/faa liyet-raporlari
THY (2016) Annual report. https://investor.turkishairlines.com/tr/mali-ve-operasyonel-veriler/faa liyet-raporlari
THY (2017) Annual report. https://investor.turkishairlines.com/tr/mali-ve-operasyonel-veriler/faa liyet-raporlari
THY (2018) Annual report. https://investor.turkishairlines.com/documents/ThyInvestorRelations/ THY_FRAT_2018.pdf
THY (2019) Annual report. https://investor.turkishairlines.com/documents/thy_frat_2019_tr.pdf
THY (2020) Board of directors. https://investor.turkishairlines.com/tr/kurumsal-yonetim/yonetim-kurulu
Topçu H (2019) 'Yerel'den 'Global'e THY'nin Yükseliş Dönemi. Remzi Kitabevi
Üsdiken B (2010) Çevresel Baskı ve Talepler Karşısında Örgütler: Kaynak Bağımlılığı Yakaşımı. In: Özen ASSŞ (ed) Örgüt Kuramı. İmge Kitabevi
Üsdiken B, Leblebici H (2009) Örgüt teorisi. In: Anderson N, Öneş D, Sinangil H, Viswesvaran (eds) Endüstri, İş ve Örgüt Psikolojisi El kitabı, vol 2
Yabanci B (2019) Work for the nation, obey the state, praise the ummah: Turkey's government-oriented youth organizations in cultivating a new nation. Ethnopolitics, 1–33
Yalçınkaya A (2020) Devlet, Aktör ve Değişim: 1983–2013 Yılları Arası Türk Hava Yolu Taşımacılığı Alanında Kurumsal Değişim. Beta
Yalçınkaya A, Adiloğlu L (2012) Türkiye'de Lisans Düzeyindeki Sivil Hava Ulaştırma İşletmeciliği (SHUİ) Eğitim Sisteminin Yapısı ve Analizi. In: 3rd international conference on new trends in education and their implications' ta sunulan bildiri
Yalçınkaya A, Nergiz A (2020) I'm an Englishman in the Board… Gayrı Millîleştirme ile Yeniden Millîleştirme Arasında British Overseas Airways Corporation'ın Türk Hava Yolları Ortaklığı (1957–1977) 3. İşletme Tarihi Konferansı, İstanbul, Türkiye
Yin R (2016) Qualitative research from start to finish, 2nd ed. The Guılford Press

Dr. Leyla Adiloğlu-Yalçınkaya received her B.Sc. degree in Aviation Management from Anadolu University in 2009. She pursued her academic career in management and organization and obtained her M.Sc. degree in International Business from Marmara University. She completed her Ph.D. in management and organization at Anadolu University in 2019. Dr. Adiloğlu-Yalçınkaya joined the Faculty of Aviation and Aeronautical Sciences, Özyeğin University in 2011. She is currently working as an Assistant Professor and Safety & Compliance Monitoring Manager at Özyeğin University. Her research interests include organizational theory, strategic management and aviation management.

Dr. Akansel Yalçınkaya received his first B.Sc. degree in Aviation Management in 2009 and second B.Sc. degree in Labor Economics and Industrial Relations in 2012 from Anadolu University. He obtained his M.Sc. degree in Business Administration from Istanbul University. He completed his integrated Ph.D. in Aviation Management with the dissertation titled "State, Agency and Change: Institutional Change in the Turkish Air Transportation Field from 1983 to 2013" at the Anadolu University, Eskişehir in 2018. Dr. Yalçınkaya joined Istanbul Medeniyet University in 2011. In 2012–2015 he worked as a research asistant at the Faculty of Aviation and Aeronautical Sciences, Anadolu University in Eskişehir. Dr. Yalçınkaya is currently working as a Assoc. Prof. Dr. at Istanbul Medeniyet University, Istanbul. His research interests include business history, organizational theory and aviation management.

Part V
Covid-19 Crisis in Airlines

Chapter 14
The Malaysian Aviation Companies' Responsiveness Plan in Mitigating the COVID-19 Crisis

Corina Joseph

Abstract The COVID-19 pandemic which has affected worldwide economic shut-down with the risk of recession in some way offers an opportunity for Malaysia to reset its aviation industry and become profitable. The government has emphasized the importance of restructuring as the industry is wringing badly, which, in turn, may lead to a collapse and requires a broader discussion of Malaysia's overall national aviation strategy. With the implementation of the Movement Control Order by the Malaysian government, this paper aims to investigate the extent of responsiveness plans and actions undertaken by three Malaysian aviation companies in battling the COVID-19 pandemic. The analysis includes the examination of online news which was captured from a Google search from a period 18 March to early October 2020. There were 82 online news captured involving three aviation companies, i.e., AirAsia, Malaysian Airline System, and Malindo. Several responsiveness plans have been revealed in the online news, i.e., capacity management, preventive measures, talent management, and financial recovery plans. Some strategies are compiled to improve the aviation industry.

Keywords Online news · COVID-19 · Responsiveness plan · Aviation

14.1 Introduction

As made known to the public eye, the COVID-19 pandemic has left a huge blow on air travel as one of the hardest hit industries. The global airline industry had to face an overwhelming effect due to the cancellation of flights, countries worldwide imposing travel bans or international flight restrictions. If not technical insolvency, this impact led airlines to be driven into substantial defaults of their contractual obligations (Lexology 2020). To airlines, the COVID-19 crisis that is still going on becomes a treacherous threat. The way people fly will change. After the virus is contained, the way on how people would fly will also be different (Bloomberg 2020). The companies

C. Joseph (✉)
UiTM Cawangan Sarawak, Samarahan, Sarawak, Malaysia
e-mail: corina@uitm.edu.my

© The Author(s), under exclusive license to Springer Nature Singapore Pte Ltd. 2022
K. Kiracı and K. T. Çalıyurt (eds.), *Corporate Governance, Sustainability, and Information Systems in the Aviation Sector, Volume I*, Accounting, Finance, Sustainability, Governance & Fraud: Theory and Application, https://doi.org/10.1007/978-981-16-9276-5_14

that have access to funds show the highest chance of surviving in the industry with the slow recovery. Conspicuous survivors of any industry shakeout could be long stretch ease transporters, for example, AirAsia X. Minimal effort transporters rely upon higher resource usage to separate them from contenders. That is simpler to do on short bounces by just expanding the number of flights a plane that works in a solitary day, yet it is harder for significant distance runs that last a lot more hours (Bloomberg 2020). Faus (2020) has highlighted a few examples on how the travel and tourism industry can be affected due to the coronavirus:

- It is warned by The World Travel and Tourism Council that 50 million jobs worldwide in the travel and tourism industry could be cut due to the pandemic.
- The worst affected continent is expected to be Asia.
- For the restoration of the industry, it could take around 10 months the moment the epidemic has died down.
- As of right now, 10% of global GDP is accounted by the tourism industry.

Malaysian airlines are not without exception being badly affected by the COVID-19. Financial support is now being requested by Malaysia-based airlines from the government due to the lurching of the coronavirus crisis (Leong and Samarathisa 2020). Despite most airlines undertaking extreme cost-cutting approaches because of the quick depletion of finances by local airlines, to pull them through, it may not be enough. To take things in a more serious note, the Visit Malaysia 2020 had been cancelled by the government due to the outbreak of COVID-19 (Leong and Samarathisa 2020). It is predicted by The International Air Transport Association (IATA) that the passenger demand in 2020 reducing by 39% could be seen by Malaysia based on a scenario in which after 3 months, critical travel restrictions are lifted, and then gradual recovery, there will be a possible reduction of over 25 million passengers demand, some US$3.3 billion (RM14.36 billion) revenue effect, approximately 170,000 job losses and a US$3.8 billion potential GDP effect (Yunus 2020).

To ensure the survival, Malaysian carriers have announced the increase in air fares. Passengers are warned by Malaysian carriers that they are most likely to have to pay over 50% for the air fares if there is imposition on social distancing when boarding the aircraft in light of COVID-19 (Yusof 2020a). There will be a very intense competition between the airline companies as the demand gradually increases, just with less seats in the local market that are available (Yusof 2020a).

The Malaysian's commitment in enforcing the travel rules somehow adversely affect the operation of carriers. It is decided by the government on March 16, 2020 that a movement control order is implemented in the nation, in which amid other matters, movement restriction and assembly in all states as well as a strict restriction on travel for foreigners and tourists into Malaysia are encompassed.

- March 16: The enforcement of the movement control order (MCO) is announced from March 18 to 31 in order to stop COVID-19 from spreading any further, including the closing of the country's borders to foreign tourists and visitors.
- April 10: Another MCO extension is announced for a fortnight until April 28, with several selected economic sectors granted access to reopen in phases under

strict healthcare guidelines as well as restricting movements throughout the MCO period.

- May 1: The conditional MCO implementation is announced, where most economic and social events have access to resume starting from May 4.
- June 7: The recovery MCO implementation is announced and commenced from June 10 to August 31 due to the success of the country in changing the curve of the COVID-19 infection by flattening it and is currently in the recovery phase (Bernama 2020a).

Another imposition is carried out by the Malaysian Aviation Commission (MAVCOM). MAVCOM had released a Notice to Airlines Relating to the Malaysian Aviation Consumer Protection Code 2016, (14) on April 8, 2020, to provide aid to airlines that are presently overwhelmed with an unparalleled volume of consumers who want to make changes on seeking refunds or their bookings, which lessened the obligations the Malaysian Aviation Consumer Protection Code 2016 (MACPC) has imposed on airlines:

- Extended timeline for resolution of criticisms and also refunds payment—an extended timeline of 60 days (from thirty days) is currently given to airlines to solve all complaints and the refund remittance from the date of receipt of the feedback and also the refund claim. From February 1, 2020 to September 30, 2020, to any or all complaints and refund requests received, this is applicable.
- The change in flight status communication—the requirements for airlines to interact with passengers and provide public information regarding any flight status change has been waived by MAVCOM. From February 1, 2020 to September 30, 2020, this is applicable to all flights that are affected (Lexology 2020).

Due to the rapid changes in the external environment as a result of the COVID-19, it is interesting to explore on the extent of responsiveness plans and actions undertaken by three Malaysian aviation companies in battling the COVID-19 pandemic. The investigation of the online news involving the three Malaysian carriers is carried out in this paper. The chronological events based on the news reported from March to early October are analyzed.

14.2 Methodology

The analysis includes the examination of online news which was captured from a Google search from a period 18 March to early October 2020. There were 82 online news captured involving three aviation companies, i.e., AirAsia, Malaysian Airline System, and Malindo. All the online news is grouped based on the respective airline companies. Next, the online news was read and the search for responsiveness plans is carried out. The themes for responsiveness are identified and explained based on the chronological events. The explanation on responsiveness plans for each airline company is explained next.

14.3 AirAsia

It appears that AirAsia is the most reported airline in the online news. The responsiveness plan to COVID-19 includes precautionary measures, capacity management, talent management, and financial recovery plans. Introducing a few precautionary measures including, i.e., advice on travel, aircraft cleaning, pre-flight, in-flight as well as arrival. Regulators, civil aviation, local governments, and health authorities, including the World Health Organization (WHO) and the International Civil Aviation Organization (ICAO) are all whom AirAsia is working with closely and are following their expert assistance (www.airasia.com).

Reaching the end of February 2020, it is announced by AirAsia Group Bhd that there is a second consecutive quarterly loss and cautioned that the organization has the possibility in missing its own internal objective for 2020 due to the COVID-19 crisis. As AirAsia started 2020 with numerous provokes by the coronavirus, it left some parts of its businesses disrupted because of movement limitations, flight postponements, and cancellations. Regardless of the vulnerabilities, as the company strive to proceed, the "spirits are high" at AirAsia.

The business is active in checking the developments and has set up proactive mitigating activities to put a limit on the disadvantage effect of COVID-19. From early February 2020, the responsiveness strategies include active capacity management, reorganizing the capacity to intra-Asean as well as domestic aircrafts and the pursuing of market share with an aggressive marketing push are all included. Incentives, discounts, and rebates are also strategies the company is engaging with, alongside authorities and the industry stakeholders (The Star 2020a). In times of crisis, the CEO of AirAsia, i.e., Tony Fernandez had a strong belief in marketing and advertising. It seemed that AirAsia had tripled its advertising instead of cutting down on marketing and flights. This shows that when someone is a level-headed leader with clear directions, they are capable in turning an almost bankrupt company into one of the most successful budget airlines worldwide despite having to face a series of crises and outbreaks (Kenas 2020).

In relation to the capacity management, the company has reduced the frequency of flights across the network and suspended certain routes temporarily as part of the airline's mitigation strategy. Flights to China have been reduced since February 2020, and AirAsia X's capacity had been reduced to 90% of its network which included Japan, Australia, South Korea, and India in March. Other than cities in China, Tianjin and Lanzhou, and Jaipur in India, the airline also put a halt in flying to Wuhan, the epicenter of the novel coronavirus outbreak (Free Malaysia, Today 2020). Not only AirAsia is continuing to monitor the traits, they are also employing aggressive education approaches to instill self-assurance of flying returned within the traveling network. It is informed by the airline that its turnaround strategy includes using an aggressive method so that expenses can be reduced and demand is stimulated, inclusive of the non-profitable routes postponement, the ability control that is ongoing to have flights fit with demand, postponing deliveries to destinations using the Airbus a330 neo aircraft, as well as engaging moist-rent arrangements in a short period of time (Leong and Samarathisa 2020). It is in mid-June 2020 where AirAsia Group Bhd plans to add more flights through July. The CEO stated the call for air journey has shown improvement with the implementation of the Recovery Movement Control Order (RMCO), where interstate journey is permitted (Malay Mail 2020).

There are changes in the responsiveness plans relating to talent management as reported in online news starting March to October 2020. In early March 2020, employees are reassured by AirAsia that they would not have to be dismissed of their job despite a lot of airlines worldwide do because of the pandemic. Zero or minimal impact was assured to be felt by employees from the epidemic which left a huge effect on air travel worldwide where the airline has put in place a variety of cost control measures across all its business operations, emphasizing on decreasing the impact on employees, especially the ones categorized in the lower income bracket. The AirAsia Group CEO Tony Fernandes made a statement at the end of April 2020 that he and the Executive Chairman, i.e., Datuk Kamarudin Meranun had refused to get their salaries during that time. Meanwhile, depending on seniority, temporary pay reductions have been accepted by the staff between 15 and 75% (Yunus 2020).

In early June 2020, 111 cabin crew individuals, 172 pilots, and 50 engineers are reported to be involved in the layoffs. Those who are in a state of worry regarding the process cuts would be notified via e-mail. All those who are concerned were promised to receive reimbursement by AirAsia. As there are no other choices in ensuring that the budget carrier continues to run like earlier, there is a possible opportunity in restructuring the agency due to the difficult-international financial situation. It was also stated in the record that long-haul finance service AirAsia X might even lessen its staff, including employees from overseas (Bernama 2020a, b).

During the second week of June 2020, AirAsia scaled back the last group of workers' salaries through up to 75% in a try to keep the airline. The retrenchment includes reducing 60% of AirAsia's cabin team and pilots for each AirAsia and its medium-haul affiliate AirAsia X. The carrier institution operates through Malaysia, Thailand, Indonesia, Japan, India, and the Philippines. Almost all of the agency's 20,000 personnel had been personally re-evaluated on the grounds depending on the income scale and overall performance, with the lay-off expected to make it through to the end of July. In early October 2020, it is confirmed by AirAsia Bhd and AirAsia x Bhd regarding the cutback of 10% of their 24,000 employees. The CEO stated the move to end off the workforce services needed to be undertaken given the fact that the aviation sector which was facing problems in recovering in the future. The institution was put under expectations to run on a smaller scope while AirAsia could make an attempt to rescue most of its personnel that was laid off as viable. According to the CEO, "AirAsia is all about its human beings. Anything human beings need to mention, we constructed a splendid organisation because of notable human beings, 24,000 top-notch humans, and we lost a few even though it turned into via no fault of theirs. So, my duty is to try and get the airline lower back (on its feet)" (Bernama 2020b).

There are several financial recovery plans continuously conducted by AirAsia. From July until September 2020, it is intended by AirAsia Group Bhd to raise RM2.4 billion to ease its tight cash flow that is threatening the budget airline's ongoing challenge. AirAsia, which had triggered the agreed criteria of Practice Note 17 status, stated that it might acquire up to RM1 billion from monetary institutions. The group and the organization propose to elevate the capital as much as RM1.4 billion while required to toughen its equity base and liquidity besides anticipating the hit implementation on these capital elevating proposals. It was found out by the budget carrier that a few "financial institutions in Malaysia have shown the willingness to assist it and grant as much as RM1 billion in subsidy" (Shankar 2020).

A proposal of restructuring RM63.5 billion of debt and having its share capital reduced by 90% was made by AirAsia X Bhd (AAX), the long-haul arm of AirAsia Group Bhd, in order to continue as a going concern in early October 2020. The COVID-19 pandemic had harshly left a huge hit upon the group as most of its planes were not available due to the closed borders for months, thus seeking to reconstruct the RM63.5 billion debt into a principal amount of RM200 million so that waive is made for the balance (Reuters 2020a).

AirAsia Group Bhd is also putting an end with its Japanese operations immediately in early October. It has been said by the airline that has been notified regarding the AirAsia Japan (AAJ) board of directors in making a decision to stop all operations with immediate effect. It has been agreed by the group to the decision made by AAJ so that the cash burn of AAJ and the Company would be reduced within the extremely challenging operating conditions in Japan in which the COVID-19 pandemic has worsened (Reuters 2020b).

As a result of excellent responsiveness plans perceived by stakeholders, AirAsia arose as the most reputable company in the list of "Most Reputable Brands of Malaysia struggle the Covid-19" in March 2020. "Strategy", "Culture", and "Delivery" were three pillars of Reputation Framework which became a topic unraveled by the recognition rankings file among Malaysian netizens who became aware of the manufacturers. The honorable "culture" has been proved by AirAsia by being truthful and ethical in offering credit account redemption and flight refunds to countries including China, Hong Kong, and Macao, followed by flight suspensions to the countries. As the budget airline was praised for its pioneer move of sending volunteers in a different flight as part of the Humanitarian Assistance and Disaster Relief (HADR) mission to bring Malaysians who have been stranded in China back home, the "strategy" pillar was also excessive for AirAsia. AirAsia's assist in transporting medical gloves to the healthcare employees in Wuhan additionally contributed to the airline's top rank among Malaysia's most reputable brands during the coronavirus pandemic apart from supplying refunds and flying back Malaysians to their homeland again (Isentia 2020).

14.4 Malaysian Airline System

MAS offers a lot of Malaysian hospitality in its daily business activities. The company is implementing several precautionary measures as part of its strategic responsiveness plan in mitigating the COVID-19 crisis. The company has put in place various improved defensive measures along with the exceptional interests in mind as the well-being of its customers and personnel is of extreme significance to the business. Together with the Ministry of Health Malaysia (MOH) and World Health Organization (WHO), the key health and safety measures adhere to the guidelines set by way of local and global health authorities to provide a safer and healthy travel experience. Provided by means of local and worldwide aviation government together with the civil aviation authority of Malaysia (CAAM) and worldwide civil aviation corporation (ICAO), Malaysia airlines also adhere to flight safety tips. As what has been stated in the guidelines: Making Sure That Everyone Is Absolutely Secured; Physical Distancing; Face Covering and Masks; Habitual Sanitation; Health Screening; Contract Tracing; Trying Out and Customer and crew care. A special video has also been released on its social media platforms with the title "Ready to fly and rediscover Malaysia" in line with the recent announcement the government has made on traveling interstates. It is hoped by MAS that local travelers can be inspired to further explore the attractiveness of our country besides boosting local tourism efforts by the showcase of beautiful and unique domestic attractions placed across Malaysia (The Star 2020b).

As for the talent management, it was reported in early May 2020 that Malaysia Airlines Berhad (MAS) had reduced their operations because of the COVID-19 epidemic and had its 13,000 employees (including employees the subsidiaries of Malaysia Aviation Group (MAG) had employed, including MAB Kargo, MAB Engineering, Firefly and MASwings) offered the choice of accepting unpaid leave for 3 months or unpaid leave for 5 days per month for 3 months beginning in April 2020 (Lexology 2020). Furthermore, to have cash flow sustained, Malaysia Airlines recommended unpaid leave and induced salary cuts between 10 and 35%. This is done so that those in the lower income bracket are protected and possible job cuts can be avoided in the effort to have costs being trimmed any further.

In terms of the capacity management, MAS has continued to steer the dynamic pricing further varying on capacity and demand in the second week of May 2020. The promotions would definitely be determined in intervals. For both international and domestics routes, all flight schedules were expected to be normalized by MAS in early June 2020. The domestic and international connectivities would be increased in early June and July, respectively, so that essential travel within the country would be facilitated as border restrictions are starting to be lifted in other countries (The Malaysia Insight 2020).

In view of the crisis, several business investments have also been halted temporarily. In order to manage its cost-effectively, MAS has proactively removed the capacity. As the COVID-19 epidemic proceeded to hammer the national carrier, Malaysia Airlines had removed 96% of its operational capacity. This was aggravated by the global travel ban so that the further spreading of the lethal virus is prevented. Since the Movement Control Order on March 18, 75% of the carrier's aircraft fleet has been grounded as the desire for travel weakens. In mid-June, MAS reduced 15% capacity for domestic operations only. Malaysia Airlines had its production reduced drastically with an overall production for the months April to June 2020, recording only 94% below its budget.

With the government's support, airlines can also initiate several initiatives, citing that the entire aviation sector value chain needs to be coordinated. Debt instruments such as equity financing as selections should be the airlines' focus in helping them be placed on the road to recovery. As the aviation business is a high-intensity cash flow environment, it is consistent with the International Air Transport Association's prediction that the debt carried by the global airline industry could increase up to U$550 billion or could even face a staggering 28% at the end of 2020 (Yusof 2020a). A suggestion was made last August 2020 that just like Air New Zealand, MAB should just pay attention on being a national carrier and expand its business in logistics as well as being creative. In order for the bottom line to be fueled, it is not just the passengers' business (Leong 2020).

14.5   Malindo

Like AirAsia and MAS, there were numerous disinfectant measures Malindo Air underwent in providing protection for their passengers, as their safety and security is preponderating to the airline. A team of disinfectant specialists deeply cleansed all of Malindo Air's ATR and Boeing aircrafts every day including night time as measures that were implemented based on directives the authorities had issued. With medical-grade disinfectants, the intensive cleansing process is carried out to make sure that all of Malindo Air aircrafts' interiors are cleaned deeply. Other measures such as increasing the supply of water sold aboard in order for the travelers to stay hydrated have conjointly been adopted in its in-flight service. Based on Malindo Air's observation, they find it preponderant in confirming that strict precautions for the workers who are on duty should have experience by taking heed of personal hygiene, valuing the importance of carrying face masks and gloves while being attentive to passenger symptoms throughout the flight. It is Malindo Air's top priority to ensure their passengers' safety no matter the circumstance.

In terms of the talent management, based on a memo Reuters had sighted, Malindo Air has instructed its staff to accept a 50% pay cut and an unpaid leave for 2 weeks as the coronavirus pandemic affects air travel demand and the broader industry badly in March 2020. Chief Executive Officer Mushafiz Mustafa Bakri stated that a few measures have been implemented by the airlines to manage with the short-fall of revenue, which included flight suspensions, appealing to suppliers to accept payments, and pleading staff to offer for unpaid leave. "With a heavy heart, with not much of any further concrete options, we are now left with no choice but to ask each one of you to take a pay cut of your basic pay of up to 50% for the next several months until normalcy returns", as what Mushafiz had stated in the memo. Employees were requested to have their number of working days reduced by up to 15 days monthly as part of the pay cut (Flee 2020).

14.6 Discussion

It appears that AirAsia has the most coverage in the online news. This could be due to the role of media visibility which serves as the coercive factor in enforcing the airline company to communicate about its strategic responsiveness plan in mitigating the COVID-19 pandemic. The coercive factor is one of the three isomorphism mechanisms introduced by DiMaggio and Powel (1983) which is used to explain the organizational behavior and practices. Coercive isomorphism is the most cited type of institutional force. According to DiMaggio and Powell (1983, p. 149), the pressures can be "exerted by other organizations on which an organization may be dependent, as well as cultural expectations in which the organizations operate". In this manner, AirAsia wants to be seen as legitimate by stakeholders, i.e., demonstrating its commitment to the crisis management by communicating using online news. In recent years, press or media plays an important role in shaping the image and reputation of organizations through the publication of news. The spread of news about the development and administrative programs of an organization could be obtained through public media. The public media plays an important role in ensuring that the citizens keep updating and informing about recent news about the organization. Pressure exerted by press visibility on the organization could influence the amount of information disclosed.

As for MAS and Malindo Airline, it appears that the online news is not used extensively to communicate about the strategic responsiveness plan in mitigating the COVID-19. Perhaps both airlines are legitimizing their strategic responsiveness plan via other communication platforms such as websites or internal bulletins.

A Way Forward

It is time for the Malaysian aviation industry to revitalize the sector. Malaysia's aviation industry needs a more extensive conversation and techniques should be reconsidered to guarantee that it is productive and eventually developing once more (Yunus 2020). Airlines receiving soft loans first become the main priority in order to prevent their collapse from happening.

The followings are suggested by Molenaar et. al. (2020) which could be undertaken by the airlines according to the flight plan:

- The airline companies should begin by deciding the ideal size and measurements of their networks and fleets, and they ought to do so within the weeks that are yet to come. The airline companies should settle on important choices—including which type of fleets to recommission first and which routes that have higher chances to recuperate—based on a few requests and market structure situations while keeping in mind that streamlining with the expectation of complimentary income.

- Mergers and acquisition and consolidation opportunities should be considered. It is expected that options including potential divestitures and the sale or minority equity stakes' purchases could be reviewed.
- The operating model and organization should be resized and restructured using a zero-based method. These includes redesigning the process, organization size, and structure.
- The moment airports and countries are opened once again; airlines should also start preparing by ramping up. Organization update must be executed on a weekly basis.

Last but not least, it is definite that in order for cash levels to be protected, finance teams have to be deeply involved, where revenues are captured as soon as possible, and cash-outs are delayed as much as possible. A project management office should be established by the airlines so that cash can be managed until the financial processes and routines are stabilized and regulated with the environment and can be applied once again.

Airlines will assess engaging in aggressive (i) operational restructuring, which is targeted amid others, having their business plans re-aligned, cost operations besides structure, and also (ii) debt restructuring and liability management—capital structure, resolution of liabilities, as well as termination or renegotiation of arduous contracts in which a reorganization is involved. With stakeholders or within formal insolvency proceedings, pursuing these plans on a consensual basis is possible (Kikoyo 2020).

As the passenger traffic will reduce than initially predicted over the next 2–3 years, it is no doubt that purchasing plane tickets is also being adjusted by Malaysian carriers. Instead of adding more flights during this unprecedented period, airlines should pay more attention on avoiding extra cost and start preserving cash. It is flexibility that is being observed by airlines because of how difficult it is for them to expect how the traffic will recover the moment COVID-19 is gone. Local carriers seek for financial assistance in order to keep their businesses active as most of them are tight on budget. However, to maintain the cash flow, they should reschedule or cancel their aircraft orders or deliveries. As most countries have their borders closed and international flights are also banned, the capacity in the short term has also been reduced (Yusof 2020b).

References

Bernama (2020a) AirAsia may lay off hundreds to pare down operations. Accessed 5 June 2020a. https://www.nst.com.my/news/nation/2020a/06/598068/airasia-may-lay-hundreds-pare-down-operations#.XtmsDwpgKrQ.facebook

Bernama (2020b) AirAsia confirms layoffs. Accessed 9 October 2020b. https://www.nst.com.my/news/nation/2020b/10/631023/airasia-confirms-layoffs

Bloomberg (2020) Covid-19: AirAsia, like other low-cost airlines, fighting for survival. https://www.nst.com.my/news/nation/2020/04/584511/covid-19-airasia-other-low-cost-airlines-fighting-survival

DiMaggio PJ, Powell WW (1983) The iron cage revisited: institutional isomorphism and collective rationality in organizational fields. Am Sociol Rev 48(2):147–160

Faus J (2020) This is how coronavirus could affect the travel and tourism industry. https://www.weforum.org/agenda/2020/03/world-travel-coronavirus-covid19-jobs-pandemic-tourism-aviation/

Flee (2020) Covid-19: Malindo Air asks staff to take up to 50% pay cut to cushion coronavirus blow. Accessed 7 March 2020. http://www.malaysianwings.com/forum/index.php?/topic/22603-covid-19-malindo-air-asks-staff-to-take-up-to-50-pay-cut-to-cushion-coronavirus-blow/

Free Malaysia Today (2020) Virus causes 25% less passengers for AirAsia X in first quarter. Accessed 28 April 2020. https://www.freemalaysiatoday.com/category/highlight/2020/04/28/virus-causes-25-less-passengers-for-airasia-x-in-first-quarter/

Isentia (2020) Prudential 'most reputable' in Covid-19 measures. Accessed 4 March 2020. http://www.dailyexpress.com.my/news/148082/airasia-rapidkl-prudential-most-reputable-in-covid-19-measures/

Kenas T (2020) Lessons From SARS: What Can We Learn To Face Covid-19? Accessed 30 March 2020. https://www.imoney.my/articles/lessons-sars-covid-19-alibaba-airasia-jd

Kikoyo H (2020) Survival Strategies for Airlines Facing Insolvency–Fallout from the Coronavirus (COVID-19) Pandemic. Accessed 17 March 2020. http://www.brownrudnick.com/alert/survival-strategies-for-airlines-facing-insolvency-fallout-from-the-coronavirus-covid-19-pandemic/

Lexology (2020) Staying airborne during COVID-19 pandemic. Accessed 6 May 2020. https://www.lexology.com/library/detail.aspx?g=6bb569f8-3d96-4408-a83e-dc62ac24e96f

Leong D (2020) Another bid for Malaysia Airlines, really? Accessed 4 Aug 2020. https://focusmalaysia.my/opinion/another-bid-for-malaysia-airlines-really/

Leong D, Samarathisa (2020) AirAsia, MAB, Malindo, Firefly seek financial aid from MoF, say sources. Accessed 19 March 2020. https://focusmalaysia.my/mainstream/airasia-mab-malindo-firefly-seek-financial-aid-from-mof-say-sources/

Malay Mail (2020) AirAsia to add more flights by July, says Tony Fernandes. Accessed 15 June 2020. https://www.malaymail.com/news/malaysia/2020/06/15/airasia-to-add-more-flights-by-july-says-tony-fernandes/1875662

Malaysian Airlines (2020) Your safety and well-being are our priority. https://www.malaysiaairlines.com/my/en/advisory/preventive-measures-and-travel-recommendations.html

Molenaar D, Bosch F, Guggenheim J, Jhunjhunwala P, Loh H, Wade B (2020) The Post-COVID-19 Flight Plan for Airlines. Accessed 31 March 2020. https://www.bcg.com/en-sea/publications/2020/post-covid-airline-industry-strategy.aspx

Reuters (2020a) AirAsia X to stave off liquidation with RM63.5 bill debt restructure. Accessed 7 October 2020a. https://www.freemalaysiatoday.com/category/highlight/2020a/10/07/airasia-x-to-stave-off-liquidation-with-rm63-5-bil-debt-restructure/?utm_source=FCM&utm_medium=Push

Reuters (2020b) AirAsia shuts Japan operations. Accessed 5 October 2020. https://www.freemalaysiatoday.com/category/nation/2020/10/05/airasia-shuts-japan-operations/

Shankar AC (2020) AirAsia says it aims to raise RM2.4 billion via debts and equity. Accessed 9 July 2020. https://www.theedgemarkets.com/article/airasia-says-it-aims-raise-rm24-billion-debts-and-equity#.XwZxDlaANLs.facebook

The Malaysia Insight (2020) Malaysia Airlines to normalise schedule in October. Accessed 8 June 2020. https://www.themalaysianinsight.com/s/252114?utm_source=dlvr.it&utm_medium=facebook

The Star (2020a) AirAsia may miss 2020a target due to Covid-19 outbreak. Accessed 27 February 2020a. https://www.klsescreener.com/v2/news/view/643758/airasia-may-miss-2020a-target-due-to-covid-19-outbreak

The Star (2020b) Matta: Malaysian airline industry needs government help to survive Covid-19. Accessed 20 April 2020b. https://www.thestar.com.my/lifestyle/travel/2020b/04/20/local-air lines-need-help-from-government-to-survive-post-covid-19-says-matta

Yusof A (2020a) Air fares may rise by more than 50pct if social distancing is enforced. Accessed 12 May 2020a. https://www.nst.com.my/business/2020a/05/591874/air-fares-may-rise-more-50pct-if-social-distancing-enforced

Yusof A (2020b) Air travel may take up to 18 months to recover, says Malaysia Airlines. Accessed 15 June 2020b. https://www.nst.com.my/business/2020b/06/600673/air-travel-may-take-18-mon ths-recover-malaysia-airlines

Yunus R (2020) Airlines need soft loans, not merger talk. Accessed 30 April 2020. https://themal aysianreserve.com/2020/04/30/airlines-need-soft-loans-not-merger-talk/

www.airasia.com (2020) COVID-19: Flying safe with AirAsia-AirAsia's safety measures on the ground and in-flight. Accessed April 2020. https://newsroom.airasia.com/news/2020/4/27/covid-19-flying-safe-with-airasia-safety-measures-airasia-implements-on-ground-and-in-flight

Dr. Corina Joseph is a full Professor at the Faculty of Accountancy, UniversitiTeknologi MARA (UiTM), Sarawak Branch. She is an Associate Member of the Malaysian Institute of Accountant. Her research interests are in CSR and Sustainability, Corporate Governance, Public Sector Accounting and Financial Criminology. Corina is an advisory member for the Social Responsibility Asia (SR Asia) Global Certification Program for Professionals in Sustainability Repotting. She is currently the Co-Chief Editor of International Journal of Service Management and Sustainability. She also serves as the Editorial Advisory Board Member of several Scopus and international refereed journals. She has received numerous national and international awards throughout her 26 years' service in UiTM.

Appendix
Airline List in Study

Airline	IATA code	Airline	IATA code
Aeroflot	SU	Japan Airlines	JL
Air Canada	AC	Korean Air	KE
Air China	CA	Latam Airlines	LA
Air France-KLM	AF/KL	Lufthansa	LH
All Nippon Airways	NH	Qantas Airways	QF
American Airlines	AA	Ryanair	RK
China Eastern Airlines	MU	Singapore Airlines	SQ
China Southern Airlines	CZ	Southwest Airlines	WN
Delta Airlines	DL	Turkish Airlines	TK
Easyjet	U2	United Continental	UA/CO

Index

Printed in the United States
by Baker & Taylor Publisher Services